Springer Series in Computational Physics

Springer Series in Computational Physics

Editors: C. A. J. Fletcher   R. Glowinski   W. Hillebrandt   M. Holt   P. Hut
H. B. Keller   J. Killeen   S. A. Orszag   V. V. Rusanov

Karl K. Sabelfeld

Monte Carlo Methods

in Boundary Value Problems

With 62 Figures

Springer-Verlag

Berlin Heidelberg New York
London Paris Tokyo
Hong Kong Barcelona
Budapest

Professor Karl K. Sabelfeld

Computing Center, Siberian Division, USSR Academy of Sciences
Prospect Lavrentyeva 6, SU-630090 Novosibirsk, USSR

Editors

C. A. J. Fletcher

Department of Mechanical Engineering
The University of Sydney
New South Wales, 2006
Australia

R. Glowinski

Institut de Recherche d'Informatique
et d'Automatique (INRIA)
Domaine de Voluceau
Rocquencourt, B. P. 105
F-78150 Le Chesnay, France

W. Hillebrandt

Max-Planck-Institut für Astrophysik
Karl-Schwarzschild-Straße 1
W-8046 Garching, Fed. Rep. of Germany

M. Holt

College of Engineering and
Mechanical Engineering
University of California
Berkeley, CA 94720, USA

P. Hut

The Institute for Advanced Study
School of Natural Sciences
Princeton, NJ 08540, USA

H. B. Keller

Applied Mathematics 101-50
Firestone Laboratory
California Institute of Technology
Pasadena, CA 91125, USA

J. Killeen

Lawrence Livermore Laboratory
P. O. Box 808
Livermore, CA 94551, USA

S. A. Orszag

Department of Mechanical and
Aerospace Engineering
Princeton University
Princeton, NJ 08544, USA

V. V. Rusanov

M. V. Keldysh Institute
of Applied Mathematics
Miusskaya pl. 4
SU-125047 Moscow, USSR

ISBN-13: 978-3-642-75979-6   e-ISBN-13: 978-3-642-75977-2
DOI: 10.1007/978-3-642-75977-2

Library of Congress Cataloging-in-Publication Data. Sabel'fel'd, K. K. (Karl Karlovich [Metody Monte-Karlo v kraevykh zadachakh. English] Monte Carlo methods in boundary value problems / K. K. Sabelfeld. p. cm.–(Springer series in computational physics) Translation of: Metody Monte-Karlo v kraevykh zadachakh. Includes bibliographical references and index. ISBN 0-387-53001-0 (U.S. : alk. paper) 1. Boundary value problems. 2. Monte Carlo method. I. Title. QA379.S2313 1991 515'.35–dc20 90-22642

© Springer-Verlag Berlin Heidelberg 1991
Softcover reprint of the hardcover 1st edition 1991

Typesetting: Springer TEX inhouse system
57/3140-543210 – Printed on acid-free paper

Preface

Methods of integral equations play an important role in investigating boundary value problems of mathematical physics. Classical examples are the solutions to the heat and electrostatic potential problems through solution of boundary integral equations. From the computational point of view the passage from a differential equation to boundary integral equations has at least two advantages: (1) the dimension of the problem is reduced, (2) in the case of an exterior boundary value problem we pass from an unbounded domain to a finite boundary of this domain. In this monograph, boundary integral equations of the potential theory (Chap. 3), elasticity (Chap. 4), and diffusion theories (Chap. 5) are used to construct walk on boundary algorithms for solving the corresponding boundary value problems.

Another important class of integral equations considered in this monograph involves local integral equations with generalized kernels. To solve these equations, straightforward probabilistic representations can be applied where conventional numerical methods fail.

Although three different classes of boundary value problems are treated here, the method of investigation based on local and global integral equations is common to all of them. Therefore, general aspects of the Monte Carlo technique utilized in the book are presented in Chap. 1. In particular, definitions and fundamental properties of the walk on spheres and walk on boundary processes are given. Simulation formulas for scalar and vector random fields are constructed. We present in Chap. 2 general approaches to constructing Monte Carlo algorithms for solving integral equations. Applications to homogeneous and coagulative formation of aerosols and clusters, transfer of these particles in turbulent flows and inertial deposition of particles on bodies (Sects. 3.3, 5.4) and stochastic problems of thin plate theory (Sect. 4.4) are presented.

Monte Carlo methods for solving boundary value problems are in a state of continuing development. This monograph is the third devoted to Monte Carlo methods in boundary value problems and the first published in English. The text by Elepov, Kronberg, Mikhailov, Sabelfeld appeared in 1980, the second by Ermakov, Nekrutkin, Sipin in 1984. Although the author dealt with certain topics that have interested him personally, he hopes that the coverage of the subject is reasonably complete.

I want to thank many colleagues: N.A. Simonov, A.E. Ultan, O.A. Kurban-muradov, S.E. Makarov, Yu.N. Kopylov, I.A. Shalimova, D.M. Rakimgulova, S.E. Pashenko, K.P. Kutsenogy for their kind help. V.P. Ilyin, I.M. Sobol, S.M. Ermakov and P.G. Skvortsov I want to thank for their valuable advice and many critical remarks, Professor G.A. Mikhailov, for encouraging me to write this book.

Novosibirsk, June 1991   *K.K. Sabelfeld*

Contents

X

List of Symbols

C_n^k	Binomial coefficients $\binom{n}{k}$
$D\xi$	Variance of a random estimate
F_{LL}, F_{NN}	Longitudinal and transversal spectra of a random field
$\int_C F[x]d_{\mathbf{w}}x$	Wiener path integral
$I_\nu(z) = \mathrm{i}^{-\nu}J_\nu(\mathrm{i}z)$	Modified Bessel function
$J_\nu(z)$	Bessel function of a complex argument z
$K[\,]$	A matrix-integral operator generated by the kernel
$K[x,y]$	A matrix kernel of an integral operator
$MF, \langle F \rangle, E[F]$	Different notations for an expectation of F
M_π	Mathematical expectation over the distribution π
$\left[\dfrac{N}{M}\right]_{R_\lambda f}$	Padé approximation of the resolvent $R_\lambda f$
$n(r,\varepsilon)$	Mean value of numbers of steps in the ε-spherical process
$n(y)$	A normal vector (exterior or interior) to the boundary at $y \in \partial G$
$N_x^r(u)$	Spherical mean value of a function $u(x)$
$Q_i^{(k)}$	Random weights
R_λ	Resolvent of an integral equation
$S(x,r)$	A sphere of radius r centered at the point x
$\mathrm{Sp}\, F(\lambda)$	Trace of $F(\lambda)$
T_ε	Computational cost of an algorithm
$u_\delta(x,y)$	Green's function
$Z_0(x,t)$	Fundamental solution of the heat equation
$\dfrac{\alpha_1}{1+}\dfrac{\alpha_2\lambda}{1+}\cdots\dfrac{\alpha_i\lambda}{1+}$	Continued fraction
Γ_ε	The ε-boundary of a domain G defined as $\Gamma_\varepsilon = \{x \in G : d(x,\partial G) < \varepsilon\}$
Δ	The Laplace operator $\dfrac{\partial^2}{\partial x_1^2} + \dfrac{\partial^2}{\partial x_2^2} + \ldots + \dfrac{\partial^2}{\partial x_m^2}$
$\lambda = \psi(\eta)$	A conformal mapping of the spectral parameter

$\mu(y), \mu^*(y)$	Densities of the simple and double layer potentials, respectively
$\mu_0(y)$	Eigenfunction of the boundary integral operator (solution of the Robin problem)
$\pi(x)$	Initial distribution density of a Markov chain
$\varrho(K)$	Spectral radius of the operator K
$\{(\xi_k, \tau_k)\}_{k=0}^{\infty}$	Walk on boundary process in the nonstationary case
σ_m, S_m, ω_m	Surfaces of a unit sphere in $\mathbf{R}^m [= 2\pi^{m/2}/\Gamma(m/2)]$
$\omega(\varepsilon)$	Continuity modulus

Introduction

Monte Carlo methods have come to be widely used to solve complicated problems in radiative transfer theory, [0.1–0.4]. Systematic development of the ideas of statistical modeling in this field is primarily due to the efforts of G.I. Marchuk, V.S. Vladimirov, N. Chentsov, G.A. Mikhailov, I.M. Sobol, J.H. Curtiss, G.E. Albert, H. Kahn, M. Kalos and J.M. Hammersley.

In the last few years Monte Carlo methods have undergone rapid development, which can be found in the original papers. This book deals with one part of this development: construction and justification of random walk alogrithms for solving boundary value problems of mathematical physics.

Why have the Monte Carlo methods recently come back into favor? The first reason is that, nowadays, statistical description of various physical processes is quite common. The method of statistical simulation is an effective and straight forward research tool exploiting a probabilistic description. On the other hand there also exists a fundamental connection between partial differential equations and random processes. This connection was was recognized long ago, and the theory of differential equations was used in probability theory. For example, *Kolmogorov* [0.5] showed that a transition function $P(t, x, G)$, the probability that a Brownian particle starting from a point x hits a set G at a moment t, satisfies a parabolic differential equation. In turn, results of probability theory were later successfully applied to investigating boundary value problems. In this connection we should mention here the works of *Wiener* [0.6], *Petrovsky* [0.7] *Khinchin* [0.8], *Courant, Friedrichs, Lewy* [0.9], *Doob* [0.10], *Dynkin* [0.11], *Cameron* [0.12], *Kovalchik* [0.13], *Ventsel* [0.14].

The probabilistic representations of solutions to boundary value problems in the form of functional (path, or continual) integrals can be used to construct Monte Carlo algorithms. However, there often exists a possibility to derive different probabilistic representations based on Markov chains whose implementation is quite simple. In addition in Monte Carlo schemes it is possible to take into account various specific features of the problem under study. For example, in the simulation of admixture transfer it is not difficult to take into account coagulation of the particles, the specific size distribution in aerosols, random character of sources, etc. When simulating the capture of aerosols by large particles by solving an exterior Neumann problem for the Laplace equation it is possible to take into account the particle's diffusion. In elasticity problems random loads can be taken into account.

One of the most impressive features of the Monte Carlo methos is the possibility to calculate probabilistic characteristics, for example, the intensity of fluctuations, correlation functions, the excess probability, etc. However, the method has a serious drawback: it is not useful when high accuracy is required, since $T(\varepsilon)$, the cost of improving the accuracy ε, is, as a rule, quite large: $T_\varepsilon = 0(\varepsilon^{-2})$. On the other hand, the Monte Carlo method does have certain advantages and is therefore often used when all other methods fail, and the time required to obtain an accurate result to within a few percent on a computer is not too long. The main advantages of the Monte Carlo method are:

1) Algorithms do not require large memory even for multidimensional problems.
2) It is possible to calculate directly the desired functionals, e.g., solution at an arbitrary point, derivatives and integrals of the solution, without finding the whole solution field.
3) Practical adaptation of the algorithms to a computer program are simple enough, allowing one to solve boundary value problems with complicated boundaries.
4) No additional difficulties arise in solving problems with discontinuous bondary functions, non-smooth boundaries, or complicated right-hand sides of equations (sources).
5) Problems with random coefficients, random boundary values and other stochastic parameters can be solved.
6) Monte Carlo algorithms are easily parallelized, i.e., calculations can be performed along the trajectories independently of one another.
7) Finally, the error of the method is estimated simultaneously at moderate computational cost.

To streamline the discussion we shall classify Monte Carlo methods for solving boundary value problems into three groups:

Approach I includes schemes based on approximate calculation of path integrals (e.g., the Wiener intergrals) representing solutions to boundary value problems.

Approach II exploits the mean value relations or, more generally, the Green's functions for standard domains such as a ball, a sphere, an ellipsoid, a cone, a half-space, etc. Monte Carlo iterations of the mean value integrals lead to so-called walk inside the domain processes, in particular, the walk on spheres processes. The theorectical basis of the method lies in the derivation of a local integral equation which is equivalent to the original boundary value problem.

Approach III is based on Monte Carlo solution of boundary integral equations. More generally, one uses so-called global integral equations whose phase space consists of the boundary, or both of the boundary and the domain. Solution to the boundary value problem is then represented as a linear functional of the solution of the global integral equation. As an example, we mention the boundary integral equations of potential theory [0.15]. We could also mention an approach based on simulation of walk on a grid processes which is a simple and relatively universal procedure. However, it can also be interpreted as a discrete variant of approach I.

Implemetation of the walk on grid process is simple but the cost of the algorithm is high because of the necessity to simulate very long trajectories. This scheme dates from about 1934 (see the work by *Petrovsky* [0.7] and the well-known work by *Courant, Friedrichs, Lewy* [0.9]).

In approach I there exist two computational schemes. The first one is based on deterministic approximate formulas for path integrals which are exact for functionals of special type. Here we refer to works by *Cameron* [0.16], *Vladimirov* [0.17], and *Sabelfeld* et al [0.18]. A detailed description of such methods can be found in [0.19]. The second scheme is based on Monte Carlo simulation of a random process which approximates the random process generated by the corresponding measure in functional space. This approach was first suggested in [0.20] (see also [0.18]). It should be noted that sometimes (for example, when Laplace' s equation is considered) the probabilistic representation involves only the exit points, and it is not necessary to simulate the Wiener process inside the domain in detail. This method (so-called walk on spheres algorithm) was investigated by *Müller* [0.21] and *Motoo* [0.22]. Its generalizations to the heat equation, several second order elliptic equations, parabolic systems, hyperbolic equations, integral and intergro-differential equations have been reported [0.18, 23, 24].

Approach II was suggested in [0.18]. Interpretation through integral equations permits one to utilize the developed Monte Carlo technique. For example, utilization of the symmetry property of the Green' s function makes it possible to construct a walk on spheres algorithm for calculating the solution simultaneously at a set of prescribed points, which is impossible in the method based on a probabilistic representation in the form of the Wiener integral. There exist many works on mean value relations and Green' s formulas for differential equations, which will be metioned in Chaps. 3–5.

The first monograph on the subject [0.18] surveys studies carried out prior to 1979. In [0.23] a slightly different approach is described where the convergence results are based on utilization of martingales.

In connection with approach III it should be noted that the usual Monte Carlo technique is based on representing the solution as a convergent Neumann series. However, the Neumann series for the boundary integral equations of the potential theory diverges, so that the standard Monte Carlo estimates are not applicable. A first attempt to construct a Monte Carlo algorithm for solving an integral equation on the basis of analytical continuation of the divergent Neumann series was made in [0.25]. Later a series of approaches to Monte Carlo construction of the resolvent of integral equations of the second kind was suggested [0.4, 0.24]. These approaches and a few techniques of convergence acceleration for Neumann series are described in detail in Chap. 2 of this book.

The main advantages of the walk on boundary algorithms are:

1) It is possible to calculate the solution and its derivatives at an arbitrary set of points simultaneously, without finding the whole solution field.
2) The algorithms permit simultaneous solution of interior and exterior Dirichlet, Neumann and third boundary value problems using a common walk on boundary process.
3) Calcultions of derivatives is simpler than the scheme based on the walk on spheres method.
4) The algorithms are free of the error due to approximations near the boundary which appears in the walk on spheres algorithms.

On the other hand, the walk inside the domain algrithms, in particular, the walk on spheres algorithms, have important advantages: practical adaptations to a computer program are simple, allowing one to solve boundary value problems with complicated boundaries. This monograph deals with random walks inside the domain and the walk on boundary algorithms for solving multidimensional boundary value problems, and contains mainly methods of approaches II and III.

The developed methods are applied to three main classes of boundary value problems in deterministic and stochastic formulations:

1) problems of the potential theory (e.g., calculation of the aerosol capture coefficient for three-dimensional bodies, evaluation of motion of charged particles in electrostatic fields)
2) problems of elasticity theory (e.g., bending of thin plates under stochastic loads)
3) diffusion problems (e.g., calculation of probalitiy distributions of random concentration, heat and mass transfer in bulk condensation, homogeneous and heterogeneous nucleations, coagulation processes, etc).

It should be noted that all calculations mentioned in the book were carried out by the Soviet computer system BESM-6. We tabulate, as the need arises, $s = 2[D\xi/N]^{1/2}$, the statistical error of calculations, and N, the number of Monte Carlo samples.

1. General Schemes for Constructing Scalar and Vector Monte Carlo Algorithms for Solving Boundary Value Problems

In this introductory chapter we formulate general approaches and basic schemes for constructing random walk algorithms (the walk on boundary and the walk on spheres algorithms) on the basis of global and local integral equations and probability representations in the form of Wiener continual integrals. Basic random walks are defined and investigated. Simulation formulas for important classes of scalar and vector random processes and fields are constructed.

1.1 Random Walks on Boundary and Inside the Domain Algorithms

The walk inside the domain algorithms are based on the use of local integral equations. To solve these equations, standard Monte Carlo methods are used, provided the corresponding Neumann series converges. These methods are not applicable to boundary integral equations of the potential theory, since the corresponding Neumann series diverges. Therefore, we introduce in Sect. 1.1.1 ε-biased estimates based on analytical continuation of the resolvent with respect to the spectral parameter of the integral equation under study. This method is described in detail in Chap. 2 where we also present different approaches for solving integral equations by the Monte Carlo methods.

1.1.1 Monte Carlo-Algorithms

Consider an equation

$$\varphi(x) = \lambda \int_G k(x, x')\varphi(x')dx' + f(x) \tag{1.1}$$

in the operator form

$$\varphi = \lambda K\varphi + f$$

where K is defined on a Banach space $X(G)$ of integrable in G functions, G is a domain of Euclidian space $\mathbb{R}^n$.

Without loss of generality we assume that it is desired to obtain the solution of (1.1) at $\lambda = \lambda_* = 1$. Suppose that K is a compact operator. For example, let

$$k(x, x') = 0(|x - x'|^{-n+r}) \quad \text{as} \quad |x - x'| \to 0$$

where r is a positive constant. Then [1.1] $\lambda_1, \lambda_2, \ldots$, the set of characteristic numbers of (1.1), is at most countable and separated from zero. The solution of (1.1) is represented in the form of a series absolute convergent in

$$\Lambda_1 = \{\lambda : |\lambda| < |\lambda|_1\} :$$

$$\varphi(x) = \sum_{i=0}^{\infty} \lambda^i K^i f(x) = f(x) + \lambda R_\lambda f(x) \tag{1.2}$$

where $R_\lambda = K + \lambda K^2 + \ldots$ is the resolvent operator. If $|\lambda_1| > 1$, solution of (1.1) at $\lambda_* = 1$ is given by (1.2), a basis of standard Monte Carlo methods [1.2, 3].

Assume that $|\lambda_1| \leq 1$ and suppose that $\lambda = \lambda_* = 1$ is not a characteristic number of (1.1). In this case there exists a unique solution of (1.1) but the Neumann series diverges. However the function $R_\lambda f$ is analytic in Λ_1, and it is possible to construct an analytical continuation of $R_\lambda f$ on all the complex plane except at the points $\lambda_1, \lambda_2, \ldots$. In particular, it can be analytically continued to the point $\lambda = \lambda_1 = 1$. One of the methods of analytical continuation is based on the change of spectral parameter λ in (1.1) [1.4–6].

Consider a simple connected domain D in the complex plane such that $\{\lambda_k\}_{k=1}^\infty \notin D, 1 \in D, 0 \in D$ and choose a conformal mapping $\lambda = \psi(\eta), \eta(0) = 0$ which maps the unit circle $\Delta = \{\eta : |\eta| < 1\}$ to D. Then $\eta_* = \psi^{-1}(1) \in \Delta$ and all characteristic values go to the exterior of Δ. Substituting $\lambda = \psi(\eta)$ to $\lambda R_\lambda f$ and expanding the latter function in powers of η yields the following representation of $F(\eta) = \psi(\eta) R_{\psi(\eta)} f$:

$$F(\eta) = \sum_{i=1}^{\infty} b_i \eta^i .$$

Here

$$b_i = \sum_{k=1}^{i} d_k^{(i)} K^k f , \quad d_k^{(i)} = \frac{1}{1!} \left\{ \frac{\partial}{\partial \eta^i} [\psi(\eta)]^k \right\}_{\eta=0} .$$

Thus the solution of (1.1) can be represented in the form

$$\varphi(x) = f(x) + \sum_{i=1}^{\infty} \left(\sum_{k=1}^{i} d_k^{(i)} K^k f(x) \right) \eta_*^i . \tag{1.3}$$

It is possible to suggest various methods of Monte Carlo calculation of the series (1.3). In the first one, we take a finite partial sum in (1.3):

$$\varphi(x) = f(x) + \sum_{i=1}^{m} l_i^{(m)} K^k f + \varepsilon(m) , \tag{1.4}$$

6

where

$$l_i^{(m)} = \sum_{k=i}^{m} d_i^{(k)} \eta_*^k .$$

Note that the remainder $\varepsilon(m)$ can be easily estimated since the series (1.3) converges like a geometric progression.

The second approach is based on a transition to a new integral equation of the form

$$\varphi_1 = K_1 \varphi_1 + f_1$$

such that the Neumann series for it can be written as $f_2 + R_1 f_1$. Note that this method is generated by the transformation $\lambda = \alpha \eta / (1 - \beta \eta)$ if $f_2 = 0$. Here α, β are complex parameters. This function maps Δ on the domain

$$D = \left\{ z : \text{Re} \left[-\frac{\alpha}{\beta}(\bar{z} - \bar{z}_0) \right] < 0 \right\} , \qquad z_0 = -\frac{\alpha|\beta|}{\beta(1 + |\beta|)} , \qquad |\beta| = 1 .$$

Indeed, in this case

$$d_k^{(i)} = \alpha^k C_{i-1}^{k-1} \beta^{i-k} , \qquad \eta_* = (\alpha + \beta)^{-1} ,$$

and (1.3) takes the form

$$\varphi(x) = f(x) + \sum_{i=0}^{\infty} K_1^i f_1 , \qquad (1.5)$$

where

$$K_1 = \frac{\beta}{\alpha + \beta} I + \frac{\alpha}{\alpha + \beta} K , \qquad f_1 = \frac{\alpha}{\alpha + \beta} K f .$$

Consequently, the series (1.5) converges if the characteristic number of K lies outside of D.

Assume that it is desired to calculate the integral

$$I_h = (\varphi, h) = \int_G \varphi(x) h(x) dx , \qquad (1.6)$$

where $h \in X^*(G)$, and $\varphi(x)$ is the solution of (1.1). It is sufficient to calculate the quantities $(K^i f, h)$, $i = 1, \ldots$, since it is possible to integrate the series (1.3–5) termwise. Let ξ_i be an unbiased estimate of $(K^i f, h)$, that is, $M\xi_i = (K^i f, h)$. This estimate can be constructed as follows [1.2].

Let $\{x_k\}_{k=0}^{\infty}$ be a stationary Markov chain of points $x_k \in G$ such that the initial point x_0 is distributed in G according to a density $\pi(x)$, and the transition $x_{k-1} \to x_k$, $k \geq 1$ is defined by the density

$$p(x_{k-1}, x_k) = r(x_{k-1}, x_k)[1 - g(x_{k-1})] ,$$

where $r(x_{k-1}, x_k)$ is the conditional distribution density provided that x_{k-1} is fixed, and $g(x_{k-1})$ is the termination probability in the transition $x_{k-1} \rightarrow x_k$.

Introduce the notations

$$Q_0 = \frac{f(x_0)}{\pi(x_0)} , \quad Q_k = Q_{k-1} \frac{k(x_k, x_{k-1})}{p(x_{k-1}, x_k)} , \quad k \geq 1 , \tag{1.7}$$

$$Q_0^* = \frac{h(x_0)}{\pi(x_0)} , \quad Q_k^* = Q_{k-1}^* \frac{k(x_{k-1}, x_k)}{p(x_{k-1}, x_k)} , \quad k \geq 1 , \tag{1.8}$$

and let N be the random length of the Markov chain $\{x_k\}$. It is assumed that the densities π and p are properly chosen, i.e., $\pi(x) \neq 0$ if $f(x) \neq 0$, and $p(x, y) \neq 0$ if $k(y, x) \neq 0$ in (1.7), and $\pi(x) \neq 0$ if $h(x) \neq 0$, and $p(x, y) \neq 0$ if $k(x, y) \neq 0$ in (1.8). Under these conditions the random variables

$$\xi_k = Q_k h(x_k) , \quad \xi_k^* = Q_k^* f(x_k) \tag{1.9}$$

are unbiased estimates for $(K^k f, h) = (f, K^{*k} h)$. If it is additionally assumed that the Neumann series for the operator $\bar{K}$ with the kernel $|k|$ converges, then [1.2]

$$\xi = \sum_{k=0}^{N} Q_k h(x_k) , \quad \xi^* = \sum_{k=0}^{N} Q_k^* f(x_k) , \tag{1.10}$$

are unbiased estimates for I_h. Using (1.3) we obtain that

$$\xi = \sum_{k=0}^{m} l_k^{(m)} Q_k h(x_k) , \quad \xi^* = \sum_{k=0}^{m} l_k^{(m)} Q_k^* f(x_k) , \tag{1.11}$$

are δ-biased estimates for I_h, where $\delta = (\varepsilon(m), h)$, $l_0^{(m)} = 1$; $g(x_k) = 0$ if $k < m$, $g(x_m) = 1$. In the second type of estimates for I_h

$$\begin{aligned} \xi &= \sum_{k=0}^{N} Q_k h(x_k) + \frac{[f(x_0) - f_2(x_0)] h(x_0)}{\pi(x_0)} \\ \xi^* &= \sum_{k=0}^{N} Q_k^* f_1(x_k) + \frac{[f(x_0) - f_2(x_0)] h(x_0)}{\pi(x_0)} \end{aligned} \tag{1.12}$$

where Q_k, Q_k^* are determined from (1.7) and (1.8), respectively, by exchanging k with k_1, and f with f_1. In (1.12) we can also take $g(x_k) = 0$ for $k < N$, $g(x_N) = 1$ where N is a fixed integer number, so the estimates ξ and ξ^* are then δ-biased. If the Neumann series for $\bar{K}_1$ converges, then $g(x_k) > 0$, $k \leq N$, and the estimates (1.12) are unbiased; however, it may appear that their variances become infinite. The random quantities ξ and ξ^* are called direct and adjoint estimates, respectively.

In the next section we describe a general scheme of vector algorithms for solving high-order elliptic equations and systems. It is based on vector random estimates for solving systems of integral equations

$$\varphi_i(x) = \sum_{j=1}^{s} \int_G k_{ij}(x,y)\varphi_j(y)dy + f_i(x) , \quad i = 1, \ldots, s . \qquad (1.13)$$

We rewrite (1.13) in the matrix form

$$\Phi(x) = \int_G K[x,y]\Phi(y)dy + F(x) , \qquad (1.14)$$

or in short form

$$\Phi = K[\,]\Phi + F .$$

Here and henceforth, Φ is a column-vector with components φ_i, $K[x,y]$ is an $s \times s$ matrix with entries $\{k_{ij}\}_{i,j=1}^{s}$, F is a column-vector with components f_i. The integral of a vector is understood to be a vector whose components are the corresponding integrals; $K[\,]$ is a matrix-integral operator with the kernel $K[x,y]$.

Assume that it is desired to calculate a vector function

$$J(x) = \int P[x,y]\Phi(y)dy = (P, \Phi) \qquad (1.15)$$

at a fixed point x.

Let $\{y_i\}$ be a Markov chain with transition density $p(x,y)$ and initial density $p_0(x)$ such that $p_0(x) \neq 0$ if $P[x,y]\Phi(y) \not\equiv 0$, and $p(x,y) \neq 0$ if $K[x,y]\Phi(y) \not\equiv 0$. Then standard arguments show [1.2] that the funtionals

$$\xi = P[x, y_i]Q_i , \quad \xi^* = Q_i^* F(y_i) \qquad (1.16)$$

are unbiased estimates for $(P, K^i[\,]F)$. In (1.16)

$$Q_i = \frac{K[y_i, y_{i-1}]Q_{i-1}}{p(y_{i-1}, y_i)} , \quad Q_0 = \frac{F(y_0)}{p_0(y_0)} \qquad (1.17)$$

are the vector weights, and

$$Q^* = \frac{Q_{i-1}^* K[y_{i-1}, y_i]}{p(y_{i-1}, y_i)} , \quad Q_0^* = \frac{P[x, y_0]}{p_0(y_0)} \qquad (1.18)$$

are the matrix weights.

Standard matrix estimates for $J(x)$ are obtained on the basis of (1.16–18) by summing the estimates for $(P, K^i[\,]F)$ as in the scalar case described above.

1.1.2 Scalar and Vector Walk Inside the Domain Algorithms

The general scheme for constructing random walk algorithms includes following points:

1) Passing from the original boundary value problem to an integral equation of the form (1.1).

2) Analysis of existence and uniqueness of the solution of the integral equation.
3) Construction of unbiased or δ-biased estimates.
4) Investigations of the variances and the cost of the random walk algorithms.
5) Construction of simulation formulas.

It is clear that it is possible to pass from the original problem to various integral equations. We shall consider two main types of integral equations: the local and the global. In the case of local integral equations the kernel is a generalized function, and the solution to this equation is represented in the form of a convergent Neumann series. In the case of global integral equations we apply one of the methods of analytical continuation described in Sect. 1.1.1 and Chap. 2.

The scalar walk on spheres algorithm for a simple diffusion equation was first constructed on the basis of a local integral equation in [1.7, 8]. We describe now the general scheme of the walk on spheres algorithm for high-order equations and systems.

Let L be an elliptic partial differential operator in $\mathbf{R}^n$ of the order $2m$ with real coefficients

$$L(x, D) = \sum_{|\alpha| \le 2m} a_\alpha D^\alpha . \tag{1.19}$$

Suppose that m differential operators $B_j(x, D)$ are given and the order $(B_j) = m_j \le 2m - 1$. Consider the boundary value problem

$$L(x, D)u = 0 , \quad x \in G \subset \mathbf{R}^n \tag{1.20}$$

$$B_j(x, D)u = g_j , \quad j = 1, 2, \ldots, m , \quad x \in \partial G = \Gamma , \tag{1.21}$$

and the system of elliptic equations of second order

$$\sum_{j=1}^m L_{ij}(x, D)u_j = 0 , \quad i = 1, \ldots, m \tag{1.22}$$

with Dirichlet boundary conditions

$$R_k u = \sum_{j=1}^m b_{kj} u_j \Big|_\Gamma = g_k , \quad k = 1, \ldots, m . \tag{1.23}$$

It is assumed that the statement of the problem (1.20, 21) and (1.22, 23) is correct in the sense that there exist unique solutions to these problems; later we assume also that the solutions are sufficiently smooth in $\bar{G}$.

We suppose that Green formulas for a ball $B(x, r) \subset G$ with the boundary $\partial B(x, r) = S(x, r)$ can be obtained for the solutions of (1.20, 21) and (1.22, 23) (mean value relations). Information concerning existence, uniqueness and Green formulas can be found, for example, in [1.8–11]. The mean value relations have the form

$$u(x) = \int_{S(x,r)} Au(x + r\omega) ds(\omega)$$

where $u = (u_1, \ldots, u_m)^T$, the kernal $A(\omega)$ is an $m \times m$ matrix, and the integral in the right-hand side is unerstood to be a vector whose components are the integrals of the components of the vector $Au(x+r\omega)$. In general, the mean value relation can be considered at an arbitrary point y not coinciding with x, the center of the ball $B(x,r)$.

Let us consider a few simple examples.

1) Suppose that $u(x)$ is a scalar function harmonic in a domain G, then

$$u(x) = \int_{S(x,r)} u(x + r\omega)ds(\omega)$$

holds for an arbitrary sphere $S(x,r) \subset G$. Note that if a point $y \in B(x,r) \subset \mathbf{R}^n$ is not coincident with the center x then the mean value relation is given by the Poisson formula

$$u(y) = \int_{S(x,r)} \frac{r^2 - |x - y|^2}{|x - z|^n} u(z)dS(z) \ .$$

2) Let $u(x)$ be the solution of the diffusion equation

$$\Delta u(x) - cu(x) = -g \ , \quad x \in G \subset \mathbf{R}^3 \ ; \quad c \geq 0,$$

then

$$u(x) = \frac{rc^{1/2}}{\sinh{(rc^{1/2})}} \int_{S(x,r)} u(x + r\omega)ds(\omega) \ .$$

3) Let $u(x)$ be a biharmonic function:

$$\Delta^2 u(x) = 0 \ , \quad x \in \mathbf{R}^m \ .$$

Denote v by $(u(x), \Delta u(x))^T$. Then

$$v_1(x) = \int v_1(x + r\omega)ds(\omega) - \frac{r^2}{2m} \int_{S(x,r)} v_2(x + r\omega)ds(\omega)$$

$$v_2(x) = \int_{S(x,r)} v_2(x + r\omega)ds(\omega) \ ,$$

i.e., we have here

$$A = \begin{pmatrix} 1 & \dfrac{-r^2}{2m} \\ 0 & 1 \end{pmatrix}$$

4) Let $u(x) = (u_1, \ldots, u_m)^T$ be a function satisfying the Lamé equation

$$\mu \Delta u(x) + (\lambda + \mu)\text{grad div } u(x) = 0 \ , \quad x \in \mathbf{R}^m \ .$$

Then

$$u_i(x) = \frac{m}{2(m+\alpha)\omega_m} \int_{S(0,1)} \left[(2-\alpha)\delta_{ij} + (m+2)\alpha w_i w_j\right] u_j ds(\omega),$$

$$i = 1,\ldots,m.$$

Here ω_m is the area of the sphere $S(0,1)$ in $\mathbf{R}^m$, $w_i = (x_i - y_i)/r$ is the direction cosine, δ_{ij} is the Kronecker delta-function, $\alpha = 1 + \lambda/\mu$.

Note that above we considered only homogeneous equations. If an inhomogeneous equation is considered, e.g., the equation $\Delta u(x) - cu(x) = -g$, then the mean value relation takes the form

$$u(x) = \int_{B(x,r)} G(x,y)g(y)dy + \int_{S(x,r)} -\frac{\partial G}{\partial n}(x,y)u(y)dS(y)$$

where $G(x,y)$ is the Green function for the ball $B(x,r)$. For example, if the function $u(x)$ satisfies the equation

$$\Delta^2 u(x) = g(x) \quad (x \in \mathbf{R}^2),$$

then the mean value relation takes the form

$$u(x) = \int_{S(x,y)} u(x+r\omega)ds(\omega) - \frac{r^2}{4}\int_{S(x,y)} \Delta u(x+r\omega)ds(\omega)$$

$$+ \int_{B(x,y)} G_1(x,y)g(y)d_B(y)$$

$$\Delta u(x) = \int_{S(x,r)} \Delta u(x+r\omega)ds(\omega) - \int_{B(x,r)} G_2(x,y)g(y)d_B(y),$$

where $G_1(x,y) = [r^2 - r_1^2(\ln(r/r_1)+1)]/4$, $G_2(x,y) = \ln(r/r_1)$, $y = x + r_1\omega$ and $d_B(y)$ shows that the integration is carried out over the disk $B(x,r)$, i.e.,

$$\int_{B(x,r)} G_1(x,y)g(y)d_B(y)$$

$$\equiv \int_0^{2\pi} \frac{d\varphi}{2\pi} \int_0^r \left[\ln(r/r_1)r_1 g(x + r_1 \cos\varphi, x + r_1 \sin\varphi)\right] dr_1.$$

If $u(x)$ satisfies the equation

$$\Delta u(x) - cu(x) = -g \quad (x \in \mathbf{R}^3)$$

then

$$u(x) = \int_{S(x,r)} \frac{rc^{1/2}}{\sinh(rc^{1/2})} u(y)ds(y)$$

$$+ \int_{B(x,r)} \frac{\sinh[(r-|x-y|)c^{1/2}]g(y)}{4\pi|x-y|\sinh(rc^{1/2})} dy.$$

Having this kind of mean value relation we can pass from the original differential problem to an integral equation of the second kind with a generalized kernel

$$u(x) = \int_G k(x,y)u(y)dy + f(x).$$

We demonstrate the procedure of passing from the mean value relation for the diffusion equation to the integral equation. To this end, let us introduce a set Γ_ε, an ε-boundary of the domain G where the problem is solved:

$$\Gamma_\varepsilon = \{x \in G : \ d(x) < \varepsilon\} \ .$$

Here $d(x)$ is the distance from the point x to the boundary (Fig. 1.1). Now, define the kernel as follows:

$$k(x,y) = \begin{cases} \dfrac{d(x)c^{1/2}}{\sinh[d(x)c^{1/2}]}\,\delta_x(y) & \text{if } x \notin \Gamma_\varepsilon \\ 0 & \text{if } x \in \Gamma_\varepsilon \end{cases}$$

where $\delta(y)$ is a generalized density defined on $S(x, d(x))$. Using the Dirac δ-function and the Heaviside θ-function we can rewrite $k(x,y)$ as follows

$$k(x,y) = \frac{d(x)c^{1/2}}{\sinh[d(x)c^{1/2}]}\,\delta(|x - y| - d(x))\theta(d(x) - \varepsilon) \ .$$

Let

$$f(x) = \begin{cases} \dfrac{1}{4\pi}\displaystyle\int_{B(x,d(x))} \dfrac{\sinh[d(x) - |x - y|)c^{1/2}]g(y)dy}{|x - y|\sinh(dc^{1/2})} & \text{if } x \notin \Gamma_\varepsilon \\ u(x) & \text{if } x \in \Gamma_\varepsilon \end{cases} \ .$$

Then it is clear that any solution of the diffusion equation satisfies the integral equation

$$u(x) = \int_G k(x,y)u(y)dy + f(x) \ .$$

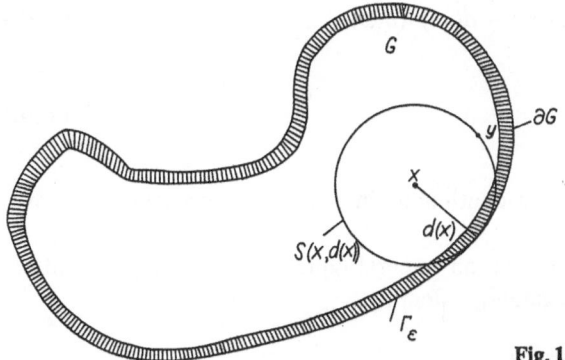

Fig. 1.1. Definition of Γ_ε

13

To be sure that the unique solution of the Dirichlet problem for the diffusion equation is coincident with the unique solution of this integral equation, it is sufficient to prove that the Neumann series converges. We denote by K the integral operator with the kernel $k(x, y)$ and show that $\|K^2\|_{L_1} < \infty$. Let $d_* = \sup_x d(x)$, where $d(x)$ are the radii of spheres lying inside G, and let $\nu(\varepsilon) = \varepsilon^2/4d_*^2$. Since $dc^{1/2} \sinh(dc^{1/2}) < 1$ for $c \geq 0$, we have

$$\int_G \int_G k(x, y)k(y, y')dy dy' \leq \int_{G \backslash \Gamma_\varepsilon} \delta_x(y) \left(\int_G \delta_y(y')dy' \right) dy$$

$$= \int_{G \backslash \Gamma_\varepsilon} \delta_x(y)dy \leq 1 - \nu(\varepsilon) < 1 .$$

Thus, $\|K^2\|_{L_1} \leq 1 - \varepsilon < 1$, and we conclude that the unique solutions of the Dirichlet problem for the diffusion equation and the derived integral equation coincide.

Note that the right-hand side of the integral equation is known only for $x \notin \Gamma_\varepsilon$. For small ε and continuous in Γ_ε solutions $u(x)$ we can approximately take in Γ_ε: $f(x) \cong \varphi(x_*(x))$ where $\varphi = u|_\Gamma$, and $x_*(x)$ is the point of Γ nearest to x.

Thus the integral equation $u = Ku + f$ is well determined, and $\|K^2\|_{L_1} < 1$, so that the Monte Carlo method can be adopted to solve it. The corresponding walk on spheres algorithms will be described for the diffusion equation and other equations of potential theory in Chap. 3. The vector walk on spheres algorithms will be considered in Chap. 4.

Now we write the local integral equation in the general case. We denote $v_j = B_j(x, D)u$, $j = 1, \ldots, m$ where u is the solution of the problem (1.20, 21), and $v_j = R_j u$ $j = 1, \ldots, m$ for (1.22, 23). Suppose that a mean value relation for v exists with a matrix kernel $A = \{a_{ij}\}_{i,j=1}^m$. Then we can write

$$v_i(x) = \sum_{j=1}^m \int_G k_{ij}(x, x')v_j(x')dx' + v_i^{(0)}(x) \tag{1.24}$$

where

$$k_{ij}(x, x') = a_{ij}(x, x')\delta(|x - x'| - d(x))\theta(d(x) - \varepsilon) , \tag{1.25}$$

$$v_i^{(0)} = \begin{cases} v_i(x) , & x \in \Gamma_\varepsilon \\ 0 & x \notin \Gamma_\varepsilon \end{cases} \tag{1.26}$$

$$i, j = 1, \ldots, m .$$

It is assumed that the vector v is continuous in Γ_ε, so we can take, as above, $v_i \cong g_i(x_*(x))$.

Note that it is possible to derive an integral equation which slightly differs from (1.24). Indeed, using the obvious identity

$$1 \equiv \theta(d(x') - \varepsilon) + \theta(\varepsilon - d(x')) ,$$

we can write

$$k_{ij}(x, x') = a_{ij}(x, x')\delta(|x - x'| - d(x))\theta(d(x') - \varepsilon) ,$$

and

$$v_i^{(0)}(x) = \sum_{j=1}^{m} \int_G a_{ij}(x, x')\delta(|x - x'| - d(x))\theta(\varepsilon - d(x'))v_j(x')dx' .$$

Sometimes, this form is preferable. For example, in the case of the diffusion equation it can be proved that $\|K\|_{L_1} < 1$. In both cases, we call integral equations of this type local integral equations. It should be noted that in the above scheme it is not necessary to take only the functions $v_j = B_j(x, D)u$, $j = 1, \ldots, m$. It is possible, if necessary, to also introduce additional components $v_{m+1}, \ldots, v_{m_1}$ such that a corresponding system of type (1.24) can be derived. The important point is only the question of equivalence of the original boundary value problem and the local integral equation.

1.1.3 Walk on Boundary Algorithms

In the walk on boundary algorithms a solution of a boundary value problem is sought in the form of an integral functional

$$u(x) = \int_\Gamma p(x, y)\Phi(y)d\sigma(y) \tag{1.27}$$

where $p(x, y)$ is a given function, and $\Phi(y)$ satisfies a boundary integral equation

$$\Phi(y) = \int_\Gamma k(y', y)\Phi(y')d\sigma(y') + F(y) , \tag{1.28}$$

where $d\sigma$ is an elemental surface on Γ. For example, in the case of the interior Dirichlet problem

$$\Delta u(x) = 0 , \quad x \in G ,$$
$$u|_\Gamma = \psi ,$$

it is known that (1.27) is the double-layer potential where

$$p(x, y) = \frac{\partial 1}{\partial n(y)|x - y|} , \quad k(y, y') = -\frac{\cos \varphi_{y'y}}{2\pi|y - y'|^2} ,$$

where $\varphi_{yy'}$ is the angle between $y - y'$ and the interior normal vector $n(y')$ to the boundary at $y' \in \Gamma$, $F(y) = \psi(y)/2\pi$. The relations may take vectorial form. For example, in the case of the interior Dirichlet problem for the Lamé equation

$$\mu\Delta u + (\lambda + \mu)\text{grad div } u = 0 , \quad u|_\Gamma = \psi$$

15

the functions $k(y, y')$ and $p(x, y)$ are the matrices $K[y', y]$ and $P[x, y]$, respectively. In nonstationary problems of diffusion and heat transfer one uses heat potentials of the form

$$u(x, t) = \int_0^t d\tau \int_\Gamma Z_0(x - y, t - \tau) \Phi(y, \tau) d\sigma_y$$

where Z_0 is a known function, and the function $\Phi(y, \tau)$ satisfies the Volterra integral equation

$$\Phi(y, \tau) = \int_0^\tau d\tau' \int_\Gamma k(y', y) \Phi(y', \tau') d\sigma_{y'} + \psi(y, \tau) . \qquad (1.29)$$

In the last case the Neumann series converges, because the spectral radius of the integral operator for Volterra equations is equal to zero. However, solutions of the stationary boundary integral equations cannot be represented in the form of a Neumann series. In this case we can use the methods described in Chap. 2, because spectral properties of these integral equations are well known. The walk on boundary algorithms are constructed for interior and exterior Dirichlet, Neumann and third boundary value problems of the potential theory in Chap. 3. The spatial problems of elasticity theory are treated in Sect. 4.3.1. Nonstationary walk on boundary algorithms for diffusion and heat equations are presented in Chap. 5.

1.1.4 Probabilistic Representations in the Form of Continual Integrals

The representation of solutions of boundary value problems in the form of a mathematical expectation is

$$u(x) = \langle F[w_x] \rangle \qquad (1.30)$$

where w_x is a random process (random sequence, Markov chain, random field, etc.), F is a functional defined on w_x, the brackets $\langle \ldots \rangle$ mean averaging over all the trajectories of w_x. The simple Monte Carlo scheme of calculation of the expectation (crude Monte Carlo) is given by the formula

$$\langle F[w_x^{(k)}] \rangle \approx \frac{1}{N} \sum_{k=1}^N F[w_x^{(k)}] ,$$

where $w_x^{(k)}$ are independent realizations of w_x, $k = 1, \ldots, N$. The probabilistic error is proportional to $[DF[w_x]/N]^{1/2}$, if the variance $DF[w_x]$ is finite. In practice, the variance is evaluated simultaneously with the expectation

$$DF[w_x] \approx \frac{1}{N} \sum_{k=1}^N F^2[w_x^{(k)}] - \left(\frac{1}{N} \sum_{k=1}^N F[w_x^{(k)}] \right)^2 .$$

It follows that the calculation of the error does not require large additional cost. In the walk on spheres and in the walk on boundary algorithms $w_x^{(k)}$ are Markov chains which "solve" corresponding local and global integral equations, respectively.

Sometimes it is possible to use direct probabilistic representation in the form of expectation of a functional defined on a continuous random process. Important functional integrals are the Wiener continual integrals [1.8, 12, 15]

$$\langle F[x(\tau)]\rangle = \int_C F[x]d_w x \ .$$

Here $d_w x$ denotes the Wiener measure, C is the space of continuous functions such that $x(0) = 0$. We now give some examples of probabilistic representations.

1) Let G be a domain with a regular (in the Wiener sense) boundary, and let $x(\tau)$ be a canonical diffusion process with the generating differential operator L. Denote by τ the first passage time: $x(\tau) \in \partial G$, $x(0) = x_0$. Let φ be a continuous function on ∂G and assume that $V \geq 0$ and g are Hölder continuous functions. Then the unique solution of the boundary value problem

$$Lu(x) - V(x)u(x) = -g(x) \ , \quad x \in G$$
$$\lim_{x \to a} u(x) = \varphi(a) \ , \quad a \in \partial G$$

at a point $x_0 \in G$ is given by the expectation

$$u(x_0) = \left\langle \int_0^\tau \exp\left\{ - \int_0^t V(x(s))ds \right\} g(x(t))dt \right\rangle$$
$$+ \left\langle \exp\left\{ - \int_0^\tau V(x(s))ds \right\} \varphi(x(\tau)) \right\rangle \ .$$

2) The solution of the Cauchy problem

$$\frac{\partial u_i}{\partial t} = \Delta u_i(x, t) + (P[x, t]u(x, t))_i u(x, 0) = \varphi(x) \ ,$$

where $x = (x_1, \ldots, x_m)$, $i = 1, \ldots, n$, $u = (u_1, \ldots, u_n)^{\mathrm{T}}$, $P[x, t] = \{p_{ij}(x, t)\}_{i,j=1}^n$, is represented in the form of an m-dimensional Wiener integral

$$u(x, t) = \int_{C^m} \exp\left\{ t \int_0^1 P\left[2\sqrt{t}y_1(\tau) + x_1, \ldots, 2\sqrt{t}y_m(\tau) + x_m, t(1 - \tau) \right] d\tau \right\}$$
$$\times \varphi\left(2\sqrt{t}y_1(1) + x_1, \ldots, 2\sqrt{t}y_m(1) + x_m \right) d_w y_1 \ldots d_w y_m \ .$$

Here exp is the exponential. It is assumed that $\{p_{ij}\}_{i,j=1}^n$ are continuous and bounded functions in the domain $\mathbf{R}^m \times (0, \infty)$ and $\varphi_i(x)$, $i = 1, \ldots, n$ are continuous and bounded functions in $\mathbf{R}^m$.

3) In the constructions of finite velocity models for diffusion problems, the following probabilistic representation of the solution of the Cauchy problem for the telegraph equation can be used [1.16]:

$$u(x,t) = \left\langle v\left(\int_0^1 (-1)^{\xi_s}\,ds\right)\right\rangle_{\xi_t} ,$$

where u solves the equation

$$u_{tt} = \Delta u - \alpha u_t , \quad \alpha > 0 , \quad u(0) = f , \quad u_t(0) = g .$$

In this representation ξ_t is the Poisson random process with the parameter $\alpha/2$; $\xi_0 = 0$, $m(t) = \alpha t/2$, and $v(\tau)$ is the solution of the Cauchy problem

$$\frac{\partial^2 v}{\partial t^2} = \Delta v , \quad v|_{t=0} = f , \quad \left.\frac{\partial v}{\partial t}\right|_{t=0} = g ;$$

hence it can be represented as

$$v(t) = \frac{t}{4\pi}\int_{S(x,t)} g(\xi)d\Omega_\xi + \frac{\partial}{4\pi\partial t}\left[t\int_{S(x,t)} f(\xi)d\Omega_\xi\right] .$$

Here $S(x,t)$ is a sphere of radius t centered at the point x, $d\Omega_\xi = \sin\theta d\theta d\varphi$.

Probabilistic representation is a convenient tool in studying asymptotics of solutions, in analytical evaluations and in constructing numerical methods. However, often it is more efficient to construct another probabilistic representation, based on Markov chains. Indeed, let us consider the following simple example of the boundary value problem

$$\Delta u(x) = 0 , \quad x \in G , \quad u|_\Gamma = \varphi .$$

Then the probabilistic representation takes the very simple form

$$u(x_0) = \langle \varphi\left(w_{x_0}(\tau)\right)\rangle .$$

It is clear from this representation that it is not necessary to construct detailed Wiener trajectories inside the domain G. Indeed, it is sufficient to simulate the random first passage points $w_{x_0}^{(i)}(\tau) \in \partial G$. This can be done by simulating the Markov chain called the walk on spheres process. Numerical realization of this process is very simple, and $n(\varepsilon)$, the average number of steps until the trajectory hits the Γ_ε-boundary is on order of $O(|\ln(\varepsilon)|)$ as $\varepsilon \to 0$. Note that $n(\varepsilon)$ for the walk on a grid with a mesh size ε is given by the asymptotics $n(\varepsilon) \approx O(\varepsilon^{-2})$ as $\varepsilon \to 0$.

1.2 Random Walks and Approximations of Random Processes

1.2.1 Walk Inside the Domain Processes

Let $G \subset \mathbf{R}^n$ be a domain with the boundary ∂G. Define a Markov chain $\{\xi_i\}_{i=0}^\infty$ called a walk on subdomains $\{G_i\} \subset G$ inside G as follows.

Let $r_0(x)$ be the density of a starting point ξ_0. When ξ_{i-1} is fixed, the next point ξ_i is sampled according to a transition density $p_i(x,y)$ on the surface ∂G_i; the termination probability in the transition $x_{i-1} \to x_i$ is denoted by $g(x_{i-1})$. The most important walk inside the domain is the walk on spheres process.

We define the walk on spheres process $\{\xi_n\}_{n=0}^{\infty}$ starting at a fixed point $x_0 \in G$ as follows

$$\xi_n = \xi_{n-1} + \omega_n d(\xi_{n-1}) , \quad \xi_0 = x_0 , \quad n = 1, 2, \ldots \quad (1.31)$$

where ω_n is a sequence of random independent unit isotropic vectors, $r_0(x) = \delta(x - x_0)$, $d(x)$ is the distance from a point x to the boundary ∂G. It is known [1.17] that this process converges to the boundary, i.e., $\xi_\infty \in \partial G$. If we introduce a termination probability $g(x) = 0$ for $x \notin \Gamma_\epsilon$ and $g(x) = 1$ if $x \in \Gamma_\epsilon$ then, obviously, the number of steps N will be finite, with probability equal to one. The terminating walk on spheres process is also called an ε-spherical process [1.8]. In this case $\{\partial G_i\}$ is a set of spheres $S(\xi_k, d(\xi_k)) \subset G$.

There exist also walk on ellipsoids [1.18], on squares [1.19], cones, half-spaces [1.20] and other processes. The transition density $p(x,y)$ can be constructed if one is able to obtain an explicit form of the Green function for the subdomain under study. It should be noted that it is possible to construct the transition $\xi_{i-1} \to \xi_i$ numerically, using inside the subdomain G_i the standard walk on spheres process. This approach will be used in Chap. 3 to construct Schwartz's alternating iterative methods.

Now we introduce the following variants of the walk on spheres processes:

I) $\quad \xi_n = \xi_{n-1} + rd(\xi_{n-1})\omega_n , \quad n = 1, 2, \ldots; \quad r \in (0, 1] \quad (1.32)$

II) $\quad \xi_n = \xi_{n-1} + \min\{r, d(\xi_{n-1})\}\omega_n , \quad n = 1, 2, \ldots \quad (1.33)$

which are also terminating after they hit Γ_ϵ. In process I, r is a fixed real parameter which could be changed from step to step. Process II permits one to organize a walk on spheres process with a fixed radius $r : \varepsilon < r < r_*$ where r_* is the radius of the maximum sphere which can be inscribed in the domain G. Near the boundary it behaves like the standard walk on spheres process. It is clear that at $r = 1$ in (1.32) and for $r > r_*$ in (1.33) we have the standard walk on spheres process.

The standard walk on spheres algorithm was studied in [1.8, 17, 21] where the logarithmic behavior of $n(\varepsilon)$, the average number of steps of the ε-spherical process inside the domain $G \backslash \Gamma_\epsilon$ was described under various assumptions about the boundary structure. From these results, however, it is not possible to obtain the dependence of $n(\varepsilon)$ on the parameter r of the processes I and II. Therefore, we derive here $n(\varepsilon, r)$ as a function of ε and r. In the limiting cases, $n(\varepsilon, r)$ coincides with the known expressions for standard ε-spherical process.

Lemma. Let $\{\xi_i(r)\}_{i=1}^{\infty}$ be a set of sequences of independent equally distributed random variables with a finite expectation (r belongs to a set of parameters X).

Let

$$v(t,r) = \min \left\{ k : \sum_{i=1}^{k} \xi_i(s) \geq t \right\} , \quad n(t,r) = Mv(t,r) .$$

Suppose that

1) $\inf_{r \in X} M\xi_i(r) = a > 0$,

2) $\lim_{N \to \infty} \sup_{r \in X} M\{\xi_i(r); \xi_i(r) \geq N\} = 0$.

Then

$$n(t,r) \geq t/M\xi_1(r) , \tag{1.34}$$
$$n(t,r) \leq t/M\xi_1(r) + \varepsilon(t) , \tag{1.35}$$

where $\lim_{t \to \infty} \frac{\varepsilon(t)}{t} = 0$.

Proof. To prove the *Lemma*, we use the method presented in [1.22]. Let

$$\xi_i(r, N) = \min(\xi_i(r), N); \quad \delta(n) = \sup_{r \in X} M\{\xi_i(r); \xi_i(r) \geq N\} ;$$

$$S_k(r) = \sum_{i=1}^{k} \xi_i(r), \quad S_0 = 0, v(t, r, N) = \min\{v(t, r), N\} ;$$

$$S_k(r, N) = \sum_{i=1}^{k} \xi_i(r, N); \quad n(t, r, N) = Mv(t, r, N) .$$

Here, r is a parameter, t is the renewal level, and N is the cut level.

By definition, $\{\xi_i(r, N)\}_{i=1}^{\infty}$ is a sequence of independent identically distributed random variables for each fixed r and N. Now we show that

$$M\xi_1(r, N) \geq M\xi_1(r) - \delta(N) . \tag{1.36}$$

Indeed

$$M\xi_1(r, N) \geq M\{\xi_1(r, N); \xi_1(r, N) < N\} = M\xi_1(r) - M\{\xi_1(r, N);$$
$$\xi_1(r, N) \geq N\} \geq M\xi_1(r) - \sup_{r \in X} M\{\xi_1(r, N); \xi_1(r, N) \geq N\}$$

$$\geq M\xi_1(r) - \delta(N) .$$

Suppose that N_0 is chosen so that $\delta(N_0) \leq a/2$. Then by Theorem 8 of [1.22] and by the Wild identity we obtain for $N_0 \geq N$

$$n(t, r, N) = Mv(t, r, N) = \frac{M S_v(r, t, N)}{M\xi_1(r, N)} < \infty ; \tag{1.37}$$

$$n(t, r) = Mv(t, r) = \frac{M S_{v(t,r)}}{M\xi_1(r)} < \infty . \tag{1.38}$$

From (1.38) and by the definition of $v(t, r)$ we get (1.34). The definition of $v(t, r, N)$ implies that

$$S_k(r, N) = S_{k-1}(r, N) + \xi_k(r, N) \leq t + N \qquad (1.39)$$

holds for the event $v(t, r, N) = k$. Substituting (1.39) into (1.37) we obtain

$$n(t, r, N) \leq \frac{t + N}{M\xi_1(r) - \delta(N)} . \qquad (1.40)$$

Now we will prove that

$$v(t, r, N) \geq v(t, r) . \qquad (1.41)$$

Indeed

$$t \leq S_k(r, N) = \sum_{i=1}^{k} \xi_i(r, N) = \sum_{i=1}^{k} \min(\xi_1(r), N) \leq \sum_{i=1}^{k} \xi_i(r) ,$$

is true for the event $v(t, r, N) = k$. Hence,

$$v(t, r) = \min \left\{ m : \sum_{i=1}^{m} \xi_i(r) \geq t \right\} \leq k .$$

Therefore,

$$n(t, r) = Mv(t, r) \leq Mv(t, r, N) = n(t, r, N) .$$

Using (1.40) we now obtain

$$n(t, r) \leq \frac{t + N}{M\xi_1(r) - \delta(N)} . \qquad (1.42)$$

Taking into account hypothesis 1 of the *Lemma* and the inequality (1.42), we get

$$n(t, r) \leq t[M\xi_1(r)]^{-1} + (t + N)[M\xi_1(r) - \delta(N)]^{-1} - t[M\xi_1(r)]^{-1}$$

$$= t[M\xi_1(r)]^{-1} + [NM\xi_1(r) + t\delta(N)][M\xi_1(r)(M\xi_1(r) - \delta(N))]^{-1}$$

$$\leq \frac{t}{M\xi_1(r)} + \frac{N + t\delta(N)/a}{a - \delta(N)} .$$

Therefore,

$$n(t, r) \leq \frac{t}{M\xi_1(r)} + \frac{N + t\delta(N)/a}{a - \delta(N)} .$$

Let $N(t) = t^{1/2}$, and

$$\varepsilon(t) = \frac{N + t\delta(N(t))/a}{a - \delta(N)} .$$

For $t \geq N_0^2$ we find

$$N(t) \geq N_0; \quad \delta(N(t)) \leq \delta(N_0) \leq \frac{a}{2} .$$

Hence,

$$0 \leq \frac{\varepsilon(t)}{t} \leq \frac{2t^{-1/2}}{a} + \frac{2\delta(t^{1/2})}{a^2} \to 0 \quad \text{as} \quad t \to \infty .$$

This completes the proof of (1.35). □

Remark. Hypothesis 2 of the *Lemma* is necessary (otherwise the statement is not true). We construct now an example which shows that hypothesis 1 is not sufficient to provide (1.35). In particular, it is not possible to obtain (1.35) from the known renewal theorem (for one sequence of independent identically distributed random variables).

Let

$$P(\xi(r) = r) = \frac{1}{r} ; \quad P(\xi(r) = 0) = 1 - \frac{1}{r} ; \quad r \in [1, \infty) \equiv X .$$

In this case $M \xi_1(r) = 1$ and

$$\lim_{N \to \infty} \sup_{r \in X} M \{\xi_1(r); \xi_1(r) \geq N\} = 1 ; \quad \sup_{r \in X} n(t, r) = \infty , \tag{1.43}$$

i.e., (1.35) is not true although hypothesis 1 is satisfied (but hypothesis 2 is not true). Now we prove (1.43). In the case of non-negative $\xi_i (i = 1, 2, \ldots)$, the following relation holds (Chap. 8 of [1.22]):

$$n(t, r) = 1 + \sum_{k=1}^{\infty} P \left(\sum_{i=1}^{\infty} \xi_i(r) < t \right) .$$

In our case

$$P \left(\sum_{i=1}^{\infty} \xi_1(r) < t \right) \geq P(\xi_1(r) = 0, \ldots, \xi_k(r) = 0) = P^k(\xi_1(r) = 0) .$$

Hence

$$n(t, r) \geq 1 + \sum_{k=1}^{\infty} P^k(\xi_1(r) = 0) = \frac{1}{1 - P(\xi_1(r) = 0)} = r ,$$

therefore $\sup_X n(t, r) = \infty$, $X = [1, \infty)$.

The average number of steps of the ε-spherical process I inside the domain $G \backslash \Gamma_\varepsilon$ can now be obtained on the basis of inequality (1.35).

Theorem 1.1. Let $\{P_n^r\}$ be the ε-spherical process of type I in $G = \{(x, y, z) : z \geq 0\}$, i.e.,

$$P_n^r = P_{n-1}^r + rd\left(P_{n-1}^r\right)\omega_n \; ; \quad r \in [\delta, 1], \delta > 0 \; ; \quad d\left(P_n^r\right) = z_n^r \, ,$$
$$n = 0, 1, \dots, N \, . \tag{1.44}$$

Then for average number of steps of the process $\{P_n^r\}$ inside the domain $G\backslash \Gamma_\varepsilon$ the following representation holds

$$n(r, \varepsilon) = \frac{c}{g(r)} + o(c) \quad \text{as} \quad \varepsilon \to 0 \, . \tag{1.45}$$

Here $c = \ln[d(P_0)/\varepsilon]$, and

$$g(r) = \frac{1}{2r}\{2r + (1 - r)\ln(1 - r) - (1 + r)\ln(1 + r)\} \, .$$

Proof. Rewrite (1.44) for each of the coordinates

$$x_n^r = x_{n-1}^r + r\omega_n^1 z_{n-1}^r \, ,$$
$$y_n^r = y_{n-1}^r + r\omega_n^2 z_{n-1}^r \, ,$$
$$z_n^r = z_{n-1}^r + r\omega_n^3 z_{n-1}^r \, ,$$

where $\omega_n^3 = \zeta_n$ is a random variable uniformly distributed on $(0, 1)$. Now,

$$d(P_n^r) = z_n^r = z_{n-1}^r(1 + r\zeta_n) = d(P_{n-1}^r)(1 + r\zeta_n) \, .$$

Consequently,

$$d(P_n^r) = \prod_{k=1}^{n}(1 + r\zeta_k)d(P_0) \, , \tag{1.46}$$

where $\{\zeta_k\}_{k=1}^{n}$ is a sequence of independent random variables uniformly distributed on $(-1, 1)$. Therefore,

$$N(r) = \min\left(n : d(P_n^r) \le \varepsilon\right) = \min\left(n : \ln\left[d\left(P_n^r\right)\right] \le \ln(\varepsilon)\right)$$
$$= \min\left(n : \ln\left[d(P_0)\right] + \sum_{k=1}^{n}\ln\left[1 + r\zeta_k\right] \le \ln(\varepsilon)\right) \, .$$

We put $\xi_k(r) = -\ln[1 + r\zeta_k]$ and $t = \ln[d(P_0)/\varepsilon]$. Then

$$N(r) = \min\left\{n : \sum_{k=1}^{n}\xi_k(r) \ge t\right\} \, .$$

Let us now show that the sequence $\{\xi_k(r)\}_{k=1}^{\infty}$, $r \in (0, 1]$ satisfies the hypothesis of the *Lemma*. First, we verify hypothesis 1:

$$M\xi_1(r) =) - M\ln(1 + r\zeta_1) = -\frac{1}{2}\int_{-1}^{1}\ln(1 + rx)dx$$

$$= -\frac{1}{2r} \int_{1-r}^{1+r} \ln(z)dz = -\frac{1}{2r}[z\ln(z) - z]\Big|_{1-r}^{1+r}$$

$$= \frac{1}{2r}\{2r + (1-r)\ln(1-r) - (1+r)\ln(1+r)\} \equiv g(r) .$$

Now we analyze the behavior of $g(r)$ as a function of r for $r \in [\delta, 1]$, $\delta > 0$. We have

$$\frac{\partial g}{\partial r} = -\frac{1}{2r^2}[2r - \ln(1+r) + \ln(1-r)] \geq 0 ; \quad g(0) = \lim_{r \to 0} g(r) = 0 .$$

Hence, $g(r)$ increases as r increases and reaches the maximum $\ln(e/2)$ at $r = 1$. Consequently,

$$M\xi_1(r) = g(r) \geq g(\delta) > 0 , \quad r \in [\delta, 1] , \quad \delta > 0 ,$$

and hypothesis 1 is satisfied. Let us now verify hypothesis 2, that is, we have to show that

$$\lim_{N \to \infty} \sup_{r \in [\delta, 1]} M\{\xi_1(r); \xi_1(r) \geq N\} = 0 . \tag{1.47}$$

Indeed,

$$0 \leq M[\xi_1(r); \xi_1(r) \geq N] \leq \frac{M\xi_1^2(r)}{N} = \frac{1}{2N} \int_{-1}^{1} \ln^2(1+rx)dx$$

$$= \frac{1}{2Nr}\{(1+r)\ln^2(1+r) - (1-r)\ln^2(1-r) - 2(1+r)\ln(1+r)$$

$$- 2(1-r)\ln(1-r) + 4r\} \leq \frac{c(\delta)}{N} .$$

This gives (1.47). Theorem 1.1 is proved. $\qquad\qquad\square$

Note that in two dimensions the same result is true with

$$g(r) = -\ln\left\{\left[1 + (1-r^2)^{1/2}\right]/2\right\} .$$

If we consider a walk on spheres process $\{\xi_k\}$ with a fixed radius r, then the following estimation of $n(\varepsilon, r)$ can be obtained. Let d_* be the radius of the maximum sphere lying inside G. We choose the parameter $r_0 \in [\delta, 1]$ as $r_0 = r/d_*$. Then $r_0 d(\xi_k) = [r/d]d(\xi_k) \leq r$. Hence,

$$n(\varepsilon, r) \leq \frac{|\ln(\varepsilon)|}{g(r_0)} \leq \frac{|\ln(\varepsilon)|}{g(r/d_*)} .$$

For small r

$$n(\varepsilon, r) = \frac{4d_*^2|\ln(\varepsilon)|}{r^2} + 0(r^4) \quad \text{as} \quad r \to 0 .$$

Thus, we have obtained the estimation (1.45) for a half-space. This estimation holds for arbitrary domains, which could be proved using the scheme described in [1.8].

1.2.2 Walk on Boundary Processes

Consider a domain $G \subset \mathbf{R}^m$ with the boundary $\partial G = \Gamma$. Let us first suppose that G is a bounded convex domain. An isotropic walk on boundary process starting from x_0 is defined as a Markov chain $\{\xi_k\}_{k=1}^{\infty} \in \Gamma$ such that the first state ξ_1 is chosen on Γ according to a uniform view angle from the point x_0. This means that the initial density has the form [1.23, 24]:

$$p_0(y) = \cos \angle (n(y), x_0 - y) \left[\sigma_m |y - x_0|^{m-1}\right]^{-1} , \qquad (1.48)$$

where $\sigma_m = 2\pi^{m/2}/\Gamma(m/2)$ is the surface area of a sphere $S(0,1) \subset R^m$, $n(y)$ is the normal vector interior to Γ at $y \in \Gamma$, $\angle(n(y), x_0 - y)$ is the angle between the vectors $n(y)$ and $x_0 - y$. The transition from a point $y_{i-1} \in \Gamma$ to $y_i \in \Gamma$ is defined analogously:

$$p\left(y_{i-1}, y_i\right) = 2 \cos \angle \left(n(y_i), y_{i-1} - y_i\right) \left[\sigma_m |y_{i-1} - y_i|^{m-1}\right]^{-1} . \qquad (1.49)$$

Suppose now that G is not convex. Then the ray x_0, y with isotropic direction may intersect Γ at $q(y)$ points. Then we choose one of them, say y, with probability $1/q$, so that

$$p_0(y) = |\cos \angle (n(y), x_0 - y)| \left[\sigma_m q(y)|y - x_0|^{m-1}\right]^{-1} .$$

The transition from y_{i-1} to y_i is carried out as follows. Let $q(y_i)$ be the number of intersections of Γ with the line having isotropic direction $y_{i-1} - y_i$ (not counting the point y_{i-1}). Simple arguments show that the density of the transition from y_{i-1} to y_i can be written as

$$p\left(y_{i-1}, y_i\right) = 2 \left|\cos \angle \left(n(y_i), y_{i-1} - y_i\right)\right| \left[\sigma_m q(y_i)|y_{i-1} - y_i|^{m-1}\right]^{-1} .$$

Note that if we had chosen the point y_i (y_{i-1} fixed) according to the density $cp_0(y_i)$ then we would have had to carry out a difficult calculation of the constant c. Note also that for brevity we used in the densities the same variables as chosen for the random points.

Generally, the sequence $\xi_i \in \Gamma$, $i = 1, 2, \ldots$ can be chosen using different laws. For example, we shall also use the following transition densities

$$p\left(y_{i-1}, y_i\right) \sim [\cos(\theta_i)]^{-\beta} \quad \text{and} \quad p\left(y_{i-1}, y_i\right) \sim |y_{i-1} - y_i|^{-1-\beta}$$

where $y_i = (r, \theta_i, \varphi)$, i.e., the angle φ is uniformly distributed on $[0, 2\pi]$.

In nonstationary problems, we shall use nonstationary walk on boundary processes. Let G be an arbitrary convex domain, and let $Q_t = G \times (0, t)$,

$\Gamma_t = \partial G \times (0, t)$, $\Gamma = \partial G \times (-\infty, \infty)$. Introduce a homogeneous Markov chain $\{(\xi_k, \tau_k)\}_{k=0}^{\infty}$ with the phase space Γ. Let $\pi(\xi_0, \tau_0)$ be the density of the initial point ξ_0, τ_0. Define the transition density as

$$p(\xi, \tau \rightarrow \xi', \tau') = p_\xi(\xi \rightarrow \xi') p_\tau(\tau \rightarrow \tau' | \xi, \xi') \qquad (1.50)$$

where p_ξ is the spatial transition density, i.e., p_ξ is given by (1.49),

$$p_\xi(\xi \rightarrow \xi') = p(\xi, \xi') \, ,$$

and $p_\tau(\tau \rightarrow \tau' | \xi, \xi')$ is the time probability density defined as

$$p_\tau(\tau \rightarrow \tau' | \xi, \xi') = \frac{\pi^{m/2} |\xi - \xi'|^m}{\Gamma(m/2)(\tau - \tau')} Z_0(\xi - \xi', \tau - \tau') \, , \qquad (1.51)$$

$$t \geq 0 \, , \quad |x| \neq 0 \, ,$$

where $Z_0(x, t)$ is the fundamental solution of the heat equation

$$Z_0(x, t) = \begin{cases} (4\pi a^2 t)^{-m/2} \exp\left\{ \dfrac{|x|^2}{4a^2 t} \right\} \, , & t \geq 0 \, , \quad |x| \neq 0 \, , \\ 0 \, , & t < 0 \, . \end{cases}$$

It is not difficult to show that $p_\tau(\tau \rightarrow \tau' | \xi, \xi')$ is indeed a transition density on $(-\infty, \tau)$ with respect to τ' since this function is positive and

$$\int_{-\infty}^{\tau} p_\tau(\tau \rightarrow \tau' | \xi, \xi') d\tau' \equiv 1 \quad \text{for all} \quad \tau \, .$$

Thus, the nonstationary isotropic walk on boundary process is constructed, according to (1.50), as follows. The first point (ξ_0, τ_0) is sampled according to an initial density $\pi(\xi, \tau)$, for example, ξ_0 can be sampled as in the stationary case. Then ξ_k is simulated also as in the stationary case, according to p_ξ, i.e., the vector $\xi_k - \xi_{k-1}$ is uniformly distributed in the space; next, the new time is calculated,

$$\tau_k = \tau_{k-1} - |\xi_{k-1} - \xi_k|^2 / (4a^2 \gamma_{m/2}) \, , \qquad (1.52)$$

where $\gamma_{m/2}$ is a random variable with a gamma distribution whose parameter is $m/2$.

Note that various modifications of this process can be considered by changing the transition densities or by varying the termination probability. For example, we shall consider a process $\{(\xi_k, \tau_k)\}$ with a backward time.

As in the walk on spheres process, it is necessary to obtain an estimation of the mean number of steps.

Theorem 1.2. Let $K_\xi(\alpha_0, h_0)$ be a cone of angle α_0 with the top at ξ_0 whose height is h_0. Suppose that the boundary ∂G satisfies the interior-cone condition, i.e., there exist α_0 and $h_0 > 0$ such that for any point $\xi \in \partial G$ it is possible to construct a cone $K_\xi(\alpha_0, h_0) \subset G$ (α_0 and h_0 do not depend on ξ). Let

$$N_t = \max\{k : \tau_0 + \tau_1 + \ldots + \tau_k < t\},$$

$$\tau_i = |\xi_i - \xi_{i+1}|^2 / 4\gamma_{m/2}^{(i)}.$$

Then

$$n_t = MN_t = 0(t - M\tau_0) \quad \text{as} \quad t \to \infty.$$

Proof. Let $\varrho_i = |\xi_i - \xi_{i-1}|$ and let

$$\varrho_i' = \begin{cases} h_0, & \text{if the ray } \xi_{i-1}, \xi_i \text{ intersects the base of } K_{\xi_{i-1}} \\ 0, & \text{otherwise} \end{cases}.$$

Then ϱ_i' are independent and $\varrho_i \geq \varrho_i'$. In addition, $\varrho_i' = h$ with the probability $2\alpha_0/\sigma_m$, where σ_m is the surface area of $S(0,1) \subset \mathbb{R}^m$.

Let $\delta > 0$ be a fixed number. We introduce the random variables

$$\gamma_i' = \begin{cases} \gamma_{m/2}^{(i)} & \text{if } \gamma_{m/2}^{(i)} \geq \delta \\ \delta & \text{if } \gamma_{m/2}^{(i)} < \delta, \end{cases}$$

and put $\tau_i' = (\varrho_i')^2 / 4\gamma_i'$, $i = 1, 2, \ldots$. Since $\tau_i \geq \tau_i'$ we get

$$N_t \leq N_t' = \max\{k : \tau_1' + \ldots + \tau_k' < t\}$$

almost everywhere.

The random variables $\tau_1', \tau_2', \ldots$ are independent, equally distributed, with expectation μ and variance σ^2. Therefore, by the renewal theory the following asymptotics holds

$$MN_t' = \frac{t}{\mu} + o(t) \quad \text{as} \quad t \to \infty.$$

Using the inequality $N_t \leq N_t'$ we complete the proof. $\qquad\qquad\square$

1.2.3 Approximation of Wiener Processes

To calculate the Wiener continual integrals by the Monte Carlo methods it is necessary to construct some approximations of the Wiener process $x(\tau)$ in (1.30).

Definition. *A set of random variables* $\{\xi(t), t \in T \subset \mathbb{R}\}$ *is called a gaussian random process if for any set of points* $t_1, \ldots, t_n \in T$ *the distribution of the vector* $(\xi(t_1), \ldots, \xi(t_n))$ *is gaussian. Let* $B(t_1, t_2)$ *be the correlation function of* $\xi(t)$*, i.e.,*

$$B(t_1, t_2) = M\left\{[\bar{\xi}(t_1) - \bar{A}(t_1)][\xi(t_2 - A(t_2))]\right\},$$

where

$$A(t) = M\operatorname{Re}\xi(t) + i\,M\operatorname{Im}\xi(t)$$

is the expectation value, and the bar means that the complex conjugate value is taken.

Let B be an $n \times n$ matrix with elements $\{B(t_i, t_j)\}$ and let b_{ij} be the elements of the matrix B^{-1}. Then the probability distribution of the vector $(\xi(t_1), \ldots, \xi(t_n))$ is given by

$$f_{t_1, \ldots t_n}(x_1, \ldots, x_n) = \frac{1}{(2\pi)^{n/2}|B|^{1/2}} \exp \left\{ -\frac{1}{2} \sum_{i,j=1}^{n} b_{ij} R_{ij} \right\} .$$

Thus the gaussian processes are uniquely defined by the functions $A(t)$ and $B(t_1, t_2)$.

Definition. A gaussian random function $x(\tau)$ is called a Wiener process if

1) $x(0) = 0 , \quad M x(\tau) = 0$

2) $M x(t_1) x(t_2) = D \min(t_1, t_2) .$

In the case $D = 1/2$, $x(t)$ is called a standard Wiener process. A piecewise linear approximation is constructed as follows [1.25].

For simplicity let us consider the Wiener process on $[0, 1]$ with $D = 1$. We divide the segment $[0, 1]$ into n parts, $0 = t_0 < t_1 < t_2 < \ldots < t_{n-1} < t_n = 1$. Since

$$W(t_i) = W(t_{i-1}) + (t_i + t_{i-1})^{1/2} \zeta_i ,$$

where $\{\zeta_i\}$ are independent gaussian random variables with zero expectation and unit variance, the linear approximation of $W(t)$ for $t \in [t_{i-1}, t_i]$ has the form

$$W(t) = W(t_{i-1}) + \frac{W(t_i) - W(t_{i-1})}{t_i - t_{i-1}} (t - t_{i-1})$$

or

$$W(t) = W(t_{i-1}) + \frac{t - t_{i-1}}{[t_i - t_{i-1}]^{1/2}} \zeta_i ,$$

$$W(0) = 0 , \quad t \in [t_i - t_{i-1}] , \quad i = 1, \ldots, n .$$

Generalizations to arbitrary gaussian processes are considered in [1.15]. Special approximations are presented in [1.8, 26].

We present here the optimal orthogonal expansions of the Wiener processes such that $M x(t_1) x(t_2) = 2D \min(t_1, t_2)$.

Let $\{\varphi_k(\tau)\}_{k=1}^{\infty}$ be a complete system of functions orthonormal on $(0, t)$ with a weight function $g(\tau)$. Then

$$x(\tau) = \sum_{k=1}^{\infty} \alpha_k \varphi_k(\tau) , \quad \alpha_k = \int_0^t x(\tau) \varphi_k(\tau) g(\tau) d\tau .$$

Let

$$x_n(\tau) = \sum_{k=1}^{n} \alpha_k \varphi_k(\tau) , \quad \bar{x}_n = x - x_n .$$

The set $\{\varphi_k(\tau)\}$ will be found from the criterion

$$\Delta_n \equiv \int_0^t \left\langle [x(\tau) - x_n(\tau)]^2 \right\rangle g(\tau) d\tau \to \min . \tag{1.53}$$

From (1.53) and the condition

$$\int_0^t \varphi_k^2(\tau) d\tau = 1$$

it can be shown that φ_k satisfy the equation [1.8]

$$\int_0^t B(\tau, s) \varphi_k(s) g(s) ds = \lambda_k \varphi_k(\tau) .$$

Substituting into this integral equation $B(\tau, s) = 2D \min(\tau, s)$ yields

$$\varphi_k''(\tau) + \frac{2D}{\lambda_k} g(\tau) \varphi_k(\tau) = 0 , \quad \varphi_k(0) = \varphi_k'(t) = 0 . \tag{1.54}$$

This boundary value problem uniquely defines the constants λ_k and the functions $\varphi_k(\tau)$.

The function $x(\tau)$ is gaussian, therefore $\xi_k = \alpha_k \lambda_k^{-1/2}$, $k = 1, 2, \ldots$ is a sequence of independent gaussian variables with zero expectations and unit variances; consequently, $M[x_n(\tau)\bar{x}_n(\tau)] = 0$. Thus, the optimal orthogonal expansion of the Wiener process has the form

$$x_n(\xi, \tau) = \sum_{k=1}^{n} \lambda_k^{1/2} \xi_k \varphi_k(\tau) . \tag{1.55}$$

For example, if $g = 1$,

$$\varphi_k(\tau) = \left(\frac{2}{t}\right)^{1/2} \sin\left[\pi\left(k - \frac{1}{2}\right)\frac{\tau}{t}\right] , \quad \lambda_k = \frac{16Dt^2}{\pi^2(2k-1)^2} .$$

If $g(\tau) = 3\tau^2/t^2$, then

$$\varphi_k(\tau) = \frac{A_k}{t} \tau^{1/2} J_{1/4}\left(\zeta_k \tau^2/t^2\right) , \quad \lambda_k = \frac{3Dt^2}{2\zeta_k^2} ,$$

where ζ_k is the k-th root of the equation $J_{-3/4}(\zeta_k) = 0$: $\zeta_1 = 1.0585$, $\zeta_2 = 4.2840, \ldots$; $J_{-3/4}$ is the Bessel function, and $A_k = 2\{3^{1/2}|J_{1/4}(\zeta_k)|\}^{-1}$. The relative error of the approximation for $g = 1$ is given by

$$\delta_n = \Delta_n / \int_0^t B(\tau, \tau) g(\tau) d\tau = 1 - \frac{8}{\pi^2} \sum_{k=1}^{n} (2k-1)^{-2} ,$$

so $\delta_1 = 0.19$, $\delta_2 = 0.099$, $\delta_3 = 0.067$, $\delta_4 = 0.050$, $\delta_5 = 0.031$. For $g(\tau) = 3\tau^2/t^2$

$$\delta_n = 1 - \left(\zeta_1^{-2} + \zeta_2^{-2} + \ldots + \zeta_n^{-2} \right),$$

thus $\delta_1 = 0.107$, $\delta_2 = 0.053$, $\delta_3 = 0.035$, $\delta_4 = 0.026$, $\delta_5 = 0.021$. Note that another expansion could be obtained on the basis of spectral representations of local homogeneous random fields described in the final part of the next section. In one dimension the formulas result in following expansion of the Wiener process ($\tau \in [0, 2\pi]$):

$$x(\tau) = \frac{\tau}{(2\pi)^{1/2}} \xi_0 + \frac{1}{(2\pi)^{1/2}} \sum_{k \neq 0} \left[\xi_k \sin(k\tau) + \eta_k(\cos(k\tau) - 1) \right],$$

where $\xi_0, \xi_{\pm 1}, \xi_{\pm 2}, \ldots, \eta_{\pm 1}, \eta_{\pm 2}, \ldots$ are independent standard gaussian random variables. This representation can be simplified to

$$x(\tau) = \frac{\tau}{(2\pi)^{1/2}} \xi_0 + \sum_{k \neq 0} (2/\pi)^{1/2} \xi_k \sin\left(\frac{k\tau}{2} \right).$$

Consider now approximations in the form of step-functions. Let

$$\varphi(u, \tau) = \begin{cases} (2D)^{1/2}, & \text{if } 0 \leq u \leq \tau \leq 1 \\ 0, & \text{if } 0 \leq \tau \leq u \leq 1 \end{cases}.$$

Assume that u is a random variable uniformly distributed on $[0, 1]$. Then φ is a random function $\varphi(u, \tau) = (2D)^{1/2} \theta(\tau - u)$ where $\theta(\tau)$ is the Heaviside step function. Let

$$y_m(u, \tau) = \left(\frac{2D}{m} \right)^{1/2} \sum_{k=1}^{m} \mu_k \theta(u_k, \tau), \tag{1.56}$$

where $u = (u_1, \ldots, u_m)$, $\{\mu_k\}$ are independent equally distributed random variables with zero expectations and unit variances (it is assumed that the distribution is symmetric with respect to zero); $\{u_k\}$ are mutually independent (and independent of $\{\mu_k\}$) random variables uniformly distributed on $[0, 1]$. It is clear that the moments $My_m(\tau)$ and $\langle y_m(\tau_1) y_m(\tau_2) \rangle$ coincide with those of the Wiener process for all $m = 1, 2, \ldots$. The convergence $y_m(\tau) \to x(\tau)$ is studied in [1.8].

The optimal orthogonal expansions and the approximations of type (1.56) can be combined. Let $\bar{H}_n$ be a subspace orthogonal to the linear span of the functions $\varphi_1(\tau), \ldots, \varphi_n(\tau)$. Since $\bar{x}_n \in \bar{H}_n$, we can take, instead of $\bar{x}_n$, the projection of $y_m(\tau)$ on $\bar{H}_n$. To this end we expand the function $\theta(\tau - s)$:

$$\theta(\tau - u) = \sum_{k=1}^{\infty} \theta_k(u) \varphi_k(\tau),$$

where

$$\theta_k(u) = \int_u^t \varphi_k(\tau) g(\tau) d\tau.$$

30

The projection of $\theta(\tau - s)$ on the first n functions $\varphi_k(\tau)$ has the form

$$\theta_n(\tau, u) = \sum_{k=1}^{n} \theta_k(u)\varphi_k(\tau) .$$

Therefore,

$$\bar{\theta}_n(u, \tau) = \theta(\tau - u) - \sum_{k=1}^{n} \theta_k(u)\varphi_k(\tau) .$$

Substituting $\theta_n(u_k, \tau)$ into (1.56) instead of $\theta(\tau - u_k)$ yields

$$z_{m,n}(u, \tau) = \left(\frac{2Dt}{m}\right)^{1/2} \sum_{k=1}^{m} \mu_k \left[\theta(\tau - u_k) - \sum_{i=1}^{n} \theta_i(u_k)\varphi_i(\tau)\right] .$$

Thus we obtain a new approximation

$$z_{m,n} = x_n + y_m - y_{m,n} ,$$

where

$$y_{m,n}(u, \tau) = \left(\frac{2Dt}{m}\right)^{1/2} \sum_{l=1}^{m} \mu_l \sum_{k=1}^{n} [\theta_k(u_l)\varphi_k(\tau)] .$$

Using these approximations it is possible to evaluate the Wiener integrals by the Monte Carlo method on the basis of the approximate formula

$$\int_C F[x]d_w x \cong \langle F[\varphi]\rangle_\varphi \tag{1.57}$$

where φ is one of the approximations $x_n(\tau)$, $y_m(u, \tau)$ or $z_{m,n}(u, \tau)$. Weak convergence of these approximations to the Wiener process is proved in [1.8].

In conclusion, we give an example of a limit theorem. Suppose that $|F[x]| \leq H(\|x\|^2)$, where

$$\int_C H(\|x\|^2)d_w x < \infty .$$

Then

$$\int_C F[x]d_w x = \lim_{n,m \to \infty} \langle F[z_{m,n}]\rangle_{z_{m,n}} . \tag{1.58}$$

The equality (1.58) remains true if the following weakened assumptions are made: (1) the functional $F[x]$ is ϱ_{L_p}-continuous for $p \geq 1$; (2) $F[x]$ is uniformly integrable with respect to the measure generated by $z_{n,m}$.

1.2.4 Simulation of Random Fields

In this section, we consider the problem of simulation of random fields with a given statistical structure. We are interested in wind velocity fields, therefore we shall construct simulation formulas for vector random fields of the following types: homogeneous, partially homogeneous, quasi-homogeneous, local homogeneous. Sometimes, we additionally require that incompressibility, isotropy and other properties hold for the random fields studied. The structure of these fields is described in [1.27].

Various methods of simulation of general-type random processes and fields are known [1.2, 28, 29]. Depending on the experimental data available, one uses, as the input information, the correlation tensor or the spectral tensor. We assume that spectral data are given. The simulation formulas presented below generalize the formulas based on the spectral representation [1.25, 30]. A survey of the spectral characteristics of fluctuations of wind velocity has been published [1.31].

Homogeneous Random Fields. As is known [1.30], simulation of a scalar homogeneous random field $v(x)$, $(\bar{v}(x) = 0)$ with a given spectral density $F(\lambda)$

$$F(\lambda) = \frac{1}{(2\pi)^n} \int_{\mathbf{R}_n} B(\varrho) \exp\{-(\lambda, \varrho)\} d\varrho \qquad (1.59)$$

where $B(\varrho)$ is the correlation function of the field $v(x)$, can be carried out by the formula

$$v(x) = u_* \left[\xi_\lambda \cos(\lambda, x) + \eta_\lambda(\lambda, x) \right] . \qquad (1.60)$$

Here

$$u_*^2 = \int_{\mathbf{R}^n} F(\lambda) d\lambda$$

is the total energy of the field $v(x)$, λ is the n-dimensional random vector with distribution density $p(\lambda) = F(\lambda)/u_*^2$, ξ_λ and η_λ are independent random variables (λ is fixed) such that

$$\bar{\xi}_\lambda = \bar{\eta}_\lambda = 0 , \quad \overline{\xi_\lambda^2} = \overline{\eta_\xi^2} = 1 .$$

Remark. Assume that

$$F(\lambda) = F_1(\lambda) + \ldots + F_N(\lambda) \qquad (1.61)$$

where $F_i(\lambda) \geq 0$, $F_i(-\lambda) = F_i(\lambda)$. To simulate (1.61), it is possible to modify the algorithm as follows. First, we simulate N independent random fields $v_i(x)$ according to $F_i(\lambda)$, then we set

$$v^{(N)}(x) = v_1(x) + \ldots + v_N(x) . \qquad (1.62)$$

Now we have described the simulation of a random field with zero mean $\bar{v}(x) = 0$ and with a given spectral density $F(\lambda)$, and no assumption about the multidimensional distributions of the random field was made (i.e., we have described the simulation of a homogeneous random field in a broad sense). It is often necessary to simulate gaussian random fields with a given spectrum function. It is then possible to use representation (1.62). In this case, keeping in mind the central limit theorem, one may hope that with an appropriate choice of the representation (1.61) $v^{(N)}(x)$ will converge to a gaussian random field as $N \to \infty$. Note that the convergence of $v^{(N)}$ to v can also be understood in a different sense. In this connection, we refer to [1.2] where the expansion (1.61) was constructed as

$$F_i(\lambda) = \begin{cases} F(\lambda), & \text{if } \lambda \in D_i \,, \\ 0 & , \quad \text{if } \lambda \notin D_i \,, \end{cases} \tag{1.63}$$

$$\bigcup_{i=1}^{N} D_i = \mathbf{R}^n \,, \quad D_i \bigcap D_j = \emptyset \; (i \neq j) \,.$$

With a few assumptions about the subdivision given in (1.63) it was proved in [1.2] that $v^{(N)}$ weakly converges to a gaussian field v.

Note that the random variables in (1.60) may have arbitrary distributions. For example, in [1.2] they were sampled from a gaussian distribution, and in [1.29] they were taken as uniform random variables. As mentioned in [1.2], the absolute one-point distribution of the random field given by (1.60) will be gaussian if ξ_λ and η_λ are gaussian (multidimensional distributions are generally not gaussian).

In the particular case when the field is isotropic, F depends only on $|\lambda|$, therefore we take in (1.60) $\lambda = r\omega$, where ω is a unit isotropic vector, r is a random variable distributed according to

$$\sigma_n p(r) r^{n-1} \,, \quad r \in (0, \infty) \quad \left(\sigma_n = \frac{2\pi^{n/2}}{\Gamma(/2)} \right) \,.$$

Now we consider the case of vector homogeneous random fields. It is convenient here to consider the complex- and real-valued random fields separately. Let

$$V(x) = (v_1(x), \dots, v_l(x))^{\mathrm{T}} \quad x \in \mathbf{R}^n$$

be a complex-valued homogeneous random field with the spectral tensor

$$F(\lambda) = \frac{1}{(2\pi)^n} \int_{\mathbf{R}^n} B(\varrho) \exp\{-(\lambda, \varrho)\} d\varrho \,, \tag{1.64}$$

where $B(\varrho) = \langle V(x+\varrho) V^*(x) \rangle$ is the correlation tensor of the random field $V(x)$. Let $p(\lambda)$ be a density function in $\mathbf{R}^n$. Then the simulation formula has the form

$$V(x) = \frac{1}{[p(\lambda)]^{1/2}} \exp\{-\mathrm{i}(\lambda, x)\} \xi_\lambda \,, \tag{1.65}$$

where the vector λ has the distribution density $p(\lambda)$, and ξ_λ is a complex random l-dimensional column vector with zero mean and the following covariance matrix (λ is fixed):

$$\langle \xi_\lambda \xi_\lambda^* \rangle = F(\lambda) .$$

The vector ξ_λ is constructed as follows. We put $F(\lambda) = G(\lambda)G^*(\lambda)$, where $G(\lambda)$ is a left triangular matrix [1.2]. Then we set

$$\xi_\lambda = G(\lambda)\xi_\lambda'$$

where ξ_λ' is an l-dimensional random vector with zero mean and the unit covariance matrix. For $p(\lambda)$, it is natural to take the function

$$p(\lambda) = \frac{1}{u_*^2} \operatorname{Sp} F(\lambda) = \frac{1}{u_*^2} \sum_{i=1}^{l} F_i(\lambda) ,$$

where

$$u_*^2 = \int_{\mathbf{R}^n} \operatorname{Sp} F(\lambda)$$

is the total field energy.

Now consider the case of the real-valued homogeneous random field $V(x)$. The generalization of the simulation formula (1.60) has the form

$$V(x) = \frac{1}{[p(\lambda)]^{1/2}} \left[\xi_\lambda \cos(\lambda, x) + \eta_\lambda(\lambda, x) \right] . \tag{1.66}$$

Here, λ is a random vector distributed according to a density $p(\lambda)$, and the $2l$-dimensional random vector $(\xi_\lambda, \eta_\lambda)^{\mathrm{T}}$ (λ is fixed) has zero mean and the covariance matrix

$$\begin{pmatrix} \langle \xi_\lambda \xi_\lambda^{\mathrm{T}} \rangle & \langle \xi_\lambda \eta_\lambda^{\mathrm{T}} \rangle \\ \langle \eta_\lambda \xi_\lambda^{\mathrm{T}} \rangle & \langle \eta_\lambda \eta_\lambda^{\mathrm{T}} \rangle \end{pmatrix} = \begin{pmatrix} \operatorname{Re} F(\lambda) & \operatorname{Im} F(\lambda) \\ -\operatorname{Im} F(\lambda) & \operatorname{Re} F(\lambda) \end{pmatrix} . \tag{1.67}$$

The correctness of the procedure follows from the fact that the $2l$-dimensional matrix in (1.67) is symmetric and positive definite (since the spectral tensor $F(\lambda)$ is Hermitian and positive definite). Another variant of simulation formulas for vector homogeneous random fields was proposed in [1.28, 29].

Remark. To simulate the vector $(\xi_\lambda, \eta_\lambda)^{\mathrm{T}}$ with the covariance matrix (1.67), it is possible to use two methods. The first method is based on the representation $F(\lambda) = GG^*$, where G is a triangular matrix [1.2], where

$$\begin{pmatrix} \operatorname{Re} F & \operatorname{Im} F \\ -\operatorname{Im} F & \operatorname{Re} F \end{pmatrix} = \begin{pmatrix} \operatorname{Re} G & \operatorname{Im} G \\ -\operatorname{Im} G & \operatorname{Re} G \end{pmatrix} \begin{pmatrix} \operatorname{Re} G & \operatorname{Im} G \\ -\operatorname{Im} G & \operatorname{Re} G \end{pmatrix}^{\mathrm{T}} .$$

Then,

$$\begin{pmatrix} \xi_\lambda \\ \eta_\lambda \end{pmatrix} = \begin{pmatrix} \operatorname{Re} G & \operatorname{Im} G \\ -\operatorname{Im} G & \operatorname{Re} G \end{pmatrix} \begin{pmatrix} \xi_\lambda' \\ \eta_\lambda' \end{pmatrix} ,$$

where ξ_λ' and η_λ' are independent l-dimensional vectors with zero mean and the unit covariance matrix.

The second method is based on the expansion of the real-valued matrix

$$\begin{pmatrix} \mathrm{Re}\, F & \mathrm{Im}\, F \\ -\mathrm{Im}\, F & \mathrm{Re}\, F \end{pmatrix} = HH^{\mathrm{T}}$$

where H is a $2l$-dimensional left triangular matrix. Then,

$$\begin{pmatrix} \xi_\lambda \\ \eta_\lambda \end{pmatrix} = H \begin{pmatrix} \xi'_\lambda \\ \eta'_\lambda \end{pmatrix} .$$

Consider an important special case of vector homogeneous random fields, i.e., a class of incompressible homogeneous fields (div $V(x) = 0$). As is known [1.27], the spectral tensor F of the incompressible homogeneous field ($3 - D$ case) has the form

$$F_{ij}(\lambda) = b^2 \left(\delta_{ij} - \frac{\lambda_i \lambda_j}{\lambda^2} \right) + \bar{a}_i a_j \left(1 - \frac{b^2}{a^2} \right)$$

$$= a^2 \left(\delta_{ij} - \frac{\lambda_i \lambda_j}{\lambda^2} \right) + \bar{b}_i b_j \left(1 - \frac{a^2}{b^2} \right) . \qquad (1.68)$$

Here $a = a(k) = (a_1(k), a_2(k), a_3(k))^{\mathrm{T}}$, $b = b(k) = (b_1(k), b_2(k), b_3(k))^{\mathrm{T}}$ are vector functions such that the vectors $a(k)$, $b(k)$ and k are mutually orthogonal, $(a, b) = (a, k) = (b, k) = 0$, where $(a, b) = \sum_{i=1}^{3} \bar{a}_i b_i$. Also, $a_j(-k) = \bar{a}_j(k)$ and $b_j(-k) = \bar{b}_j(k)$. A three-dimensional homogeneous incompressible random field is thus fully specified by two vector functions (in general, complex-valued) satisfying the above conditions.

Now we give simulation formulas in the case of a complex-valued random field. Put

$$p(\lambda) = \mathrm{Sp}\, F = \frac{a^2(\lambda) + b^2(\lambda)}{u_*^2}$$

where

$$u_*^2 = \int_{\mathbf{R}^3} (a^2 + b^2(\lambda)) d\lambda .$$

Then, sample λ according to $p(\lambda)$ and calculate

$$V(x) = \frac{\exp\{i(\lambda, x)\}}{[p(\lambda)]^{1/2}} \left\{ m(\lambda) \left[\xi_\lambda \times \frac{\lambda}{|\lambda|} \right] + d\, \frac{\bar{m}(\lambda)}{|M(\lambda)|} \xi_\lambda^{(1)} \right\} ,$$

where

$$d = [|a^2 - b^2|]^{1/2} ,$$
$$m(\lambda) = \min\{|a(\lambda)|, |b(\lambda)|\} , \qquad M(\lambda) = \max\{|a(\lambda)|, |b(\lambda)|\} ,$$

ξ_λ is a standard three-dimensional gaussian vector, $\xi_\lambda^{(1)}$ is a standard gaussian random variable independent on the vector ξ_λ. We now verify that the random field constructed has the desired spectral tensor.

We set

$$\Lambda_1 = \{\lambda : a^2(\lambda) \geq b^2(\lambda)\} \ , \quad \Lambda_2 = \mathbf{R}^3 \backslash \Lambda_1$$

and define a matrix

$$\Omega_\lambda = \frac{1}{|\lambda|} \begin{pmatrix} 0 & \lambda_3 & -\lambda_2 \\ -\lambda_3 & 0 & \lambda_1 \\ \lambda_2 & -\lambda_1 & 0 \end{pmatrix} .$$

Then,

$$\xi_\lambda \times \frac{\lambda}{|\lambda|} = \Omega_\lambda \xi_\lambda \ , \quad (\Omega_\lambda \Omega_\lambda^{\mathrm{T}})_{ij} = \delta_{ij} - \frac{\lambda_i \lambda_j}{|\lambda|^2} \ .$$

From this, using (1.68) we obtain

$$\begin{aligned}
\langle V(x+\varrho)V(x)|\lambda \in \Lambda_1\rangle &= \frac{\exp\{\mathrm{i}(\lambda,\varrho)\}}{p(\lambda)} \left\langle \left[|b(\lambda)|\Omega_\lambda \xi_\lambda + d\frac{\bar{a}(\lambda)}{|a(\lambda)|}\xi_\lambda^{(1)}\right] \right. \\
&\quad \left. \times \left[|b(\lambda)|\,\xi_\lambda^{\mathrm{T}}\Omega_\lambda^{\mathrm{T}} + d\frac{a^{\mathrm{T}}(\lambda)}{|a(\lambda)|}\xi_\lambda^{(1)}\right] \right|\lambda \in \Lambda_1 \right\rangle \\
&= \frac{\exp\{\mathrm{i}(\lambda,\varrho)\}}{p(\lambda)}\left[b^2(\lambda)\Omega_\lambda \Omega_\lambda^{\mathrm{T}} + d^2\frac{\bar{a}(\lambda)a^{\mathrm{T}}(\lambda)}{|a^2(\lambda)|}\right] \\
&= \frac{\exp\{\mathrm{i}(\lambda,\varrho)\}}{p(\lambda)}\,F(\lambda) \ .
\end{aligned}$$

Likewise,

$$\langle V(x+\varrho)V(x)|\lambda \in \Lambda_2\rangle = \frac{\exp\{\mathrm{i}(\lambda,\varrho)\}}{p(\lambda)}\,F(\lambda) \ .$$

Thus, the spectral tensor of the random field constructed is really $F(\lambda)$.

Now we present simulation formulas in the case of real-valued random fields. Let ξ_λ and η_λ be independent three-dimensional random vectors with zero mean and unit covariance matrix, and let $\xi_\lambda^{(1)}$ and $\eta_\lambda^{(1)}$ be mutually independent (and independent of ξ_λ and η_λ) random variables with zero mean and unit variance. We set

$$\begin{aligned}
&\xi_\lambda' = \xi_\lambda \times \frac{\lambda}{|\lambda|} \ , \quad \eta_\lambda' = \eta_\lambda \times \frac{\lambda}{|\lambda|} \\
&\xi_\lambda^{(2)}(a) = \xi_\lambda^{(1)}\mathrm{Re}\,\{a\} - \eta_\lambda^{(1)}\mathrm{Im}\,\{a\} \\
&\eta_\lambda^{(2)}(a) = \xi_\lambda^{(1)}\mathrm{Im}\,\{a\} + \eta_\lambda^{(1)}\mathrm{Re}\,\{a\} \ .
\end{aligned}$$

It is easy to verify that

$$\begin{aligned}
\left\langle \xi_\lambda' {\eta_\lambda'}^{\mathrm{T}}|\lambda \right\rangle &= \left\langle \eta_\lambda' {\xi_\lambda'}^{\mathrm{T}}|\lambda \right\rangle = 0 \\
\left\langle \xi_\lambda' {\xi_\lambda'}^{\mathrm{T}}|\lambda \right\rangle &= \left\langle \eta_\lambda' {\eta_\lambda'}^{\mathrm{T}}|\lambda \right\rangle = \Omega_\lambda \Omega_\lambda^{\mathrm{T}}
\end{aligned}$$

and

$$\left\langle \xi_\lambda^{(2)}(a)\xi_\lambda^{(2)^T}(a)|\lambda \right\rangle = \left\langle \eta_\lambda^{(2)}(a)\xi_\lambda^{(2)^T}(a)|\lambda \right\rangle = \text{Re} \left\{ F^{(1)}(\lambda) \right\}$$

$$\left\langle \xi_\lambda^{(2)}(a)\eta_\lambda^{(2)^T}(a)|\lambda \right\rangle = -\left\langle \eta_\lambda^{(2)}(a)\xi_\lambda^{(2)^T}(a)|\lambda \right\rangle = \text{Im} \left\{ F^{(1)}(\lambda) \right\} ,$$

where

$$F_{js}^{(1)}(\lambda) = \bar{a}_j(\lambda)a_s(\lambda) , \quad s, j = 1, \ldots, l .$$

Using the above properties and (1.66), we conclude that the spectral tensor of the random field

$$V(x) = \frac{1}{[p(\lambda)]^{1/2}} \left\{ m(\lambda) \left[\xi_\lambda' \cos(\lambda, x) + \eta_\lambda' \sin(\lambda, x) \right] \right.$$

$$\left. + \frac{d}{|M(\lambda)|} \left[\xi_\lambda^{(2)}(M) \cos(\lambda, x) + \eta_\lambda^{(2)}(M) \sin(\lambda, x) \right] \right\} ,$$

where λ is sampled according to the density

$$p(\lambda) = \frac{a^2(\lambda) + b^2(\lambda)}{u_*^2} ,$$

is $F(\lambda)$.

From the practical point of view it is interesting to construct simulation formulas for isotropic random vector fields.

Let $V(x) = (v_1(x), v_2(x), v_3(x))^T$ be a real-valued isotropic random field. As is known [1.27], its spectral tensor has the form

$$F_{ij}(\lambda) = \left[F_{LL}(|\lambda|) - F_{NN}(|\lambda|) \right] \frac{\lambda_i \lambda_j}{\lambda^2} + F_{NN}(|\lambda|)\delta_{ij} , \tag{1.69}$$

$$i, j = 1, 2, 3 .$$

Here, F_{LL} and F_{NN} are longitudinal and transverse three-dimensional spectra, respectively (they are positive functions).

Let

$$\left[u_*^{(L)} \right]^2 = \int_{\mathbf{R}^3} F_{LL}(|\lambda|)d\lambda ,$$

$$\left[u_*^{(N)} \right]^2 = \int_{\mathbf{R}^3} F_{NN}(|\lambda|)d\lambda .$$

Assume that the random vector $\lambda^{(L)} = (\lambda_1^{(L)}, \lambda_2^{(L)}, \lambda_3^{(L)})$ is distributed according to the density $F_{LL}(|\lambda|)/[u_*^{(L)}]^2$, and the vector $\lambda^{(N)}$ is independently sampled from the density $F_{NN}(|\lambda|)/[u_*^{(N)}]^2$. Then the simulation formula for the isotropic field with spectral tensor (1.69) takes the form

$$V(x) = u_*^{(N)} \left[\left(\xi^{(N)} \times \frac{\lambda^{(N)}}{|\lambda^{(N)}|} \right) \cos\left(\lambda^{(N)}, x \right) + \left(\eta^{(N)} \times \frac{\lambda^{(N)}}{|\lambda^{(N)}|} \right) \sin\left(\lambda^{(N)}, x \right) \right]$$

$$+ u_*^{(L)} \left[\xi^{(L)} \frac{\lambda^{(L)}}{|\lambda^{(L)}|} \cos\left(\lambda^{(L)}, x \right) + \eta^{(L)} \frac{\lambda^{(L)}}{|\lambda^{(L)}|} \sin\left(\lambda^{(L)}, x \right) \right] , \tag{1.70}$$

37

where $\xi^{(N)}$ and $\eta^{(N)}$ are independent three-dimensional random vectors such that $\bar{\xi}^{(N)} = \bar{\eta}^{(N)} = 0$, and

$$\overline{\xi^{(N)}\xi^{(N)T}} = \overline{\eta^{(N)}\eta^{(N)T}} = E$$

(E is the identity matrix); $\xi^{(L)}$ and $\eta^{(L)}$ are mutually independent (and independent of $\xi^{(N)}$ and $\eta^{(N)}$) scalar random variables such that

$$\bar{\xi}^{(L)} = \bar{\eta}^{(L)} = 0 , \quad \overline{\left[\xi^{(L)}\right]^2} = \overline{\left[\eta^{(N)}\right]^2} = 1 .$$

Note that $F_{LL} = 0$, if the field is incompressible (div $V = 0$) [1.27] and $F_{NN} = 0$, if the field is potential (rot $V = 0$). Thus, in the case of an incompressible field, we take in the simulation formula only the first term (with the index N). The second term (with the index L) is the simulation formula for the potential random field.

Partially Homogeneous Random Fields. Let $x = (y, z)$, $y \in \mathbf{R}^n$, $z \in \mathbf{R}^m$, and let

$$V(x) = (v_1(x), \ldots, v_l(x))^T .$$

Assume that the random field $V(y, z)$ is homogeneous with respect to the variable y, i.e.,

$$\langle V(y_1, z_1)V^*(y_2, z_2)\rangle = B(y_1 - y_2, z_1, z_2) .$$

Random fields with this property are called partially homogeneous random fields. Consider the partial spectral tensor

$$\psi(\lambda, z_1 z_2) = \frac{1}{(2\pi)^n} \int_{\mathbf{R}^n} B(\varrho, z_1, z_2)\exp\{-(\lambda, \varrho)\}d\varrho . \tag{1.71}$$

Let us first consider the case of a complex-valued random field $V(x)$. It is not difficult to verify that the correlation tensor of the field

$$V(y, z) = \frac{1}{[p(\lambda)]^{1/2}} \exp\{i(\lambda, y)\}\xi_\lambda(z) , \tag{1.72}$$

where λ is distributed according to $p(\lambda)$, and $\xi_\lambda(z)$ (λ is fixed) is a homogeneous l-dimensional complex-valued random field with correlation tensor (1.71)

$$\langle \xi_\lambda(z_1)\xi_\lambda^*(z_2)\rangle = \psi(\lambda, z_1, z_2) ,$$

is equal to $B(\varrho, z_1, z_2)$.

To justify this scheme, it is necessary to prove that $\psi(\lambda, z_1, z_2)$ (λ is fixed) is a correlation tensor of a complex-valued random field. To do this, it is sufficient to show that $\psi(\lambda, z_1, z_2)$ is positive definite, i.e.,

$$\sum_{r,s=1}^{k} \alpha_r\psi(\lambda, z_r, z_s)\alpha_s^* \geq 0 \tag{1.73}$$

for arbitrary $z_1, z_2, \ldots, z_k \in \mathbf{R}^m$ and complex vectors

$$\alpha_1 = (\alpha_{11}, \ldots, \alpha_{1l}), \ldots, \alpha_k = (\alpha_{k1}, \ldots, \alpha_{kl}) \, .$$

Inequality (1.73) follows from the fact that the left-hand side of (1.73) coincides with the spectrum of the complex-valued homogeneous scalar field

$$v(y) = \sum_{r=1}^{k} \alpha_r v(y, z_r) \, .$$

Assume now that $V(y, z)$ is a real-valued random field. Then the representation (1.72) takes the form

$$V(y, z) = \frac{1}{[p(\lambda)]^{1/2}} \, [\xi_\lambda(z) \cos(\lambda, y) + \eta_\lambda(z) \sin(\lambda, y)] \tag{1.74}$$

where the $2l$-dimensional real-valued field $(\xi_\lambda(z), \eta(z))^{\mathrm{T}}$ (λ is fixed) has the following correlation tensor

$$
\begin{aligned}
&\begin{pmatrix} \overline{\xi_\lambda(z_1)\xi_\lambda^{\mathrm{T}}(z_2)} & \overline{\xi_\lambda(z_1)\eta_\lambda^{\mathrm{T}}(z_2)} \\ \overline{\eta_\lambda(z_1)\xi_\lambda^{\mathrm{T}}(z_2)} & \overline{\eta_\lambda(z_1)\eta_\lambda^{\mathrm{T}}(z_2)} \end{pmatrix} \\
&= \begin{pmatrix} \mathrm{Re}\,\{\psi(\lambda, z_1, z_2)\} & \mathrm{Im}\,\{\psi(\lambda, z_1, z_2)\} \\ -\mathrm{Im}\,\{\psi(\lambda, z_1, z_2)\} & \mathrm{Re}\,\{\psi(\lambda, z_1, z_2)\} \end{pmatrix} \, .
\end{aligned}
\tag{1.75}
$$

The matrix on the right-hand side of (1.75) is symmetric and positive definite, therefore representation (1.74) is justified. The symmetry follows directly from the fact that the tensor $\psi(\lambda, z_1, z_2)$ is Hermitian. To prove that the above mentioned matrix is positive definite, note that

$$\sum_{r,s=1}^{k} \alpha_r \psi(\lambda, z_r, z_s) \alpha_s^* \geq 0 \tag{1.76}$$

for arbitrary $z_1, z_2, \ldots, z_k \in \mathbf{R}^m$ and complex vectors $\alpha_1, \ldots, \alpha_k$ as in the case of (1.73). It should be noted that the real part of (1.73) is coincident with

$$\sum_{r,s=1}^{k} (\alpha_r^{\mathrm{R}}, \alpha_r^{\mathrm{I}}) \begin{pmatrix} \psi_{r,s}^{\mathrm{R}} & \psi_{r,s}^{\mathrm{I}} \\ -\psi_{r,s}^{\mathrm{I}} & \psi_{r,s}^{\mathrm{R}} \end{pmatrix} \begin{pmatrix} [\alpha_r^{\mathrm{R}}]^{\mathrm{T}} \\ [\alpha_s^{\mathrm{I}}]^{\mathrm{T}} \end{pmatrix} \, ,$$

where

$$\alpha_r = \alpha_r^{\mathrm{R}} + i\alpha_r^{\mathrm{I}} \, , \quad \psi_{r,s}^{\mathrm{R}} = \mathrm{Re}\,\{\psi(\lambda, z_r, z_s)\} \, , \quad \psi_{r,s}^{\mathrm{I}} = \mathrm{Im}\,\{\psi(\lambda, z_r, z_s)\} \, .$$

Consequently, the tensor in (1.75) is positive definite.

The method described thus enables us to reduce the construction of a random field of $n + m$ variables to simulation of a random field of m variables, which is very important from a practical point of view.

Quasi-Homogeneous Random Fields. Let $V(x)$ be a scalar random field with zero mean. Consider a normalized correlation function

$$\frac{B(x_1, x_2)}{[B(x_1, x_2)B(x_1, x_2)]^{1/2}} . \tag{1.77}$$

We introduce new variables: $r = x_1 - x_2$, $R = (x_1 + x_2)/2$. A random field $V(x)$ is called quasi-homogeneous if (1.77), as a function of variables r and R, is weakly dependent on R [1.32].

It is natural to introduce the following generalization of this notion to l-dimensional partially homogeneous vector fields $V(x, y)$, $x \in \mathbb{R}^n$, $y \in \mathbb{R}^m$ with the partial spectral tensor given by (1.71). Let

$$\tilde{\psi}_{r,s}(\lambda, z_1, z_2) = \frac{\psi_{r,s}(\lambda, z_1, z_2)}{[\psi_{r,r}(\lambda, z_1, z_2)\psi_{s,s}(\lambda, z_1, z_2)]^{1/2}} . \tag{1.78}$$

The random field $V(y, z)$ is called quasi-homogeneous if $\tilde{\psi}_{r,s}(\lambda, z_1, z_2)$ depends only on $z_1 - z_2$. Consider an extended class of random fields. Let

$$\Lambda_{z_1, z_2} \equiv \operatorname{supp} \psi(\cdot, z_1, z_2) = \bigcup_{i=1}^{l} \bigcup_{j=1}^{l} \operatorname{supp} \psi_{ij}(\cdot, z_1, z_2)$$

be the support of the tensor ψ with respect to λ (z_1 and z_2 are fixed). The following inequality holds:

$$|\psi_{ij}(\cdot, z, z)|^2 \le \psi_{ii}(\lambda, z, z)\psi_{jj}(\lambda, z, z) .$$

This follows from the fact that $\psi_{ij}(\lambda, z_1, z_2)$ is a correlation tensor of a complex-valued random vector field $\xi_\lambda(z)$ and from the Cauchy-Schwartz-Buniakowsky inequality. This yields

$$\Lambda_{z,z} \equiv \bigcup_{i=1}^{l} \operatorname{supp} \psi_{ii}(\cdot, z, z) .$$

From the more general inequality

$$|\psi_{ij}(\lambda, z_1, z_2)|^2 \le \psi_{ii}(\lambda, z_1, z_1)\psi_{jj}(\lambda, z_2, z_2)$$

it follows that

$$\Lambda_{z_1, z_2} \subset \Lambda_{z_1, z_1} \cap \Lambda_{z_2, z_2} . \tag{1.79}$$

Definition. *A complex-valued random vector field $V(y, z)$ homogeneous with respect to y, is called quasi-homogeneous, if*

$$\psi(\lambda, z_1, z_2) = A(\lambda, z_1)\tilde{\psi}(\lambda, z_1 - z_2)A^*(\lambda, z_2) \tag{1.80}$$

where the matrix valued function $A(\lambda, z)$ satisfies the following condition

$$\det A(\lambda, z) \neq 0 \quad \forall z \in \mathbf{R}^m, \quad \forall \lambda \in \Lambda_{z,z}.$$ (1.81)

Note that (1.81) is satisfied if $\eta_z(y) = V(y, z)$ has a full rank, i.e.

$$\det \psi(\lambda, z, z) \neq 0 \quad \forall z, \quad \forall \lambda \in \operatorname{supp} \psi(\cdot, z, z) = \Lambda_{zz}.$$

Indeed, if $\lambda \in \Lambda_{z,z}$, then it follows from (1.80) that $\det A(\lambda, z) \neq 0$. Thus, from (1.79) we obtain (1.81).

Note that $\tilde{\psi}_{r,s}(\lambda, z_1 - z_2)$ in equation (1.80) is the partial spectral tensor of the random field

$$\tilde{V}(y, z) = \int_{\mathbf{R}^n} \tilde{A}_1(y - y', z)V(y', z)dy'$$ (1.82)

where

$$\tilde{A}_1(y, z) = \int_{\mathbf{R}^n} \exp\{i(y, \lambda)\} A^{-1}(\lambda, z)d\lambda$$

and $V(y, z)$ is the given partially homogeneous random field.

The formulation of a quasi-homogeneous random field in terms of the partial spectral tensor is convenient in constructing the simulation formulas given below. It is possible to present other equivalent formulations of this by using the correlation tensor or the realizations of the random field.

Assume that the correlation tensor of the random field $V(y, z)$ satisfies the relation

$$\begin{aligned} B(y_1 - y_2, z_1, z_2) &= \langle V(y_1, z_1)V^*(y_2, z_2) \rangle \\ &= \int_{\mathbf{R}^n} dy_1' \int_{\mathbf{R}^n} dy_2' \tilde{A}(y_1 - y_1', z_1)B(y_1' - y_2', z_1 - z_2) \\ &\quad \times \tilde{A}^*(y_2 - y_2', z_2) \end{aligned}$$ (1.83)

where

$$\tilde{A}(y, z) = \int_{\mathbf{R}^n} \exp\{i(y, \lambda)\} A(\lambda, z)d\lambda$$

and

$$\tilde{B}(y, z) = \int_{\mathbf{R}^n} \exp\{i(y, \lambda)\} \tilde{\psi}(\lambda, z)d\lambda.$$

The equivalence of (1.80) and (1.83) is obvious. Another equivalent definition of a quasi-homogeneous random field can be derived from the relation

$$V(y, z) = \int_{\mathbf{R}^n} \tilde{A}(y - y', z)\tilde{V}(y', z)dy'$$

where $\tilde{V}(y, z)$ is a random field homogeneous in (y, z).

We now present the simulation formulas. First, let us consider a complex-valued quasi-homogeneous random field. The general form of the simulation formula coincides with (1.72) where $\xi_\lambda(z)$ is chosen as follows:

$$\xi_\lambda(z) = A(\lambda, z)\tilde{\xi}_\lambda(z)$$

where $\tilde{\xi}_\lambda(z)$ is a homogeneous l-dimensional complex-valued random field with the correlation matrix $\psi(\lambda, z_1 - z_2)$ which is simulated according to the standard formula (1.65).

Let us consider now a real-valued random field. The simulation formula coincides now with (1.74) where the pair $\xi_\lambda(z)$, $\eta_\lambda(z)$ is calculated as follows:

$$\begin{pmatrix} \xi_\lambda(z) \\ \eta_\lambda(z) \end{pmatrix} = \begin{pmatrix} \mathrm{Re}\,\{A(\lambda, z)\} & \mathrm{Im}\,\{A(\lambda, z)\} \\ -\mathrm{Im}\,\{A(\lambda, z)\} & \mathrm{Re}\,\{A(\lambda, z)\} \end{pmatrix} \begin{pmatrix} \tilde{\xi}_\lambda(z) \\ \tilde{\eta}_\lambda(z) \end{pmatrix}. \tag{1.84}$$

Here, a pair $\tilde{\xi}_\lambda(z)$, $\tilde{\eta}_\lambda(z)$ constitutes a $2l$-dimensional real-valued homogeneous random field with the correlation tensor

$$\begin{pmatrix} \mathrm{Re}\,\tilde{\psi}(\lambda, z_1 - z_2) & \mathrm{Im}\,\tilde{\psi}(\lambda, z_1 - z_2) \\ -\mathrm{Im}\,\tilde{\psi}(\lambda, z_1 - z_2) & \mathrm{Re}\,\tilde{\psi}(\lambda, z_1 - z_2) \end{pmatrix}. \tag{1.85}$$

It remains to show that the correlation tensor of the pair $\xi_\lambda(z)$, $\eta_\lambda(z)$ is equal to $\psi(\lambda, z_1, z_2)$. This is implied by the following equality which is equivalent to (1.80):

$$\begin{pmatrix} \psi_{12}^R & \psi_{12}^I \\ -\psi_{12}^I & \psi_{12}^R \end{pmatrix} = \begin{pmatrix} A_1^R & A_1^I \\ -A_1^I & A_1^R \end{pmatrix} \begin{pmatrix} \tilde{\psi}_{12}^R & \tilde{\psi}_{12}^I \\ -\tilde{\psi}_{12}^I & \tilde{\psi}_{12}^R \end{pmatrix} \begin{pmatrix} A_2^R & A_2^I \\ -A_2^I & A_2^R \end{pmatrix} \tag{1.86}$$

where

$$\psi(\lambda, z_1, z_2) = \psi_{12}^R + \mathrm{i}\psi_{12}^I, \quad \tilde{\psi}(\lambda, z_1, z_2) = \tilde{\psi}_{12}^R + \mathrm{i}\tilde{\psi}_{12}^I$$

$$A(\lambda, z_1) = A_1^R + \mathrm{i}A_1^I, \qquad A(\lambda, z_2) = A_2^R + \mathrm{i}A_2^I.$$

Remark. It is not difficult to construct a version of the expansion in (1.80) where $\tilde{\psi}(\lambda, z)$ is not a partial spectral tensor provided that (1.81) is not satisfied. However it may appear that the condition (1.81) is violated but $\tilde{\psi}(\lambda, z)$ in (1.80) is a partial spectral tensor. Consequently, in this case the simulation formulas presented are valid.

Simulation of Local Homogeneous Random Vector Fields. A special class of non-homogeneous random fields which play an important role in practical problems of turbulence theory is that of local homogeneous random fields [1.27]. The local homogeneous random field is defined as a random field in which all the distributions of the differences of the field components on a set of pairs of points do not change under arbitrary transfers of these pairs [1.27]. We use the spectral representation of a local homogeneous random field

$$V(x) = \int_{\mathbf{R}^n} [\exp\{\mathrm{i}(\lambda, x)\} - 1]dZ(\lambda) + v_0 \ (x \in \mathbf{R}^n) \tag{1.87}$$

where $v_0 = V(0)$ is a constant l-dimensional random vector, $Z(\Delta\lambda)$ is a vector complex-valued measure in $\mathbf{R}^n$ with non-correlated increments such that $dZ_j(-\lambda) = dZ_j^*(\lambda)$:

$$\langle dZ_j(\lambda)d\bar{Z}_s(\lambda)\rangle = F_{js}(\lambda)d\lambda , \qquad (1.88)$$

where $F_{js}(\lambda)$ are the components of the spectral tensor $F(\lambda)$. In addition, $F(\lambda)$ is positive definite for any $\lambda \neq 0$, $F_{js}(\lambda) = F_{js}^*(-\lambda) = F_{sj}^*(\lambda)$ and

$$\int_{|\lambda|<\varepsilon} \lambda^2 F_{ii}(\lambda)d\lambda < \infty , \quad \int_{|\lambda|>\varepsilon} F_{ii}(\lambda)d\lambda < \infty , \quad i = 1,\dots,l \qquad (1.89)$$

for arbitrary $\varepsilon > 0$.

As is known [1.27], if a spectral tensor $F(\lambda)$ satisfies the above conditions, then there exists a local homogeneous random field with the spectral tensor $F(\lambda)$. Moreover, if $F(\lambda)$ satisfies the condition [stronger than (1.89)]

$$\int_{\mathbf{R}^n} F_{ii}(\lambda)d\lambda < \infty , \quad i = 1,\dots l$$

then the field $V(x)$ is homogeneous.

Note that it is not possible to obtain the distributions of $V(x)$ on the basis of $F(\lambda)$ since the random value v_0 in (1.87) and its correlation with $dZ(\lambda)$ are not known. In the framework of spectral theory, it is correct to speak about distributions of the field increments at two different points. In the gaussian case, the local homogeneous random field is fully specified by

$$D_{js}(x, r_2, r_1) = \langle [v_j(x + r_2) - v_j(x)] [\bar{v}_s(r_1) - \bar{v}_s(0)] \rangle . \qquad (1.90)$$

Consider the structure tensor [1.27]:

$$D_{js}(r) = \langle [v_j(x + r) - v_j(x)] [\bar{v}_s(x + r) - \bar{v}_s(x)] \rangle . \qquad (1.91)$$

From (1.87,91), we obtain

$$D_{js}(r) = 2 \int_{\mathbf{R}^n} [1 - \cos(\lambda, r)] F_{js}(\lambda)d\lambda . \qquad (1.92)$$

We represent F_{js} through D_{js} by using (1.92). We assume additionally that

$$\int_{\mathbf{R}^n} |\lambda| F_{js}(\lambda)d\lambda < \infty . \qquad (1.93)$$

From (1.92,93), we have

$$\nabla D_{js}(r) = 2 \int_{\mathbf{R}^n} \lambda \sin(\lambda, r) F_{js}(\lambda)d\lambda . \qquad (1.94)$$

Hence, $2\lambda F_{js}(\lambda)$ can be obtained as a sine-transformation of $\nabla D_{js}(r)$ if condition (1.93) is satisfied. From (1.87,90) we obtain

$$D_{js}(x, r_2, r_1) = \int_{\mathbf{R}^n} e^{i(\lambda, x)} \left[e^{i(\lambda, r_2)} - 1 \right] \left[e^{-i(\lambda, r_1)} - 1 \right] F_{js}(\lambda) d\lambda . \qquad (1.95)$$

Thus, (1.90–95) show that in the gaussian case the structure tensor D_{js} fully specifies the local homogeneous field. Note that if $V(x)$ and $F_{js}(\lambda)$ are real-valued, then the following simple relation holds:

$$\begin{aligned} D_{js}(x, r_2, r_1) = D_{sj}(x, r_2, r_1) = &\tfrac{1}{2} \{ D_{js}(x + r_1) + D_{js}(x - r_2) \\ &- D_{js}(x - r_2 + r_1) - D_{js}(x) \} . \end{aligned} \qquad (1.96)$$

Consider now the problem of construction of simulation formulas for the increments of a vector local homogeneous random field $V(x)$ with a given spectral tensor $F(\lambda)$. Let $h = (h_1, \ldots, h_n)$ be a vector with positive coordinates, and let

$$\Delta_{h_i} V(x) = V(x + h_i e_i) - V(x) , \quad i = 1, \ldots, n ,$$

where e_i is a unit vector. We introduce the ln-dimensional column vector

$$\Delta_{h_i} V(x) = \left((\Delta_{h_1} V)^{\mathrm{T}}, \ldots, (\Delta_{h_n} V(x)^{\mathrm{T}}) \right)^{\mathrm{T}} .$$

From (1.95), we have

$$\left\langle \Delta_{h_j} V(x + \varrho) \left(\Delta_{h_s} V(x) \right)^* \right\rangle = \int_{\mathbf{R}^n} e^{i(\lambda, \varrho)} \mu_j(\lambda) \mu_s(\bar{\lambda}) F(\lambda) d\lambda \qquad (1.97)$$

where

$$\mu_j(\lambda) = \left[\exp(i\lambda_j h_j) - 1 \right] , \quad j = 1, \ldots, n .$$

Thus, $\Delta_h V$ has the spectral tensor $\mathcal{F}(\lambda)$:

$$\mathcal{F}(\lambda) = \begin{pmatrix} \mu_1(\lambda) F \\ \vdots \\ \mu_n(\lambda) F \end{pmatrix} \left(\bar{\mu}_1(\lambda) F^*, \ldots, \bar{\mu}_n(\lambda) F^* \right) . \qquad (1.98)$$

First, we present the simulation formula for the case of complex-valued fields:

$$\Delta_{h_j} V(x) = \mu_j(\lambda) \exp \{ i(x, \lambda) \} \frac{\xi_\lambda}{[p(\lambda)]^{1/2}} , \quad j = 1, \ldots, n . \qquad (1.99)$$

Here, ξ_λ (λ is fixed) is an l-dimensional complex-valued random column vector with the zero mean and with the covariance matrix

$$\langle \xi_\lambda \xi_\lambda^* | \lambda \rangle = F(\lambda)$$

[$\lambda \in \mathbf{R}^n$ is distributed according to a density $p(\lambda)$]. It is easy to verify that component (1.99) of the field $\Delta_h V$ satisfies (1.97).

Consider now the case of real-valued fields. Let ξ_λ and η_λ be independent real-valued l-dimensional column vectors with zero mean and with the unit covariance matrix, and let Ξ_λ and H_λ be ln-dimensional column vectors:

$$\Xi_\lambda = \begin{pmatrix} G_1^R \xi_\lambda + G_1^I \eta_\lambda \\ \dots\dots\dots \\ G_n^R \xi_\lambda + G_n^I \eta_\lambda \end{pmatrix} , \quad H_\lambda = \begin{pmatrix} -G_1^I \xi_\lambda + G_1^R \eta_\lambda \\ \dots\dots\dots \\ -G_n^I \xi_\lambda + G_n^R \eta_\lambda \end{pmatrix} .$$

Here,

$$G_j^R = \mathrm{Re}\,[\mu_j(\lambda) G_j(\lambda)] , \quad G_j^I = \mathrm{Im}\,[\mu_j(\lambda) G_j(\lambda)] ,$$

where $F(\lambda) = G(\lambda) G(\lambda^*)$, and G is the left triangular matrix. It is not difficult to verify that

$$\langle \Xi_\lambda \Xi_\lambda^T | \lambda \rangle = \langle H_\lambda H_\lambda^T | \lambda \rangle = \mathrm{Re}\,\mathcal{F}(\lambda)$$
$$\langle \Xi_\lambda H_\lambda^T | \lambda \rangle = -\langle H_\lambda \Xi_\lambda^T | \lambda \rangle = \mathrm{Im}\,\mathcal{F}(\lambda) .$$

Using the general form of the simulation formula for homogeneous real-valued random fields (1.66), we find out that

$$\Delta_h V(x) = \frac{1}{[p(\lambda)]^{1/2}} \left[\Xi_\lambda \cos(\lambda, x) + H_\lambda \sin(\lambda, x) \right] \tag{1.100}$$

has the spectral tensor $\mathcal{F}(\lambda)$.

We rewrite (1.100) in the coordinate form

$$\Delta_{h_i} V(x) = \big[(G_j^R \xi_\lambda + G_j^I \eta_\lambda) \cos(\lambda, x)$$
$$+ (-G_j^I \xi_\lambda + G_j^R \eta_\lambda) \sin(\lambda, x) \big]\, p(\lambda)^{-1/2} \quad (j = 1,\dots,n) . \tag{1.101}$$

It is sometimes useful to simulate derivatives of the random field $V(x)$. It is then necessary to assume that

$$\int \lambda^2 \mathrm{Sp}\, F(\lambda) d\lambda < \infty . \tag{1.102}$$

This condition ensures the existence of the spectral representation of the homogeneous random field $\nabla V(x)$:

$$\nabla V(x) = \begin{pmatrix} \dfrac{\partial V}{\partial x_1} \\ \vdots \\ \dfrac{\partial V}{\partial x_n} \end{pmatrix} . \tag{1.103}$$

We assume that condition (1.102) is satisfied; in turbulence theory, (1.102) holds since F decreases rapidly as λ increases [1.27]. Note that if (1.102) is violated, then it is possible to choose $\lambda_0 > 0$ and put $F(\lambda) = 0$ for $|\lambda| > \lambda_0$. Differentiating (1.87) yields [1.27]:

$$\left\langle \frac{\partial V(x+\varrho)}{\partial x_j} \left(\frac{\partial V(x)}{\partial x_s} \right)^* \right\rangle = \int_{\mathbf{R}^n} e^{i(\lambda, \varrho)} \lambda_j \lambda_s F(\lambda) d\lambda . \tag{1.104}$$

Consequently, the spectral tensor $F(\lambda)$ of the ln-dimensional random field (1.103) has the following representation:

$$\mathcal{F}(\lambda) = \begin{pmatrix} \lambda_1 G \\ \vdots \\ \lambda_n G \end{pmatrix} (\lambda_1 G^*, \dots, \lambda_n G^*) \tag{1.105}$$

where $G(\lambda)$ is the l-dimensional matrix determined from $F(\lambda) = G(\lambda)G^*(\lambda)$.

First, let us consider the general case of a complex-valued field $V(x)$. Let $\xi_\lambda = G\xi_\lambda'$, where ξ_λ' is a random l-dimensional column-vector with zero mean and the unit covariance matrix. It is not difficult to verify that the random field with components

$$\frac{\partial V(x)}{\partial x_j} = \lambda_j \frac{\exp\{i(\lambda, x)\}}{[p(\lambda)]^{1/2}} \xi_\lambda , \quad j = 1, \dots, n$$

in fact has the desired spectral tensor (1.105).

Consider a real-valued field. Let ξ_λ' and η_λ' be independent real-valued l-dimensional random vectors with zero mean and the unit covariance matrix. Now, we have

$$\langle \xi_\lambda \eta_\lambda^T | \lambda \rangle = \langle \eta_\lambda \eta_\lambda^T | \lambda \rangle = \operatorname{Re} F(\lambda) ,$$
$$\langle \xi_\lambda \eta_\lambda^T | \lambda \rangle = - \langle \eta_\lambda \xi_\lambda^T | \lambda \rangle .$$

Assume that

$$G^R(\lambda) = \operatorname{Re} G(\lambda) , \quad G^I(\lambda) = \operatorname{Im} G(\lambda) .$$

Define a pair of ln-dimensional random column vectors

$$\Xi_\lambda = \begin{pmatrix} \lambda_1 \left(G^R \xi_\lambda' + G^I \eta_\lambda' \right) \\ \lambda_n \left(G^R \xi_\lambda' + G^I \eta_\lambda' \right) \end{pmatrix} , \quad H_\lambda = \begin{pmatrix} \lambda_1 \left(-G^I \xi_\lambda' + G^R \eta_\lambda' \right) \\ \lambda_n \left(-G^I \xi_\lambda' + G^R \eta_\lambda' \right) \end{pmatrix} .$$

Then, the covariance matrix of $\begin{pmatrix} \Xi_\lambda \\ H_\lambda \end{pmatrix}$ (λ is fixed) is given by

$$\begin{pmatrix} \operatorname{Re} \mathcal{F} & \operatorname{Im} \mathcal{F} \\ -\operatorname{Im} \mathcal{F} & \operatorname{Re} \mathcal{F} \end{pmatrix}$$

From this, we obtain

$$\nabla V(x) = \frac{1}{[p(\lambda)]^{1/2}} \left[\Xi_\lambda \cos(\lambda, x) + H_\lambda \sin(\lambda, x) \right]$$

or in the coordinate form

$$\frac{\partial V}{\partial x_j}(x) = \lambda_j \left\{ \left(G^R \xi_\lambda' + G^I \eta_\lambda' \right) \cos(\lambda, x) \right.$$
$$\left. + \left(-G^I \xi_\lambda' + G^R \eta_\lambda' \right) \cos(\lambda, x) \right\} [p(\lambda)]^{-1/2} , \quad j = 1, \dots, n .$$

1.2.5 Stochastic Problems and Double Randomization

Let us first consider a simple example. Suppose that the following diffusion equation

$$\Delta u(x) - c(x)u(x) = -g(x,\omega) , \quad x \in G$$
$$u(y) = \varphi(y) , \qquad\qquad y \in \Gamma = \partial G$$

describes a diffusion process where $g(x,\omega)$ is a stochastic source generated by a random element $\omega \in \Omega$. Generally, it may appear that the absorption coefficient $c(x)$ is a random field ($x \in G$), or the function $\varphi(y)$, $y \in \Gamma$ is random. Moreover, there exist situations where the boundary is described as a random surface. Consequently, $u(x)$ is also random, and it is interesting to obtain the statistical characteristics of $u(x)$, for example, the expectation $\langle u(x) \rangle$, the variance $Du(x)$, the distributions $P\{u(x) > u_0\}$, etc. Direct solution of the boundary value problem for thousands of samples of the random element ω is practically impossible. Therefore, one usually uses the double randomization scheme. In this scheme we first choose a realization of ω, say ω_1, and then do not solve the boundary value problem exactly but construct only a Monte Carlo estimate of the functional to be found. Then the process is repeated for a series of independent samples of $\omega : \omega_1, \omega_2, \ldots, \omega_N$, so that the total number of realizations is sufficient to calculate the desired statistical characteristics exactly enough. Keeping this in mind let us desribe the double randomization scheme more generally [1.8].

Suppose that it is desired to calculate the expectation of a random quantity

$$\xi = \sum_{n=0}^{\infty} \prod_{k=0}^{k_n} \eta_{nk}(\omega_1) , \quad \omega_1 \in \Omega_1 ,$$

where the realizations of random quantities η_{nk} are integrals. The randomization principle consists of changing η_{nk} with their random estimates $\tilde{\eta}_{nk}$. Let $\tilde{\eta}_{nk} = \tilde{\eta}_{nk}(\omega_1, \omega_2)$ where $\omega_2 \in \Omega_2$. The probability measure on $\Omega_1 \times \Omega_2$ is introduced according to the multiplication law of measures.

Assume that the estimates $\tilde{\eta}_{nk}$ are conditionally independent and unbiased, i.e.,

$$M \prod_{k=0}^{k_n} \tilde{\eta}_{nk}(\omega_1) = \prod_{k=0}^{k_n} \eta_{nk}(\omega_1) .$$

Then

$$M\tilde{\xi} = M \sum_{n=0}^{\infty} \prod_{k=0}^{k_n} \tilde{\eta}_{nk} = M \sum_{n=0}^{\infty} M \left\{ \prod_{k=0}^{k_n} \tilde{\eta}_{nk} \Big| \omega_1 \right\} = M\xi ;$$

$$D\tilde{\xi} = DM(\tilde{\xi}|\omega_1) + MD(\tilde{\xi}|\omega_1) = D\xi + M(\tilde{\xi}^2 - \xi^2) .$$

It is obvious that the double randomization principle is extended to multiple random elements $\omega_1, \omega_2, \omega_3, \ldots$.

Consider a vectorial random variable

$$\xi = \xi(\sigma, \omega) = \{\xi_1(\sigma, \omega), \ldots, \xi_n(\sigma, \omega)\} \, ,$$

where ω is a random point of an abstract space (for example, a trajectory of a Markov chain), and $\sigma = (\sigma_1, \ldots, \sigma_n)$ is a given random vector. It is assumed that the joint distribution of ω and σ is given. It is desired to calculate the statistical characteristics of ξ, e.g., the correlation function for the random vector of conditional expectations

$$J = J(\sigma) = M(\xi | \sigma)$$

and mutual correlation moments of J and σ, i.e.,

$$K\left[J_k, J_j\right] = M\left[(J_k - J_k^*)(J_j - J_j^*)\right] \, , \quad k, j = 1, \ldots, n \, ,$$
$$K\left[J_k, \sigma_j\right] = M\left[(J_k - J_k^*)(\sigma_j - \sigma_j^*)\right] \quad k = 1, \ldots, n \, ; \quad j = 1, \ldots, m \, .$$

Here $\sigma_j^* = M\sigma_j$, $J_k^* = MJ_k$. If for each value of σ the exact vector $J(\sigma)$ is known then the problem can be directly solved by the Monte Carlo method. Note that in the general case we can apply the double randomization technique using one sample (for mutual correlation moments) or two samples of ω (for autocorrelation moments). Indeed,

$$K[J_k, \sigma_i] = M\left[(J_k - J_k^*)(\sigma_i - \sigma_i^*)\right] = M\left[(\sigma_i - \sigma_i^*)M(\xi_k - J_k^* | \sigma)\right]$$
$$= M\left[(\sigma_i - \sigma_i^*)(\xi_k - J_k^*)\right] = K[\sigma_i, \xi_k] \, .$$

Let $\omega^{(1)}$, $\omega^{(2)}$ be two independent samples of ω for a given value of σ and let $\xi_k^{(1)} = \xi_k\left(\sigma, \omega^{(1)}\right)$, $\xi_j^{(2)} = \xi_j\left(\sigma, \omega^{(2)}\right)$, then

$$K\left[\xi_k^{(1)}, \xi_j^{(2)}\right] = M\left[\left(\xi_k^{(1)} - J_k^*\right)\left(\xi_j^{(2)} - J_j^*\right)\right]$$
$$M\left[M\left(\xi_k^{(1)} - J_k^* | \sigma\right) M\left(\xi_j^{(2)} - J_j^* | \sigma\right)\right] = K\left[J_k, J_j\right] \, .$$

Consider now a linear functional equation $L\varphi = f$ with a stochastic parameter, say, a random field σ. Suppose that the unique solution of this equation exists for all realizations of σ. Assume also that some statistical characteristics of the random solution $\varphi(x, \sigma)$ exist; for example, the correlation function

$$B(x_1, x_2) = \left\langle \left[\varphi(x_1) - \langle\varphi(x_1)\rangle\right]\left[\varphi(x_2) - \langle\varphi(x_2)\rangle\right]\right\rangle \, ,$$

the average $M\varphi(x) = \langle\varphi(x)\rangle$, the intensity of fluctuations

$$I_\varphi = \frac{\left\{\langle\varphi^2(x)\rangle - \langle\varphi(x)\rangle^2\right\}}{\langle\varphi(x)\rangle^2} \, ,$$

etc. Here $\langle\ldots\rangle$ denotes averaging over the ensemble of the solutions φ.

Let $J_k(\varphi)$, $k = 1, \ldots, m$ be the functionals of the solution of $L\varphi = f$ to be calculated. Assume that unbiased estimates $\xi_k(\omega)$ are obtained:

$$J_k = M\xi_k(\omega) , \quad k = 1,\ldots,m ,$$

where ω is a random process (e.g., the walk on spheres process). In fact, ξ_k depend also on σ:

$$\xi_k = \xi_k(\omega,\sigma) , \quad J_k = J_k(\sigma) , \quad M\left[\xi_k(\omega,\sigma)|\sigma\right] = J_k(\sigma) .$$

In addition, it may appear that ω and σ are dependent random elements. The problem is to calculate the quantities

$$J_k = \langle J_k(\sigma) \rangle , \quad R_{kj} = \left\langle \left[J_k(\sigma)J_j(\sigma)\right]\right\rangle , \quad j,k = 1,\ldots,m .$$

These quantities could be evaluated using the double randomization technique which is here based on the relations:

$$\langle J_k(\sigma)\rangle = \langle M\xi_k(\omega,\sigma)\rangle = M_{(\omega,\sigma)}\xi_k(\omega,\sigma) ,$$

$$\langle J_k(\sigma)J_j(\sigma)\rangle = M_{(\omega_1,\omega_2,\sigma)}\left[\xi_k(\omega_1,\sigma)\xi_j(\omega_2,\sigma)\right] .$$

Here M denotes averaging with respect to the distributions indicated by the subscripts; ω_1, ω_2 are conditionally independent trajectories constructed for a fixed realization of σ. From these representations it is clear that to calculate J_k, it is necessary to construct one trajectory ω for a fixed sample of σ, while, to calculate R_{kj}, it is necessary to construct two independent trajectories for a fixed sample of σ.

2. Monte Carlo Algorithms
for Solving Integral Equations

Monte Carlo algorithms for solving integral equations when the Neumann series for the corresponding majorant equation converges are well developed [2.1–3]. However, many integral equations which are equivalent to the boundary value problems do not statisfy this condition. Morever, often the Neumann series diverges even for the original integral equation to be solved. This situation arises often in many problems of mathematical physics [2.4–7]. Perhaps it may be found that integral equations with convergent Neumann series compose only a small class of integral equations of practical interest.

In this chapter we propose several techniques useful for solving general integral equations with compact operators by the Monte Carlo method. In Sect. 2.1 we describe Monte Carlo algorithms based on analytical continuation of the initial element of the resolvent by special transformation of the spectral parameter $\lambda = \psi(\eta)$. The method of Sect. 2.1 has not been previously used in Monte Carlo methods, though this approach is well known in numerical analysis [2.4, 8]. It should be noted, however, that the method described in Sect. 2.1.3 can be implemented practically only by the Monte Carlo method. Asymptotically unbiased estimates based on approximations of the kernel of an integral equation by a singular function are proposed in Sect. 2.2. The method of transformation is applied in Sect. 2.3 to solve the eigen-value problems for integral equations.

In the last section we give a survey of alternative methods of construction of the resolvent and some numerical results to illustrate possibilities of some modifications, for example, continuation by Mittag–Leffler method combined with the transformation $\lambda = i(i+\eta)/(i-\eta)$.

2.1 Algorithms Based on Numerical Analytical Continuation

2.1.1 Statement of the Problem and the Main Definitions

Let X be a Banach functional space of functions $\varphi(x), x \in \mathbf{R}^n$, and let U be a continuous linear operator: $U : X \to X$. Consider an equation of the second kind

$$\varphi - \lambda U\varphi = f , \qquad (2.1)$$

where λ is a complex parameter. If X is a real space then complex extension of U is taken. We use also another form of (2.1):

$$\mu\varphi - U\varphi = f .$$ (2.2)

We denote by $\rho(U)$ and $\sigma(U)$ the resolvent set of U and the spectrum of U, respectively. This means that $\rho(U)$ is the set of values of μ such that (2.2) is uniquely solvable for arbitrary f, and $\sigma(U)$ is the complementary set.

We denote by $\pi(U)$ the non-singular values of the operator U, and by $\chi(U)$ the complementary set [$\chi(U)$ is called characteristic set of the operator U]. It is clear that if $\lambda \in \chi(U)$, then $\mu = \lambda^{-1} \in \sigma(U)$ and vice versa.

Let $\lambda \neq 0$ be a non-singular value. The operator defined by

$$I + \lambda R_\lambda = (I - \lambda U)^{-1}$$ (2.3)

is called a resolvent operator.

In the case of regular value of μ the following resolvent operator can be considered:

$$B_\mu = (\mu I - U)^{-1} .$$ (2.4)

In (2.3, 4), I denotes the identity operator.

Note that the resolvent B_μ is usually used in the theory of functional analysis as an inverse operator. In the theory of integral equations [2.4, 9] the resolvent R_λ is defined as in (2.3). If $\mu \neq 0$, then

$$B_\mu = \frac{1}{\mu} I + \frac{1}{\mu^2} R_{1/\mu}$$

and vice versa; if $\lambda \neq 0$, then

$$R_\lambda = \frac{1}{\lambda} U B_{1/\lambda} = \frac{1}{\lambda} B_{1/\lambda} U .$$

Thus all the statements concerning R_λ could be reformulated through B_μ and vice versa.

We shall suppose in the following discussion that the Fredholm alternativ theorem for a pair of adjoint integral equations

$$(I - U)\varphi = f , (I - U)^* \varphi^* = g$$

is true. It is then sufficient to assume that the operator U^m is compact, m being a positive integer. Mainly we deal with integral operators of the form

$$U\varphi = K\varphi = \int_G k(x, x')\varphi(x')dx'$$ (2.5)

where the kernel $k(x, x')$ is an arbitrary function (for example, a generalized function), $G \subset R^n$ is a domain such that the following conditions hold

51

1) $\left\{ \int_G |k(x,x')|^r dx \right\}^{1/r} \leq c_1 , \quad x' \in G ,$

2) $\left\{ \int_G |k(x,x')|^\sigma dx' \right\} \leq c_2 , \quad x \in G ,$

3) $p - r(p-1) < \sigma < p$.

Note that an integral operator with this kind of kernel will transform $L^p(G)$ to $L^p(G)$ with the norm

$$\|U\| \leq c_1^{(1-\sigma)/p} c_2^{\sigma/p} .$$

In particular, we shall use the kernels of the potential theory

$$K(x,x') = \frac{b(x,x')}{|x - x'|^{m'}}$$

where $b(x,x')$ is a bounded continuous function and $m < n$. If in addition $p > n(n-m)$, then $K : L^p(G) \to C(G)$ (Theorem XI.3.7 in [2.10]).

We denote by $\lambda_1, \lambda_2, \dots$ the characteristic numbers of the integral equation with a compact operator K

$$\varphi(x) = \lambda K\varphi + f(x) \tag{2.6}$$

and assume that $|\lambda_1| \leq |\lambda_2| \leq \dots$. The Monte Carlo methods for solving (2.6) are based on the representation

$$\varphi = (I - \lambda K)^{-1} f = f + \lambda R_\lambda f \tag{2.7}$$

where

$$R_\lambda = K + \lambda K^2 + \dots \tag{2.8}$$

which converges for $\lambda : |\lambda| < |\lambda_1|$.

If n terms in (2.8) are taken, the remainder can be estimated as

$$|\Delta_n| = 0(\delta^{n+1} n^{r-1}) \tag{2.9}$$

where $\delta = |\lambda|/|\lambda_1|$, r is the multiplicity of the pole [2.8]; (2.9) shows that if λ is close to λ_1 (i.e., $|\lambda| \lesssim |\lambda_1|$) then the series (2.8) converges very slowly and diverges if $|\lambda| \geq |\lambda_1|$. Thus, the problem can now be stated. The first problem is to accelerate the convergence of the power series (2.8) for $\lambda : |\lambda| \lesssim |\lambda_1|$. The second problem is to construct $(I - \lambda K)^{-1} f$ for $\lambda : |\lambda| \geq |\lambda_1|$, using values [explicit or approximate] of the integrals $K^n f, n = 1, 2, \dots$.

Note that if the kernel $k(s,t)$ vanishes for $s < t$, then we come to the Volterra' equation

$$\varphi(s) - \lambda \int_0^s k(s,t)\varphi(t)dt = f(s) . \tag{2.10}$$

This kind of equation will be used for solving the nonstationary boundary value problems in Chap. 5. It is known that the resolvent of such equations is an entire function of λ, so the Neumann series (2.8) converges for all values of λ. Another type of Volterra' equation with singularities will be used in Chap. 5.

2.1.2 Analytical Continuation of Neumann Series Based on the Spectral Parameter Transformation

The first to apply the spectral parameter transformation to solve integral equations was perhaps E. *Goursat* [2.11]. Special transformations were considered also in [2.4, 12]. This technique was used in [2.8] to solve various problems of numerical analysis and in [2.13, 14] to investigate some applied problems in physics. Numerical analytical continuation methods and stability estimations were considered in [2.15].

In this section, we consider special spectral transformations which can be used when solving integral equations by Monte Carlo methods. The computational cost of this technique is estimated.

Thus, we consider the problem of construction of the solution to (2.6) for $\lambda \in D, \lambda \neq \lambda_k, k = 1, 2, \ldots$., where D is a simple connected domain of the complex plane lying in the domain of definition of the function $R_\lambda f$ such that all the characteristic numbers lie outside of D and the point $\lambda = 0$ belongs to D. It is assumed that in the neighborhood of the point $\lambda = 0$ the resolvent $R_\lambda f$ is defined by the power series

$$R_\lambda f = \sum_{k=0}^{\infty} c_k \lambda^k , \quad c_k = K^{k+1} f . \tag{2.11}$$

On a plane of a complex variable η consider a unit disk $\Delta = \{\eta : |\eta| < 1\}$ and construct a function

$$\lambda = \psi(\eta) = a_1 \eta + a_2 \eta^2 + \ldots , \tag{2.12}$$

a conformal mapping of Δ on D. Then

$$R_{\psi(\eta)} f = \sum_{n=0}^{\infty} b_n \eta^n \tag{2.13}$$

is obtained by substitution of (2.12) into (2.11), where

$$b_n = \sum_{k=1}^{n} d_k^{(n)} c_k , \quad d_k^{(n)} = \frac{1}{n!} \left\{ \frac{\partial^n}{\partial \eta^n} [\psi(\eta)]^k \right\}_{\eta=0} . \tag{2.14}$$

It is clear from this that it is possible to choose adequately the domain D (eliminating, for instance, all the poles λ_k), transform the value $\lambda = \lambda_*$ (at this point we seek the solution to the integral equation) to a point $\eta = \eta_* = \psi^{-1}(\lambda_*)$ lying in Δ. Then the desired solution to the integral equation is represented in the form

$$\varphi = f + \lambda_* R_{\psi(\eta_*)} f \qquad (2.15)$$

where the series $R_{\psi(\eta)} f$ converges absolutely and uniformly on Δ. Note that this approach can also be used to accelerate the convergence in the case when $|\lambda| \lesssim |\lambda_1|$. We consider a simple example which will be used later on to construct walk on boundary algorithms. Assume that all the characteristic values are real, and $\lambda_k \in (-\infty, -a)$, $a > 0$. Without loss of generality, we assume that it is desired to solve (2.6) at $\lambda = \lambda_* = 1$. Then it is convenient to choose the mapping of the disk Δ on the domain D_a = the complex plane with a cut along the real axis from $-a$ to $-\infty$:

$$\lambda = \psi(\eta) = \frac{4a\eta}{(1 - \eta)^2} \, . \qquad (2.16)$$

Then the series $R_{\psi}(\eta) f$ converges absolutely and uniformly on Δ, and the coefficients (2.14) can be accurately calculated:

$$d_k^{(n)} = (4a)^k C_{k+n-1}^{2k-1} \, .$$

To construct a biased random estimate we chosse m so that the remainder of the series is equal to a desired quantity ε. Then

$$R_{\lambda_*} f \cong \sum_{k=1}^{m} b_k \eta_*^k = \sum_{k=1}^{m} \eta_*^n \sum_{i=1}^{k} d_i^{(k)} c_i = \sum_{k=1}^{m} c_k l_k^{(m)} \, , \qquad (2.17)$$

where

$$l_k^{(m)} = \sum_{n=k}^{m} d_k^{(n)} \eta_*^n \, . \qquad (2.18)$$

The coefficients $l_k^{(m)}$ are calculated in advance, according to (2.18), and the integrals $c_k = K^{k+1} f$ are calculated by Monte Carlo methods. Let ξ_k be an unbiased Monte Carlo estimate for c_{k-1}, then an ε-biased estimate for the solution to the integral equation has the form

$$\zeta_\varepsilon^{(m)}(x) = f(x) + \sum_{k=1}^{m} \xi_k l_k^{(m)} \, . \qquad (2.19)$$

Theorem 2.1. In the case of the transformation (2.16), $l_k^{(m)} \le 1$ for all k and m. If $D\xi_k \le \sigma^2$, then the cost of the estimate $\zeta_\varepsilon^{(m)}$ needed to achieve the error of order ε has the form

$$T_\varepsilon = 0(|\ln(\varepsilon)|^3 / \varepsilon^2) \, . \qquad (2.20)$$

Proof. Since (2.13) converges as a power series, then it is sufficient to show that $l_k^{(m)} \le 1$. Note that

$$\eta_* = \psi^{-1}(\lambda_*) = \left[(a + \lambda_*)^{1/2} - a^{1/2} \right] / \left[(a + \lambda_*)^{1/2} - a^{1/2} \right] > 0$$

54

and $d_k^{(n)} = (4a)^k C_{k+n-1}^{2k-1} > 0$, $n = k, k+1, \ldots$, therefore, $l_k^{(m)}$ monotonically increases as m increases, for fixed arbitrary k. In the limit $m \to \infty$, $l_k^{(\infty)} = 1$, since (2.18) gives 1^k, $k = 1, 2, \ldots$. It follows from this that $|l_k^{(m)}| \leq 1$ for arbitrary k and m. □

The following statements include an analogous result for a class of transformations.

Theorem 2.2. Assume that the conformal mapping $\lambda = \psi(\eta)$ has only simple poles on the bound $|\eta| = 1$. Then the cost of the estimate (2.19) is of order $(|\ln(\varepsilon)|^4 / \varepsilon^2)$ if $D(k) \equiv D\xi_k \leq \sigma^2$ and if

$$q = \frac{c|\eta_*|}{1 - |\eta_*|} < 1 , \tag{2.21}$$

where c is the constant of the inequality $|a_i| \leq c$.

Proof. The estimation $|a_i| \leq c$ can be obtained from the representation [2.16]:

$$\sum_{n=0}^{\infty} a_n z^n = \sum_{i=1}^{k} \frac{c_i}{1 - z_i z} + \sum_{n=0}^{\infty} b_n z^n ,$$

where $|z_i| = 1$, $i = 1, \ldots, k$, $\lim_{n \to \infty} \sup |b_n|^{1/n} < 1$. From this we have

$$|a_n| = \left| \sum_{i=1}^{k} c_i z_i^n + b_n \right| \leq \sum_{i=1}^{k} |c_i| + B , |b_n| < B .$$

Therefore

$$|\psi(\eta_*)| \leq c \frac{|\eta_*|}{1 - |\eta_*|} .$$

Using the Cauchy inequality we obtain

$$|d_k^{(n)}| \leq c^k \frac{|\eta_*|^k}{(1 - |\eta_*|)^k |\eta_*|^n} ,$$

consequently,

$$|l_k^{(m)}| \leq \sum_{n=k}^{m} \frac{c^k |\eta_*|^k}{(1 - |\eta_*|)^k} = q^k (m - k + 1) .$$

Thus,

$$D\zeta_\varepsilon^{(m)}(x) \leq m \sum_{n=k}^{M} (m - k + 1)^2 q^{2k} D(k) < \frac{q^2 \sigma^2 m^2}{1 - q^2} .$$

The last inequality proves the theorem because $m = 0 \, (\ln |\varepsilon|)$ due to the fact that the series (2.13) converges as a power series. □

Theorem 2.3. Assume that a regular function $\lambda = \psi_B(\eta) = a_1\eta + a_2\eta^2 + \ldots$ maps the disk $|\eta| < 1$ on a convex domain. Then the condition (2.21) takes the form

$$q = \frac{|\psi(\eta_*|}{1 - |\eta_*|} < 1 \, , a_1 = 1 \, .$$

The cost has the asymptotics $0 \ (|\ln(\varepsilon)|^4/\varepsilon^2)$ if $D(k) \le \sigma^2$.

Proof. The function $\psi_B(\eta)$ has the property $|a_k| \le 1$ [2.16]. To complete the proof one only needs to use the arguments of the proof of Theorem 2.2. $\quad\square$

Remark. Another assumption about $D(k)$ can lead to other cost estimations in Theorems 2.1–3. Indeed, if $D(k) = p^k\sigma^2$, $p < 1$ (for example, if $\|K^2/r\| < 1$), then $D\zeta_e^{(m)} < c\sigma^2 m$ if $l_k^{(m)} < c_1$ and $T_\varepsilon = 0(|ln(\varepsilon)|^2/\varepsilon^2)$. If $D(k) = p^k\sigma^2$, $p > 1$, then $D\zeta_e^{(m)} < q^2\sigma^2 m^3/(1 - |pq^2|)$ if $|pq^2| < 1$, and $T_\varepsilon = 0(|\ln(\varepsilon)|^4/\varepsilon^2)$. Analogously, it is possible to extend the condition $q \le 1$ if $D(k) = p^k\sigma^2$, $p < 1$, assuming $|pq^2| < 1$ instead of (2.21). Thus, to estimate the cost of (2.19), it is necessary to have more information about $l_k^{(m)}$ and $D(k)$.

It should be noted, that sometimes it is difficult to obtain an explicit expansion of the desired mapping. For example, the mapping of the disk Δ on a strip with a vertical cut [2.8]

$$\lambda = \psi(\eta) = a - i\left[\frac{a+c}{\pi} \ln\left(1 - \frac{z_0 - \eta\bar{z}_0}{1 - \eta}\right)\right]$$
$$+ \frac{b-a}{\pi} \ln\left(1 + \frac{a+c}{b-a}\frac{z_0 - \eta\bar{z}_0}{1 - \eta}\right) \tag{2.22}$$

can be expanded as

$$\lambda = a - i\left[\frac{a+c}{\pi} \ln(1 - z_0) + \frac{b-a}{\pi} \ln\left(1 + \frac{a+c}{b-a}z_0\right)\right]$$
$$+ \frac{a+c}{\pi n}(\bar{z}_0 - z) \sum_{i=1}^{m} \left[\frac{(1 - \bar{z}_0)^{i-1}}{(1 - z_0)^i} - \frac{(1 - d\bar{z})^{i-1}}{(1 + \bar{d}z)^i}\right]\eta^n + \ldots \, ,$$
$$d = \frac{a+c}{b-a} \, .$$

However, to determine the parameter z_0 (from the condition $\psi(0) = 0$, it is necessary to solve numerically the transcendental equation

$$-a_i = \frac{a+c}{\pi} \ln(1 - z_0) + \frac{b-a}{\pi} \ln\left(1 + \frac{a+c}{b-a}z_0\right)$$

by a special algorithm.

It should be noted that in these cases it is perhaps simpler to choose another mapping preserving the main properties of the original mapping. For example, in the case of (2.22), it is convenient to use the mapping of the disk Δ on the complex plane λ with a cut along the ray Re $\{\lambda\} = b$; Im $\{\lambda\} \geq 0$. Subsequent mappings (translation along b, rotation by $\pi/2$, mapping on a half-plane, then on a disk) enables one to construct the explicit transformation

$$\eta = \frac{[b - \lambda]^{1/2} + i[b - \lambda]^{1/2} - b^{1/2}(1 + i)}{[b - \lambda]^{1/2} + i[b - \lambda]^{1/2} - b^{1/2}(1 - i)} ,$$

$$\lambda = b + i \left[1 + i \frac{\exp\{i\theta\} + \eta}{\exp\{i\theta\} - \eta} \right]^2 \frac{b}{2} .$$

There is no need then to solve a transcendental equation. The desired expansion has the form (for $\theta = 0$)

$$\lambda = \sum_{k=1}^{\infty} (-2b)(1 + ki)\eta^k . \tag{2.23}$$

Note that this simple transformation can lead to an acceleration of the convergence of the Neumann series at λ_* close to $1 (b = 1)$. As an illustration, let us now show the results of the calculation of the function $(1 - \lambda)^{-1}$ (Table 2.1).

Consider now the connection between the method of transformation of the spectral parameter and the iterative methods for solving integral equations. This enables one to carry out the calculations more effectively, and to investigate the convergence of the iterative methods.

We start with the simplest transformation, namely, translation along λ_0. Note that this kind of transformation cannot be described by (2.12). This will be considered in the next section. Translation along λ_0 leads to the simplest method of inverse iterations (the resolvent iteration).

Indeed, let us seek the solution to (2.6) in the form

$$\varphi = \sum_{n=0}^{\infty} (\lambda - \lambda_0)^n \varphi_n .$$

Table 2.1. Calculations of $(1 - \lambda)^{-1}$ on the basis of (2.23)

Number of terms	λ_*	η_*	$f(\lambda_*) = \sum_{k=1}^{n} b_k \eta^k$	$(1 - \lambda)^{-1}$ Neumann series	Exact solution
12	0.8	$-0.20 + 0.46\,i$	$5.0 + 0.061\,i$	3.7	5
12	0.75	$-0.20 + 0.40\,i$	$3.9 + 0.0021\,i$	2.9	4
13	0.7	$-0.19 + 0.34\,i$	$3.33 + 0.44 \times 10^{-4}\,i$	–	3.33
12	1.5	$0.41 + 0.58\,i$	$-1.6 - 0.48\,i$	–	-2
13	1.1	$0.24 + 0.75\,i$	$-13.36 + 1.55\,i$	–	-10
20	1.1	$0.24 + 0.76\,i$	$-10.01 + 0.01\,i$	–	-10
13	$1 - 0.2\,i$	$0.11 + 0.54\,i$	$10^{-2} - 5.017\,i$	–	$-5\,i$

Then

$$\varphi_i = \lambda_0 K\varphi_i + K\varphi_{i-1} , \quad i \geq 1 , \quad \varphi_0 = \lambda_0 K\varphi_0 + f . \tag{2.24}$$

Let

$$\varphi^{(m)} = \sum_{n=0}^{m} (\lambda - \lambda_0)^n \varphi_n . \tag{2.25}$$

Then, obviously,

$$\varphi^{(m)} = \lambda_0 K\varphi^{(m)} + (\lambda - \lambda_0) K\varphi^{(m-1)} + f , \quad m \geq 1 . \tag{2.26}$$

The parameter λ_0 must be chosen on the basis of information about the first eigen-value of the integral equation.

Another variant of the resolvent iteration which coincides with the process proposed in [2.2] can be obtained by using the following transformation

$$\lambda = \varphi(\eta) = \frac{\alpha}{1 - \beta\eta} .$$

Indeed, in this case

$$\varphi_j = \alpha K\varphi_j + \beta\varphi_{j-1} , \quad \varphi_0 = \lambda_0 K\varphi_0 + f , \quad j > 1 , \tag{2.27}$$

and consequently,

$$\varphi^{(m)} = \alpha K\varphi^{(m)} + \beta\eta\varphi^{(m-1)} + f(1 - \beta\eta) . \tag{2.28}$$

The parameters α, β are chosen on the basis of the information about the spectrum of the operator K. This is not a difficult problem because of the obvious geometrical sense of the transformation $\lambda = \alpha/(1 - \beta\eta)$.

Randomization of the processes (2.26, 28) can be carried out in various manners. One version will be considered in the next section. Another is based on the standard vectorial algorithm for solving the system of integral equations (2.24, 27). The cost of the algorithms was obtained in [2.2]:

$$T_\varepsilon \leq c\varepsilon^{-2} g(m)m \tag{2.29}$$

where

$$D \sum_{i=1}^{m} \xi_{i,x} \leq g(m) ,$$

and $\xi_{i,x}$ is the ith component of the vectorial estimation for solving the system (2.24, 27); the quantity m is determined by the formula

$$m = c\ln(\varepsilon)[\ln\{R(\lambda)\}]^{-1} ,$$

where $R(\lambda) = \varrho[(I - \lambda K)^{-1}]$ is the spectral radius of the operator $(I - \lambda K)^{-1}$.

The general transformation

$$\psi(\eta) = \frac{\alpha_1 + \alpha_2 \eta}{1 + \beta \eta}$$

generates the iterative process

$$\varphi_j = \alpha_1 K \varphi_j + \alpha_2 K \varphi_{j-1} + \beta \varphi_{j-1} , \quad j > 1 ; \quad \varphi_0 = \alpha_1 K \varphi_0 + f ;$$
$$\varphi^{(m)} = \alpha_1 K \varphi^m + \eta \alpha_2 K \varphi^{(m-1)} + \beta \eta \varphi^{(m-1)} + f(1 - \beta \eta) . \tag{2.30}$$

In the case of a general transformation of the type (2.12) we have (for known coefficients a_j)

$$\varphi_0 = f , \quad \varphi_1 = \alpha_1 K \varphi_0 , \quad \varphi_j = a_j K \varphi_0 + a_{j-1} K \varphi_1 + \ldots + a_1 K \varphi_{j-1} ,$$

and for $\varphi^{(m)}$

$$\varphi^{(m)} = f + \eta a_1 K \varphi^{(m-1)} + a_2 \eta^2 K \varphi^{(m-2)} + \ldots$$
$$+ a_{m-1} \eta^{m-1} K \varphi^{(1)} + \eta^m a_m K f .$$

This can be rewritten in a form convenient for randomization:

$$\varphi^{(m)} = \left(1 - a_1 \eta - a_2 \eta^2 - \ldots - a_{m-1} \eta^{m-1}\right) f + a_1 \eta \left(K \varphi^{(m-1)} + f\right)$$
$$+ \ldots + a_{m-1} \eta^{m-1} \left(K \varphi^{(1)} + f\right) + \eta^m a_m K f .$$

Consider now some special transformations of the type of (2.12). The case $\lambda = a\eta$ leads to the method of successive approximations. The transformation (2.16) generates the iterative process ($j > 2$)

$$\varphi_0 = f , \quad \varphi_1 = 4aK \varphi_0 , \quad \varphi_2 = 4aK \varphi_1 + 2\varphi_1 ,$$
$$\varphi_j = 4aK \varphi_{j-1} + 2\varphi_{j-1} - \varphi_{j-2} ,$$

and for $\varphi^{(m)}$ ($m > 2$)

$$\varphi^{(m)} = 4a\eta K \varphi^{(m-1)} + 2\eta \varphi^{(m-1)} - \eta^2 \varphi^{(m-2)} + f(1 - \eta^2) .$$

The two-step iterative processes are generated by transformations $\lambda = \alpha\eta/(1 - \beta\eta)$. Indeed, in this case

$$\varphi_{j-1} = \alpha K \varphi_{j-2} + \beta \varphi_{j-2} ,$$

therefore

$$\varphi^{(m)} = \alpha\eta K \varphi^{(m-1)} + \beta\eta \varphi^{(m-1)} + f(1 - \beta\eta) . \tag{2.31}$$

The three-step iterative processes are generated by transformation (2.16), and by

$$\lambda = \psi(\eta) = \frac{2\alpha\eta}{1 + \eta^2} .$$

In this case

$$\varphi_0 = f \, , \quad \varphi_1 = 2\alpha K\varphi_0 \, , \quad \varphi_2 = 2\alpha K\varphi_1 \, , \quad \varphi_j = 2\alpha K\varphi_{j-1} - \varphi_{j-2} \, ,$$

i.e.,

$$\varphi^{(m)} = 2\alpha\eta K\varphi^{(m-1)} - \eta^2\varphi^{(m-2)} + f(1 + \eta^2) \, , \quad m > 2 \, . \tag{2.32}$$

We rewrite the process (2.32) for solving (2.6) at $\lambda_* = 1$ in the form

$$\varphi^{(m)} = (1 + \eta_*^2) \left(K\varphi^{(m-1)} + f \right) - \eta^2\varphi^{(m-2)} \, , \quad m > 2 \, , \quad \text{where}$$

$$\eta_* = \alpha - [\alpha^2 - 1]^{1/2} \, .$$

Consider a generalized universal iterative algorithm

$$\varphi^{(m)} = \varphi^{(m-1)} + f^{(m)}(K) \left[K\varphi^{(m-1)} + f - \varphi^{(m-1)} \right] \, , \tag{2.33}$$

where $f^{(m)}(K)$ is a special function of the operator K which is calculated by the Danford-Cauchy formula [2.10]

$$f^{(m)}(K) = \frac{1}{2\pi i} \int_\Gamma f^{(m)}(\zeta)(\zeta - K)^{-1} d\zeta \, , \tag{2.34}$$

where $\Gamma \subset \Delta_1$ is a simple closed smooth curve with a positive orientation covering all the eigen-values of the operator K; $f^{(m)}(\zeta)$ is a holomorphic function in the domain Δ_1 containing all the eigen-values $\{\lambda_k\}$. Usually, one uses polynomials for $f^{(m)}(\zeta)$ [2.17]. Appropriate choice of $f^{(m)}(\zeta)$ can lead to a significant acceleration of the iterative process. An example of utilization of (2.34) will be given in Chap. 4.

Note that analytical continuations could be applied also to differential equations, for example, to the equation $\Delta u + \lambda u = 0$. Indeed, this equation can be solved by the Monte Carlo method for $\lambda : \mathrm{Re}\,\{\lambda\} < 0$, then one of the methods of continuation can be applied to obtain the solution for $\lambda : \mathrm{Re}\,\{\lambda\} \geq 0$. In particular, in the case of translation we come to the iteration of the resolvent of the differential equation; in turn, this is connected with solution of the equations $(\Delta - \lambda)^n u = 0$. This approach is described in Chap. 4.

2.1.3 Transformations of the Type $\lambda = \psi(\eta) = a_0 + a_1\eta + a_2\eta^2 + \ldots$

General transformation of the spectral parameter $\lambda = \psi(\eta) = \sum_{i=0}^{\infty} a_i\eta^i$ is composed of the transformation (2.12) and translation. Therefore, this transformation generates the iterative process

$$\varphi_0 = a_0 K\varphi_0 + f \, , \quad \varphi_j = \sum_{i=0}^{j} a_i K\varphi_{j-1} \, , \quad j = 1, \ldots, m \, , \tag{2.35}$$

and

$$\varphi^{(m)} = \sum_{i=0}^{m} a_i\eta^i K\varphi^{(m-1)} + f \, . \tag{2.36}$$

The coefficients b_n are calculated as follows:

$$b_0 = \sum_{k=0}^{\infty} a_0^k c_k \,, \quad b_n = \sum_{k=0}^{\infty} d_k^{(n)} c_k \,, \quad n \geq 1 \,,$$

where $d_k^{(n)}$ are determined from

$$d_k^{(n)} = \frac{1}{na} \sum_{k=1}^{n} [i(k+1) - n] a_i d_k^{(n-i)} \,,$$

$$d_n^{(0)} = a_0^n \,, \quad k, n = 1, 2, \ldots \,.$$

Consider some error estimations when biased Monte Carlo algorithms are used, i.e., when a finite number of terms of a series is taken into account. Non-trivial questions arise here in the case of translation [see the process (2.26)].

Thus, assume that it is desired to calculate a resolvent $R_{\eta_1} f$ where η_1 belongs to a new circle of convergence centered at the point η_0. Let $\xi_0 = \eta_1 - \eta_0$. Then

$$R_{\eta_1} f = \sum_{m=0}^{\infty} b_m \xi_0^m \,, \tag{2.37}$$

where

$$b_m = \sum_{n=0}^{\infty} C_n^m \eta_0^{n-m} c_n \,. \tag{2.38}$$

In (2.38) it is assumed that $C_n^m = 0$ if $n < m$. Thus, the value of the resolvent at the point η_1 which maybe lies outside of the circle of convergence of the Neumann series, is represented as a convergent series (2.37), through c_k, the coefficients of expansion of $R_\lambda f$ at the point $\lambda = 0$. However, in (2.37, 38) we use only a finite number of terms that results in an error which can be estimated according to the scheme given in [2.13].

Let R be such that $|\eta_0| < R$, and $\alpha = |\eta_0|/R$, $\beta = |\xi_0|/|\eta_0|$. We take k terms in (2.38),

$$b_m^k = \sum_{n=0}^{k} C_n^m \eta_0^{n-m} c_n \,. \tag{2.39}$$

Now we take p terms in (2.37) provided that k is fixed:

$$R_{p,k}(\xi_0) = \sum_{m=0}^{p} b_m^k \xi_0^m \,. \tag{2.40}$$

Then

$$R_{\eta_1} f = R_{p,k}(\xi_0) + \varepsilon_1 + \varepsilon_2 \,, \tag{2.41}$$

where the errors have the form

$$\varepsilon_1 = \sum_{m=p+1}^{\infty} b_m \xi_0^m , \qquad (2.42)$$

$$\varepsilon_2 = \sum_{m=0}^{p} (b_m - b_m^k) \xi_0^m . \qquad (2.43)$$

It is clear that $\varepsilon_1 \to 0$ as $k \to \infty$ since the series (2.37) converges.

Consider the error ε_2. We show that $\varepsilon_2 \to 0$ if k and p go to infinity by a special law [2.31]. Let

$$g(t) = \frac{\alpha \beta^t}{t^t (1-t)^{1-t}} , \qquad 0 \le t \le 1 , \qquad (2.44)$$

and let s be the minimal root of the equation $g(t) = 1$.

Theorem 2.4. Let $s_0 \in (0, 1)$ be chosen so that $s_0 \le s$. Then

$$R_{p,k}(\xi_0) \to R_{\eta_1} f$$

if $p = [s_0 k]$, as $k \to \infty$ ($[r]$ denotes the integer part of r).

Proof. By the Cauchy inequality

$$|c_n| \le \frac{M}{R^n} , \qquad (2.45)$$

where M denotes the supremum of $R_\eta f$ on the circle $|\eta| = R$. Then

$$|\varepsilon_2| \le \sum_{m=0}^{p} \sum_{n=k+1}^{\infty} M C_n^m \alpha^n \beta^m .$$

The case when $g(t) < 1$ for all $t \in (0, 1)$ is trivial since then $\alpha(\beta + 1) < 1$; hence, $|\eta_0| + |\eta_1 - \eta_0| < R$ and thus η_1 lies in the circle of convergence of the Neumann series.

Suppose now that there exists a t such that $g(t) = 1$. The function $g(t)$ has a maximum at $t = \beta/(1 + \beta)$ since $g'(t) = g(t) \ln[\beta(1 - t)/t]$. Therefore $s_0 < \beta/(1 + \beta)$. Taking into account that $g(s_0) < 1$ we obtain $s_0 < 1 - \alpha$. Thus

$$|\varepsilon_2| \le M \sum_{n=k+1}^{\infty} \alpha^n \sum_{m=0}^{p} C_n^m \beta^m . \qquad (2.46)$$

In the last sum the term $C_n^p \beta^p$ is maximal, so the sum can be estimated by $(p+1) C_n^p \beta^p$. Indeed, since $p = [s_0 k]$, the inequality

$$C_n^m \beta^m < C_n^{m+1} \beta^{m+1}$$

is equivalent to

$$\frac{1}{n-m} < \frac{\beta}{m+1}, \quad \text{since}$$

$$\frac{m+1}{n-m} < \frac{s_0}{1-s_0}.$$

Thus, we obtain from (2.46)

$$|\varepsilon_2| \leq M \sum_{n=k+1}^{\infty} \alpha^n (p+1) C_n^p \beta^p = M(p+1)\beta^p \sum_{n=k+1}^{\infty} C_n^p \alpha^n$$

$$= M(p+1)\beta^p \sum_{n=0}^{\infty} C_{n+k+1}^p \alpha^{n+k+1} = M(p+1)\beta^p \alpha^{k+1} \sum_{n=0}^{\infty} C_{n+k+1}^p \alpha^n$$

$$= M(p+1)\beta^p \alpha^{k+1} C_{k+1}^p \sum_{n=0}^{\infty} W_n \alpha^n,$$

where

$$W_n = \frac{C_{n+k+1}^p}{C_{k+1}^p} = \prod_{i=0}^{n-1} \frac{k+2+l}{k+2-p+l} \leq \frac{1}{(1-s_0)^n},$$

since

$$\frac{k+2+l}{k+2-p+l} < \frac{1}{1-s_0},$$

(where the inequality $p = [s_0 k] \leq s_0 k$ was taken into account). Thus,

$$|\varepsilon_2| \leq M(p+1)\beta^p \alpha^{k+1} C_{k+1}^p \sum_{n=0}^{\infty} \left(\frac{\alpha}{1-s_0}\right)^n.$$

From this we see that it is necessary to investigate the right-hand side of the estimation

$$|\varepsilon_2| \leq M(p+1)\beta^p \alpha^{k+1} C_{k+1}^p \left(1 - \frac{\alpha}{1-s_0}\right)^{-1}. \tag{2.47}$$

Exchanging $k!$ with $(2\pi k)^{1/2} k^k \exp(-k) \exp(\theta_k)$, where $|\theta_k| < 1/12k$, we obtain

$$C_{k+1}^p = \frac{(k+1)!}{p!(k+1-p)!}$$

$$= \frac{[2\pi(k+1)]^{1/2}(k+1)^{k+1}e^{-(k+1)}e^{\theta}}{(2\pi p)^{1/2}p^p e^{-p}[2\pi(k+1-p)]^{1/2}(k+1-p)^{k+1-p}e^{-(k+1-p)}}$$

$$= \frac{(k+1)^{1/2}}{[2\pi p(k+1-p)]^{1/2}} \frac{e^{\theta}(k+1)^{k+1}}{p^p(k+1-p)^{k+1-p}}$$

$$= \frac{(k+1)^{1/2}e^{\theta}}{p^p[1-(p/k+1)]^{k+1-p}(k+1)^{-p}} \frac{1}{[2\pi p(k+1-p)]^{1/2}}$$

$$= \frac{(k+1)^{1/2}e^\theta}{[2\pi p(k+1-p)]^{1/2}} \frac{1}{(p/k+1)^p[1-(p/k+1)]^{k+1-p}}$$

$$= \frac{(k+1)^{1/2}e^\theta}{[2\pi p(k+1-p)]^{1/2}} \left[\frac{1}{(p/k+1)^{(p/k+1)}[1-(p/k+1)]^{[1-(p/k+1)]}} \right]^{k+1},$$

(2.48)

where $\theta = \theta_{k+1} - \theta_p - \theta_{k+1-p}$; $|\theta_k| < 1/12k$. Substituting (2.48) into (2.47) yields

$$|\varepsilon_2| \leq \frac{M(p+1)(k+1)^{1/2}e^\theta}{[2\pi p(k+1-p)]^{1/2}[1-(\alpha/1-s_0)]} \left[g\left(\frac{p}{k+1} \right) \right]^{k+1}$$

$$\leq M_1(k)[g(s_0)]^k,$$

(2.49)

where

$$M_1(k) = \frac{M(p+1)(k+1)^{1/2}e^\theta}{[2\pi p(k+1-p)]^{1/2}[1-(\alpha/1-s_0)]} g(s_0).$$

The estimation (2.49) shows that $\varepsilon_2 \to 0$ as $k \to \infty$, since $g(s_0) < 1$. $\qquad \square$

Calculations showed that (2.49) could be improved, in particular, by using more exact estimations in (2.45–47).

2.2 Asymptotically Unbiased Estimates Based on Singular Approximation of the Kernel

We now continue consideration of the acceleration and extension of convergence of the Monte Carlo methods for solving the integral equation

$$\varphi(x) = \int k(x,y)\varphi(y)dy + f(x).$$

(2.50)

The method presented here is based on the well known approach [2.4] exploiting a singular approximation of the kernel $k(x,y)$. However, our technique differs from that of [2.4] because it is ideally explicit and uses auxiliary integral equations whose kernels are the error of the approximation. The solution to the equation to be solved is simply represented through linear functionals of these auxiliary equations. This fact is effectively exploited in the Monte Carlo method. The second difference of the approach presented consists of the possibility to use a "few-point approximation", in particular, a one-point approximation

$$k(x,y) = k(x,y_0)\gamma(y) + k_1(x,y).$$

(2.51)

It is only required that the Neumann series for the integral equation with the kernel $|k_1(x,y)|$ converges.

Note that this approach is close to Betmann's method [2.4]. However, practical application of Betmann's method is difficult since the resolvent representation is not so simple.

2.2.1 Finite-Dimensional Case and One-Point Approximation

For the sake of simplicity consider a system of linear algebraic equations, i.e., (2.50) with the form $x = Ax + b$, or in n coordinates

$$x_i = \sum_{j=1}^{n} a_{ij}x_j + b_i , \quad i = 1,\dots,n . \tag{2.52}$$

Consider two auxiliary linear systems

$$z_i = \sum_{j=1}^{n} d_{ij}z_j + b_i ,$$

$$y_i = \sum_{j=1}^{n} d_{ij}y_j + a_{ij_0} , \quad i = 1,\dots,n \tag{2.53}$$

where $d_{ij} = a_{ij} - a_{ij_0}\gamma_j$ are the entries of a matrix D, $a_{ij_0}[i = 1,\dots,n]$ is a fixed column of the matrix A, $\gamma^{\mathrm{T}} = (\gamma_1,\dots,\gamma_n)$ is an arbitrary vector.

Theorem 2.5. Assume that the systems (2.52, 53) are uniquely solvable and

$$(\gamma, y) = \sum_{j=1}^{n} \gamma_i y_i \neq 1 .$$

Then

$$x_i = z_i + y_i \frac{\sum_{j=1}^{n} \gamma_j z_j}{1 - \sum_{j=1}^{n} \gamma_j y_j} , \quad i = 1,\dots,n . \tag{2.54}$$

Proof. Let $T = (\gamma, x)$. We prove first that $x_i = z_i + y_i T$, $i = 1,\dots,n$. Indeed,

$$z_i + y_i T = \sum_{j=1}^{n} d_{ij}z_j + b_i + T \sum_{j=1}^{n} d_{ij}y_j + T a_{ij_0}$$

$$= b_i + a_{ij_0}T + \sum_{j=1}^{n} d_{ij}(z_j + y_j T) . \tag{2.55}$$

Substituting $a_{ij} = d_{ij} + a_{ij_0}\gamma_j$ into (2.52) yields

$$x_i = \sum_{j=1}^{n} d_{ij}x_j + b_i + a_{ij_0}T ,$$

so we obtain the desired equality. We use it in the following transformations:

$$x_i = z_i + y_i \sum_{j=1}^{n} \gamma_j x_j = z_i + y_i \sum_{j=1}^{n} \gamma_j (z_j + y_j T)$$

$$= z_i + y_i \sum_{j=1}^{n} \gamma_j z_j + y_i T(\gamma, y) = z_i + y_i(\gamma, z) + (x_i - z_i)(\gamma, y)$$

$$= z_i(1 - (\gamma, y)) + y_i(\gamma, z) + x_i(\gamma, y) ,$$

and arrive at the desired equality

$$x_i = z_i + \frac{y_i(\gamma, z)}{1 - (\gamma, y)}$$

which proves the theorem. $\square$

The equality (2.54) will be useful if solution of (2.53) is somehow simpler than the construction of the solution of the original equation. In particular, the situation when $\varrho(A) > 1$ but $\varrho(D) < 1$ is interesting from the point of view of Monte Carlo methods because in this case it is possible to construct estimates for z_i, y_i, (γ, z) and (γ, y) on a single Markov chain simultaneously.

We turn now to (2.50) with the kernel (2.51). In this case the two auxiliary equations have the form

$$\varphi_1(x) = \int_G k_1(x, y)\varphi_1(y)dy + f(x) ,$$

$$\varphi_2(x) = \int_G k_1(x, y)\varphi_2(y)dy + k(x, y_0) ,$$
(2.56)

where y_0 is an arbitrary fixed point of G; $\gamma(y)$ is an arbitrary function such that (γ, φ_1) and (γ_1, φ_2) exist.

Theorem 2.6. Assume that the equations (2.56) are uniquely solvable for arbitrary right-hand sides from a class of functions, and suppose that

$$(\gamma, \varphi_2) = \int_G \gamma(y)\varphi_2(y)dy \neq 1 .$$

Then

$$\varphi(x) = \varphi_1(x) + \varphi_2(x)\frac{\int_G \gamma(y)\varphi_1(y)dy}{1 - (\gamma, \varphi_2)} .$$
(2.57)

The proof is analogous to that of Theorem 2.5. Indeed, substituting (2.51) into (2.50) yields

$$\varphi(x) = f(x) + \int_G k_1(x, y)\varphi(y)dy + k(x, y_0)J ,$$
(2.58)

where $J = (\gamma, \varphi) = \int_G \gamma(y)\varphi(y)dy$. We show now that $\varphi(x) = \varphi_1(x) + \varphi_2(x)J$.

Indeed,

$$\varphi_1(x) + \varphi_2(x)J = \int_G k_1(x,y)\varphi_1(y)dy + f(x)$$

$$+ J\int_G k_1(x,y)\varphi_2(y)dy + k(x,y_0)J = f(x) + k(x,y_0)J$$

$$+ \int_G k_1(x,y)[\varphi_1(y) + \varphi_2(y)J]dy \ .$$

Taking into account (2.58) we obtain the desired equality. Like in Theorem 2.5, we use it in the following transformations:

$$\varphi(x) = \varphi_1(x) + \varphi_2(x)J = \varphi_1(x) + \varphi_2(x)\int_G \gamma(y)\varphi(y)dy$$

$$= \varphi_1(x) + \varphi_2(x)\int \gamma(y)[\varphi_1(y) + \varphi_2(y)J]dy$$

$$= \varphi_1(x) + \varphi_2(x)(\gamma,\varphi_1) + \varphi_2(x)J(\gamma,\varphi_2)$$

$$= \varphi_1(x) + \varphi_2(x)(\gamma,\varphi_1) + [\varphi(x) - \varphi_1(x)](\gamma,\varphi_2)$$

$$= \varphi_1(x) + \varphi_2(x)(\gamma,\varphi_1) + \varphi(x)(\gamma,\varphi_2) - \varphi_1(x)(\gamma,\varphi_2)$$

$$= \varphi_1(x)[1 - (\gamma,\varphi_2)] + \varphi(x)(\gamma,\varphi_2) + \varphi_2(x)(\gamma,\varphi_1)$$

and the Theorem is proved. $\qquad\qquad\qquad\qquad\qquad\qquad\qquad\qquad\square$

Using (2.57), it is possible to construct an asymptotically unbiased estimate for $\varphi(x)$ provided that $\varrho(\bar{K}_1) < 1$, where $\bar{K}_1$ is the integral operator with the kernel $|k_1|$.

Let $\xi_1(x)$, $\xi_2(x)$, and $\xi_{1\gamma}(x)$, $\xi_{1\gamma}(x)$ be unbiased estimates for $\varphi_1(x)$, $\varphi_2(x)$, and $(\varphi_1(x),\gamma)$, $(\varphi_2(x),\gamma)$, respectively. Note that due to (2.56) all these estimates could be constructed on a single Markov chain; moreover, $\xi_{1\gamma}(x)$, $\xi_{1\gamma}(x)$ will differ only by the initial weights. Then

$$\eta(x) = \xi_1(x) + \xi_2(x)\frac{\tilde{\xi}_{2\gamma}^{(m)}}{1 - \tilde{\xi}_{2\gamma}^{(m)}} \ ,$$

$$\tilde{\xi}_{l\gamma}^{(m)} = \frac{1}{m}\sum_{k=1}^{m}\xi_{l\gamma}^{(k)}, \qquad l = 1,2 \tag{2.59}$$

is the asymptotically unbiased estimate for $\varphi(x)$. Indeed, let us assume that $\tilde{\xi}_{1\gamma}^{(i)}$, $1 - \tilde{\xi}_{2\gamma}^{(i)}$, $i = 1,2,\ldots,m$ are independent realizations of $\xi_{1\gamma}$, $1 - \xi_{2\gamma}$, respectively, constructed on a Markov chain. Then by the well known theorem for the distribution of a function of random variables [2.18] the random quantity

$$\zeta = \sum_{i=1}^{m}\xi_{1\gamma}^{(i)}\left[\sum_{i=1}^{m}\left(1 - \xi_{2\gamma}^{(i)}\right)\right]^{-1}$$

is asymptotically normal with the mean $(\gamma, \varphi_1)/[1 - (\gamma, \varphi_2)]$ and the variance

$$\frac{1}{m} \left[\frac{(\gamma, \varphi_1)}{1 - (\gamma, \varphi_2)} \right]^2 \left\{ \frac{D\xi_{1\gamma}}{(\gamma, \varphi_1)^2} - 2 \frac{\text{cov}\left[\xi_{1\gamma}(1 - \xi_{2\gamma})\right]}{(\gamma, \varphi_1)[1 - (\gamma, \varphi_2)]} + \frac{D(1 - \xi_{2\gamma})}{[1 - (\gamma, \varphi_2)]^2} \right\} .$$

Analogous arguments can be used for the random variable $\xi_2(x)\tilde{\xi}_{1\gamma}^{(m)}/[1 - \tilde{\xi}_{2\gamma}^{(m)}]$, hence, η is an asymptotically unbiased estimate for $\varphi(x)$.

2.2.2 Systems of Integral Equations

The approach described above can be generalized to systems of integral equations. Systems of linear algebraic equations can also be treated. Let us consider a system of integral equations

$$\varphi(x) = \int_G K[x, y] dy + f(x) , \tag{2.60}$$

where $f(x)$ is a known vector; $K[x, y]$ is a matrix with elements $\{k(x, y)\}_{i,j=1}^m$; $\varphi(x) = (\varphi^1(x), \dots, \varphi^m(x))^{\mathsf{T}}$ is the column vector of functions to be found. We will not make special assumptions about the properties of the integral operator $K[\,]$, the class of vector functions φ, f and the domain G. We assume only that the system (2.60) is uniquely solvable, and now generalize the representation (2.54, 57).

For simplicity we first consider the case $n = 1$, i.e.,

$$M[x, y] = K[x, y] - \delta(x)\gamma^{\mathsf{T}}(y)$$

or in more detail,

$$M_{ij}(x, y) = k_{ij}(x, y) - \delta^i(x)\gamma^i(y) , \tag{2.61}$$

where $\delta(x)$ and $\gamma(y)$ are some arbitrary column-vectors with components $\delta^1(x), \dots,$ $\delta^n(x)$ and $\gamma^1(y), \dots, \gamma^m(y)$, respectively.

We introduce two auxiliary systems of integral equations

$$\psi_0(x) = \int_G M[x, y]\psi_0(y) dy + f(x) ,$$
$$\psi_1(x) = \int_G M[x, y]\psi_1(y) dy + \delta(x) . \tag{2.62}$$

Then the following representation for the solution to (2.60) holds:

$$\varphi(x) = \psi_0(x) + \psi_1(x) \frac{\int_G \gamma^{\mathsf{T}}(y)\psi_0(y) dy}{1 - \int_G \gamma^{\mathsf{T}}(y)\psi_1(y) dy} .$$

This relation is a corollary of the general Theorem 2.7 proved below.

Let

$$M[x,y] = K[x,y] - \sum_{k=1}^{n} \delta_k(x)\gamma_k^{\mathrm{T}}(y) . \qquad (2.63)$$

We introduce $n+1$ auxiliary systems

$$\psi_0(x) = \int M[x,y]\psi_0(y)dy + f(x) ,$$

$$\psi_1(x) = \int M[x,y]\psi_1(y)dy + \delta_1(x) , \qquad (2.64)$$

$$\cdots\cdots\cdots\cdots\cdots$$

$$\psi_n(x) = \int M[x,y]\psi_n(y)dy + \delta_n(x) .$$

Denote by A the matrix with elements

$$a_{ij} = \int \gamma_i^{\mathrm{T}}(y)\psi_j(y)dy$$

and denote by b the vector with components

$$b_i = \int \gamma_i^{\mathrm{T}}(y)\psi_0(y)dy , \quad i,j = 1,\ldots,n .$$

Theorem 2.7. Assume that the system (2.64) is uniquely solvable and there exists an operator $(I - A)^{-1}$. Then the solution to (2.60) is represented as

$$\varphi(x) = \psi_0 + \psi^{\mathrm{T}}(x)J ,$$

where the vector J is determined from the system of linear algebraic equations

$$J = AJ + b . \qquad (2.65)$$

Proof. Substituting $K[x,y]$ from (2.63) into (2.60) yields

$$\varphi(x) = \int M[x,y]\varphi(y)dy + \delta_1(x) \int \gamma_1^{\mathrm{T}}(y)\varphi(y)dy + \ldots$$

$$+ \delta_n(x) \int \gamma_n^{\mathrm{T}}(y)\varphi(y)dy + f(x) = \int M[x,y]\varphi(y)dy$$

$$+ \sum_{i=1}^{n} \delta_i(x)J_i + f(x) , \qquad (2.66)$$

where

$$J_i \equiv \int \gamma_i^{\mathrm{T}}(y)\varphi(y)dy = \int \sum_{j=1}^{n} \gamma_i^{\mathrm{T}}(y)\varphi^j(y)dy .$$

Now,

$$\psi_0(x) + \sum_{i=1}^{n} \psi_i(x) J_i = \int M[x,y] \left\{ \psi_0(y) + \sum_{i=1}^{n} \psi_i(y) J_i \right\} dy$$

$$+ \sum_{i=1}^{n} \delta_i(x) J_i + f(x) . \tag{2.67}$$

Comparison of (2.66) and (2.67) shows that

$$\varphi(x) = \psi_0(x) + \sum_{i=1}^{n} \psi_i(x) J_i .$$

Note that

$$J_i = \int \gamma_i^{\mathrm{T}}(y) \varphi(y) dy = \int \gamma_i^{\mathrm{T}}(y) \left\{ \psi_0(y) + \sum_{j=1}^{n} \psi_j(y) J_j \right\} dy ,$$

or

$$J_i = J_1 \int \gamma_i^{\mathrm{T}}(y) \psi_1(y) dy + \ldots + J_n \int \gamma_i^{\mathrm{T}}(y) \psi_n(y) dy$$

$$+ \int \gamma_i^{\mathrm{T}}(y) \psi_0(y) dy ,$$

$[i = 1, \ldots, n]$, i.e., the vector J satisfies the system (2.65). $\square$

Note that the approach described can be applied also to a system of linear equations $x = Ax + b$ where A is an $m \times m$ matrix.

We introduce a matrix

$$B = A - \alpha_1 \beta_1^{\mathrm{T}} - \ldots - \alpha_n \beta_n^{\mathrm{T}} ,$$

where $\alpha_1, \ldots, \alpha_n$ and $\beta_1, \ldots, \beta_n$ are arbitrary column vectors, i.e., the matrix B is obtained from A by substraction of singular matrices of the form ($n < m$)

$$\alpha_i \beta_i^{\mathrm{T}} = \begin{pmatrix} \alpha_i^1 \beta_i^1 & \alpha_i^1 \beta_i^2 & \cdots & \alpha_i^1 \beta_i^n \\ \alpha_i^2 \beta_i^1 & \alpha_i^2 \beta_i^2 & \cdots & \alpha_i^2 \beta_i^n \\ \cdots\cdots\cdots\cdots\cdots\cdots \\ \alpha_i^n \beta_i^1 & \alpha_i^n \beta_i^2 & \cdots & \alpha_i^n \beta_i^n \end{pmatrix} .$$

Consider the $n + 1$ auxiliary linear system

$$x_0 = B x_0 + b ,$$
$$x_1 = B x_1 + \alpha_1 ,$$
$$\cdots\cdots\cdots\cdots\cdots$$
$$x_n = B x_n + \alpha_n .$$

As in the cases described above

$$x = x_0 + \sum_{j=1}^{n} J_j x_j ,$$

where $J_1, \ldots, J_m$ are components of the vector J which satisfies the equation $J = TJ + t$. Here T is the matrix with elements $T_{ij} = \beta_i^{\mathrm{T}} x_j$; t is a vector with components $t_i = \beta_i^{\mathrm{T}} x_0$.

2.2.3 General Case of the Kernel Approximation

The results of Sect. 2.2.1 can be generalized to the case when

$$K(x,y) = M(x,y) + \sum_{i=1}^{n} \alpha_i(x)\beta_i(y) .$$

Consider $n+1$ auxiliar equations

$$\varphi_0(x) = \int_G M(x,y)\varphi_0(y)dy + f(x) ,$$

$$\varphi_i(x) = \int_G M(x,y)\varphi_i(y)dy + \alpha_i(x) , \quad i = 1,\ldots,n .$$

Assuming that these equations are uniquely solvable, we derive, as previously, that

$$\varphi(x) = \varphi_0(x) + \sum_{i=1}^{n} \varphi_i(x)J_i ,$$

where $J = (J_1,\ldots,J_n)^{\mathrm{T}}$ is determined from

$$J = TJ + b , \tag{2.68}$$

T being a matrix with elements

$$T_{ij} = \int_G \alpha_i(y)\beta_i(y)dy ,$$

$$b_i = \int_G \varphi_0(y)\beta_i(y)dy , \quad i,j = 1,\ldots,n ,$$

provided that $\det T \neq 0$.

Note that it is possible to use a more general approximation, namely

$$M(x,y) = \frac{1}{\Delta}\begin{vmatrix} k(x,y) & \alpha_1(x) & \cdots & \alpha_n(x) \\ \beta_1(y) & a_{11} & \cdots & a_{1n} \\ \cdots & \cdots & \cdots & \cdots \\ \beta_n(y) & a_{n1} & \cdots & a_{nn} \end{vmatrix}$$

$$\Delta = \begin{vmatrix} a_{11} & \cdots & a_{1n} \\ \cdots & \cdots & \cdots \\ a_{n1} & \cdots & a_{nn} \end{vmatrix} , \tag{2.69}$$

where the functions $\{\alpha_i(x)\}$, $\{\beta_i(y)\}$, and the constants $\{a_{ij}\}$ are chosen arbitrarily.

Theorem 2.8. Assume that $\theta = \det(A - T) \neq 0$, where A is the matrix from (2.69) and T is the matrix from (2.68), and suppose that $\Delta \neq 0$. Then the following representation for the solution to (2.60) holds:

$$\varphi(x) = \frac{1}{\theta} \begin{vmatrix} \varphi_0(x) & \varphi_1(x) & \cdots & \varphi_n(x) \\ -b_1 & a_{11} - T_{11} & \cdots & a_{1n} - T_{1n} \\ \cdots & \cdots & \cdots & \cdots \\ -b_n & a_{n1} - T_{n1} & \cdots & a_{nn} - T_{nn} \end{vmatrix} , \tag{2.70}$$

where the constants $\{b_i\}_{i=1}^n$ are determined from (2.68).

Proof. Let Δ_{ij} be the adjunct corresponding to a_{ij} in the matrix A. Then (2.69) can be written as

$$M(x, y) = k(x, y) - \sum_{i=1}^n \sum_{j=1}^n \frac{\Delta_{ij}}{\Delta} \alpha_j(x)\beta_i(y) ,$$

hence,

$$k(x, y) = M(x, y) + \sum_{i=1}^n \sum_{j=1}^n \frac{\Delta_{ij}}{\Delta} \alpha_j(x)\beta_i(y) .$$

It is necessary to show that (2.70) satisfies the equation

$$\int_G (k(x, y)\varphi(y)dy = \varphi(x) - f(x) .$$

To this end, we substitute $k(x, y)$ into the last equality and consider the first term

$$\int M(x, y)\varphi(y)dy$$

$$= \frac{1}{\theta} \begin{vmatrix} \bar{M}^{(0)}(x, y) & \bar{M}^{(1)}(x, y) & & \bar{M}^{(n)}(x, y) \\ -b_1 & a_{11} - T_{11} & \cdots & a_{1n} - T_{1n} \\ \cdots & \cdots & \cdots & \cdots \\ -b_n & a_{n1} - T_{n1} & \cdots & a_{nn} - T_{nn} \end{vmatrix} ,$$

where $\bar{M}^{(i)}(x, y) = \int M(x, y)\varphi_i(y)dy$, $i = 0, 1, \ldots, n$. Consequently,

$$\int_G M(x, y)\varphi(y)dy$$

$$= \frac{1}{\theta} \begin{vmatrix} \varphi_0(x) - f(x) & \varphi_1(x) - \alpha_1(x) & \cdots & \varphi_n(x) - \alpha_n(x) \\ -b_1 & a_{11} - T_{11} & \cdots & a_{1n} - T_{1n} \\ \cdots & \cdots & \cdots & \cdots \\ -b_n & a_{n1} - T_{n1} & \cdots & a_{nn} - T_{nn} \end{vmatrix} .$$

Consider the second summand. First, we handle the functionals (β_i, φ):

$$(\beta_i, \varphi) = \int \beta_i(y)\varphi(y)dy$$

72

$$= \frac{1}{\theta} \begin{vmatrix} (\beta_i, \varphi_0) & (\beta_i, \varphi_1) & & (\beta_i, \varphi_n) \\ -b_1 & a_{11} - T_{11} & \cdots & a_{1n} - T_{1n} \\ \cdots & \cdots & \cdots & \cdots \\ -b_n & a_{n1} - T_{n1} & \cdots & a_{nn} - T_{nn} \end{vmatrix}$$

$$= \frac{1}{\theta} \begin{vmatrix} b_i & T_{i1} & \cdots & T_{in} \\ -b_1 & a_{11} - T_{11} & \cdots & a_{1n} - T_{1n} \\ \cdots & \cdots & \cdots & \cdots \\ -b_n & a_{n1} - T_{n1} & \cdots & a_{nn} - T_{nn} \end{vmatrix}$$

$$= \frac{1}{\theta} \begin{vmatrix} 0 & a_{i1} & \cdots & a_{in} \\ -b_1 & a_{11} - T_{11} & \cdots & a_{1n} - T_{1n} \\ \cdots & \cdots & \cdots & \cdots \\ -b_n & a_{n1} - T_{n1} & \cdots & a_{nn} - T_{nn} \end{vmatrix}.$$

In the last step we added the first and the $(i + 1)$-th row. Multiplying these equalitites by Δ_{ij}/Δ and summing with respect to i yields:

$$\int \sum_{i=1}^{n} \frac{\Delta_{ij}}{\Delta} \beta_i(y) dy$$

$$= \frac{1}{\theta} \begin{vmatrix} 0 & 0 & \cdots & 1 & \cdots & 0 \\ -b_1 & a_{11} - T_{11} & \cdots & a_{1j} - T_{1j} & \cdots & a_{1n} - T_{1n} \\ \cdots & \cdots & \cdots & & & \cdots \\ -b_n & a_{n1} - T_{n1} & \cdots & a_{nj} - T_{nj} & \cdots & a_{nn} - T_{nn} \end{vmatrix}.$$

Here we used the well known relation

$$\frac{1}{\Delta} \sum_{i=1}^{n} a_{ik} \Delta_{ij} = \begin{cases} 0 & \text{if } k \neq j \\ 1 & \text{if } k = j. \end{cases}$$

Multiplying the equality obtained by $\alpha_j(x)$ and summing over j gives the following expression for the second term:

$$\int \sum_{i=1}^{n} \sum_{j=1}^{n} \frac{\Delta_{ij}}{\Delta} \alpha_j(x) \beta_i(y) = \sum_{j=1}^{n} \alpha_j(x) \int \sum_{i=1}^{n} \frac{\Delta_{ij}}{\Delta} \beta_i(y) \varphi(y) dy$$

$$= \frac{1}{\theta} \begin{vmatrix} 0 & \alpha_1(x) & \cdots & \alpha_n(x) \\ -b_1 & a_{11} - T_{11} & \cdots & a_{1n} - T_{1n} \\ \cdots & \cdots & \cdots & \cdots \\ -b_n & a_{n1} - T_{n1} & \cdots & a_{nn} - T_{nn} \end{vmatrix}.$$

Adding this equality to the expression obtained for the second term yields

$$\int_G k(x,y)\varphi(y)dy$$

$$= \frac{1}{\theta} \begin{vmatrix} \varphi_0(x) - f(x) & \varphi_1(x) & \cdots & \varphi_n(x) \\ -b_1 & a_{11} - T_{11} & \cdots & a_{1n} - T_{1n} \\ \cdots & \cdots & \cdots & \cdots \\ -b_n & a_{n1} - T_{n1} & \cdots & a_{nn} - T_{nn} \end{vmatrix}$$

$$= \frac{1}{\theta} \begin{vmatrix} \varphi_0(x) & \varphi_1(x) & \cdots & \varphi_n(x) \\ -b_1 & a_{11} - T_{11} & \cdots & a_{1n} - T_{1n} \\ \cdots & \cdots & \cdots & \cdots \\ -b_n & a_{n1} - T_{n1} & \cdots & a_{nn} - T_{nn} \end{vmatrix}$$

$$- \frac{1}{\theta} \begin{vmatrix} \varphi_0(x) & \varphi_1(x) & \cdots & \varphi_n(x) \\ 0 & a_{11} - T_{11} & \cdots & a_{1n} - T_{1n} \\ \cdots & \cdots & \cdots & \cdots \\ 0 & a_{n1} - T_{n1} & \cdots & a_{nn} - T_{nn} \end{vmatrix}$$

$$= \varphi(x) - f(x)$$

and the Theorem is proved. $\qquad\qquad\qquad\qquad\qquad\qquad\qquad\qquad\qquad\square$

2.3 The Eigen-value Problem for the Integral Operators

2.3.1 Calculation of Eigen-values
on the Basis of the Transformation $\lambda = \psi(\eta)$

The problem of determining the characteristic values $\{\lambda_i\}$ for the eigen-values $\mu_i = \{\lambda_i^{-1}\}$ from $\lambda K\varphi = \varphi$ for a compact integral operator K is equivalent to the problem of determining the poles of a meromorphic function with the initial element

$$R_\lambda f = \sum_{k=0}^{\infty} c_k \lambda^k , \quad c_k = K^{k+1} f \qquad (2.71)$$

Thus, as in Sect. 2.1.2, two problems can be formulated. The first is the acceleration of the "power series method" based on the relation

$$\lambda_1 = \lim_{n\to\infty} \frac{a_n}{a_{n+1}} , \quad a_n \left(K^{n+1} f, h \right) , \qquad (2.72)$$

since it may appear that the convergence in (2.72) will be very slow. Indeed, let $\varepsilon_n = |(a_n/a_{n+1} - \lambda_1)/\lambda_1|$. Then $\varepsilon_n = O(\delta_1^{n+1})$, where $\delta_1 = |\lambda_1/\lambda_2|$, if λ_1, λ_2 are simple poles and $\varepsilon_n = O(\delta_1^{n+1} n^{r-1})$ if the pole λ_1 is simple and the pole λ_2 has multiplicity r.

The second problem is to construct an approximation for arbitrary λ_k on the basis of the coefficients of the series (2.71). Assume that the domain D involves a single pole $\lambda_k = \lambda_* = \mu_k^{-1}$ of the function $R_\lambda f$. Then the function $F(\eta) = R_{\psi(\eta)} f$ will have a single pole $\eta_k = \eta_* = \psi^{-1}(\lambda_*)$ lying on the band $|\eta| = 1$. Therefore, to calculate η_k, it is possible to use (2.72) where c_k must be changed with b_n from (2.14); then the λ_k are calculated from $\lambda_k = \psi(\eta_k)$. Note that the explicit choice of D and the construction of the corresponding transformation are often difficult. Therefore other poles in the disk $|\eta| < 1$ are allowed, it is only required that the pole η_k is minimal.

Consider such an example assuming that all λ_k are real and $0 < \lambda_1 < \lambda_2 <$ …. Suppose that an approximate value of λ_1 was obtained to within accuracy ε and it is desired to determine λ_2. For the transformation $\lambda = \psi(\eta)$, we choose the mapping of the complex plane with a cut along the line $\Lambda = \{\mathrm{Re}\,\{\lambda\} = \lambda_{1\varepsilon}, \mathrm{Im}\,\{\lambda\} \geq 0\}$ on the circle $|\eta| < 1$:

$$\psi(\eta) = \lambda_{1\varepsilon} + i\,\frac{\lambda_{1\varepsilon}}{2}\left[1 + i\,\frac{1+\eta}{1-\eta}\right]^2 \tag{2.73}$$

where $\lambda_{1\varepsilon} = \lambda_1 - \varepsilon$. This function maps the ray $(-\infty, \lambda_{1\varepsilon})$ on a half-circle centered at $(0.5, 0.5)$ of radius $\sqrt{2}/2$, and the ray $(\lambda_{1\varepsilon}, \infty)$ goes to the segment $x + y = 1$, $x, y \geq 0$. The point $\lambda_{1\varepsilon}$ goes to $\eta_{1\varepsilon} = (0, 1)$, and the point $\eta_2 = \psi^{-1}(\lambda_2)$ lies inside the disk $|\eta| < 1$ (Fig. 2.1).

Let $\Delta = |\lambda_1 - \lambda_2|$, $\delta = |\lambda_1|$. Then $|\eta_1| > |\eta_2|$ if $\varepsilon < \varepsilon_* = \delta^2/(\Delta + 2\delta)$. Numerical experiments with a rational function $R_\lambda f$ having two poles λ_1, λ_2 give the results listed in Table 2.2.

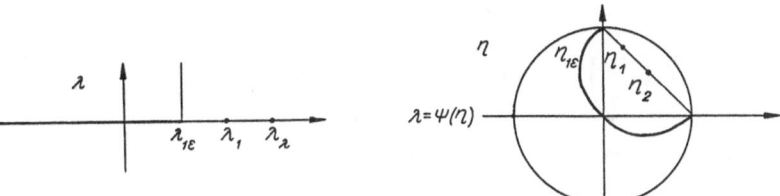

Fig. 2.1. Mapping (2.73)

Table 2.2. Calculation of the second pole of $R_\lambda f$

λ_1	λ_2	$\lambda_1 \simeq c_{11}/c_{12}$	$\lambda_2 = \psi(\eta_2)$, $\eta_3 \simeq b_{11}/b_{12}$
1	$1 - i$	$0.997 + 0.013\,i$	$1.002 - 0.998\,i$
1	$2 - i$	$1.000 + 0.004\,i$	$2.214 - 0.818\,i$

Suppose that it is desired to calculate simultaneously $R_{\lambda_*}f$, where $\lambda_1 < \lambda_* < \lambda_2$. The mapping (2.73) does not work here because $|\eta_*| > |\eta_2|$. Let

$$\lambda = \psi(\eta) = \frac{z_1 - z_2}{1 - \psi_1(\eta)} + z_2 , \tag{2.74}$$

where

$$\psi_1(\eta) = \frac{1}{[\eta - 1]^{1/2}}\left[\left(\frac{\bar{z}_1}{\bar{z}_2}\right)^2 \eta - \left(\frac{z_1}{\bar{z}_2}\right)^2\right]^{1/2} .$$

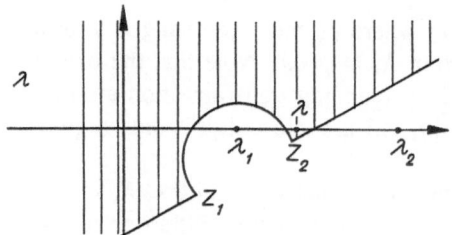

Fig. 2.2. Mapping (2.74)

The function ψ^{-1} maps the domain shown in Fig. 2.2 on the circle $|\eta| < 1$.

To calculate the coefficients b_n, it is necessary to evaluate the coefficients in the expansion

$$\psi(\eta) = \sum_{k=1}^{n} a_k \eta^k$$

and the value η_*. To this end we first expand the function $-\psi_1$ as

$$-\psi_1(\eta) = \sum_{k=1}^{n} d_k \eta^k$$

where

$$d_n = r_n + \sum_{j=1}^{n} \frac{(2j-3)!!(2n-2j-1)!!(\bar{z}_2/z_2)^{2j}}{(n-j)!j!(z_1/z_2)^{2j-1}} \,,$$

$$r_n = -\frac{(2n-1)!}{n!}\left(\frac{z_1}{z_2}\right)\,, \quad p!! \equiv 1\,, \quad \text{if} \quad p < 0\,.$$

We put now $a'_0 = -z_2$, $d'_0 = d_0 + 1$, then

$$a_1 = -\frac{a_0 d_1}{d'_0}\,, \quad a_2 = -\frac{a'_0 d_2 + a_1 d_1}{d'_0}\,,\dots,$$

$$a_n = -\frac{a'_0 d_n + a_1 d_{n-1} + \dots + a_{n-1} d_1}{d'_0}\,.$$

Thus, $R_{\lambda_*} f = \sum_{k=1}^{n} b_k \eta_*^k$, where

$$\eta_* = \psi^{-1}(\lambda_*) = \frac{(\lambda_* - z_1)^2(\lambda_* - z_2)^{-2} - (z_1/z_2)^2}{(\lambda_* - z_1)^2(\lambda_* - z_2)^{-2} - (\bar{z}_1/z_2)^2}\,.$$

Note that calculations have shown that the parameters z_1 and z_2 do not significantly affect the rate of convergence of the method.

2.3.2 Calculation of Eigen-values by Asymptotically Unbiased Estimates

There exist other methods of calculation of eigen-values [2.17, 19]. In [2.20] the resolvent is represented in the form of a continued fraction. However, the error of these methods strongly depends on the error in the calculation of λ_1. Therefore it is interesting to construct a method which enables one to improve the accuracy of calculation of λ_1. We present now a method based on the estimates of Sect. 2.2.

According to the arguments of Sect. 2.2 we can introduce a new kernel which is related to the kernel of the integral equation

$$\varphi(x) = \lambda \int_G k(x,y)\varphi(y)dy \tag{2.75}$$

through the formula

$$k_1(x,y) = k(x,y) - k(x,y_0)\gamma(y) . \tag{2.76}$$

In addition, we assume that $\varrho(\bar{K}_1) < 1$.

Substituting (2.76) into (2.75) yields

$$\varphi(x) = \lambda \int_G k_1(x,y)\varphi(y)dy + \lambda k(x,y_0) \int_G \gamma(y)\varphi(y)dy . \tag{2.77}$$

We introduce the condition

$$\int_G \gamma(y)\varphi(y)dy \equiv 1 . \tag{2.78}$$

Then (2.77) is rewritten as

$$\varphi_\lambda(x) = \lambda \int_G k_1(x,y)\varphi_\lambda(y)dy + f(x) , \tag{2.79}$$

where $f(x) = \lambda K(x,y_0)$ and we put $\varphi \equiv \varphi_\lambda$ for convenience.

Conversely, if λ is such that (2.78,79) hold, then (2.75) is true, i.e., λ is then an eigen-value and the solution $\varphi_\lambda(x)$ to (2.79) is the corresponding eigen-function.

Suppose that $\{\lambda_i\}$ are all real and $\varrho(\bar{K}_1) < 1$. Then the following algorithm of calculation of $\lambda_1 > 0$ can be formulated. Choose h and a set of test values

$$\Lambda = \{\lambda_{\min}, \lambda_{\min} + h, \ldots, \lambda_{\min} + ph\} ,$$

i.e., a partition of the interval (a, b) where λ_1 lies. By the method of dependent sampling calculate $\{\varphi_\lambda; \lambda \in \Lambda\}$ and $\{a_\lambda = (\gamma, \varphi_\lambda); \lambda \in \Lambda\}$. The set of λ for which a_λ is close to 1 composes a new segment (a', b'). Next the step h is decreased and the whole process is repeated.

Convergence of this method can be investigated only in simple cases, e.g., when (γ, φ) monotonically depends on λ.

2.4 Alternative Constructions of the Resolvent: Modifications and Numerical Experiments

2.4.1 Continuation by the Mittag–Leffler Method Combined with the Transformation $\lambda = i(i + p)/(i - p)$

The methods described in Sect. 2.1 require information about the characteristic numbers $\lambda_1, \lambda_2, \ldots$ of the integral equation. The general situation can be formulated as follows: the more detailed the information about the characteristic set $\chi(K)$ of the integral operator, the faster and the more explicit is the method that can be constructed.

However, there often arises a need to have a method which converges for all $\lambda \in \pi(K)$. Combination with the method of transformation of the spectral parameter permits to accelerate the method.

Consider such a method based on analytical continuations of Mittag–Leffler and Lindeleff [2.21, 22]. The problem is to construct the resolvent

$$R_\lambda f = \sum_{k=0}^{\infty} \lambda^k c_k , \quad c_k = K^{k+1} f ,$$

which is a meromorphic function with poles at $\lambda_1, \lambda_2, \ldots$.

The Mittag–Leffler star of the function $R_\lambda f$ is defined as follows. Let $R_\lambda f$ be the function generated by analytic continuation, along radial lines from the origin, of the initial element determined by convergence of $\sum_{k=0}^{\infty} \lambda^k c_k$. The open set in which $R_\lambda f$ is thus defined is the Mittag–Leffler star. This star consists of all points of the complex plane not of the form $\varrho\zeta$ where $\varrho \geq 1$ and ζ is a singular point of $R_\lambda f$.

The following result is known [2.22] (continuation by Lindeleff):

$$L_\delta(f) = \sum_{n=0}^{\infty} \exp\{-\delta n \ln(n)\} c_n \lambda^n \to R_\lambda f , \quad \text{as} \quad \delta \to 0 , \qquad (2.80)$$

uniformly in an arbitrary closed bounded domain lying in the Mittag–Leffler star, and

$$\varepsilon = |L_\delta(f) - R_\lambda f| \leq c\delta$$

where the constant c depends on $|\lambda - \lambda_1|$. This is a special case of the P-summation method defining the sum $\sum_{k=1}^{n} a_k$ as $\lim_{\delta \to 0} \sum_{n=1}^{\infty} A_n(\delta) a_n$, where $A_n(\delta) \to 1$ as $\delta \to 0$. In the Mittag–Leffler method $A_n(\delta) = [\Gamma(1 + \delta n)]^{-1}$.

Note that (2.80) slowly converges for small δ if $|\lambda|$ is large enough. It works well if $|\lambda|$ is small (but does not lie close to the pole).

For illustration we show numerical results where a three-dimensional boundary integral equation was solved by the Monte Carlo method ($\lambda_1 = -1$) at a point λ_*. The number of terms in (2.80) was varied, $\delta = 0.1$ was fixed (Table 2.3).

Table 2.3. Solution of a potential problem by $L_\delta(f)$-method

λ_*	N	Approximation $L_\delta(f)$	Exact solution
	30	$3.85 + 10^{-9}\,i$	
	40	$2.45 + 10^{-9}\,i$	
-1.5	60	$0.73 + 10^{-9}\,i$	0.4
	70	$0.48 + 10^{-9}\,i$	
	80	$0.398 + 10^{-9}\,i$	
$0.5 + i$	30	$0.401 + 0.85\,i$	$0.4 + 0.8\,i$
$0.3 + 0.9\,i$	30	$0.55 + 0.78\,i$	$0.54 + 0.7\,i$
1.5	20	$-2.021 + 0.731 \times 10^{-4}\,i$	-2

Suppose now that $\{\lambda_k\}$ are all real and it is desired to solve an integral equation for $\lambda \in \pi(K)$, $\lambda_k < \lambda < \lambda_{k+1}$, for some k, for example, $\lambda_1 < 1 < \lambda_2$. This case is often used in practice, but (2.80) cannot be applied because $\pi(K)$ does not lie in the Mittag–Leffler star. To overcome this difficulty, we use the transformation $\lambda = i(i+p)/(i-p)$. Then the real axis goes to the unit circle $|\eta| = 1$, and the point $\lambda = 1$ goes to the point $\eta_* = 1$ which lies in the Mittag–Leffler star. Thus, using the method described in Sect. 2.1.3 and (2.80), it is possible to calculate $R_\lambda f$ at $\lambda_* \in (\lambda_k, \lambda_{k+1})$. This gives the result shown in last row of Table 2.3 at $\lambda\psi* = 1.5$ ($\eta_* = 0.923 + 0.385\,i$).

2.4.2 Generalized Summation Methods

Generalized summation of series has been used by Euler, Markov, Poisson, Abel, Cesaro, Voronoj, etc. [2.21]. Detailed investigations in this field were pursued by *Hardy* [2.21] and *Cooke* [2.23] (see also the bibliography in [2.21]). The continuation methods can be considered from the point of view of a generalized summation of divergent series. On this basis some transformations which lead to acceleration of series are considered. In particular, a method of multiplication applied to eliminate "a pole of infinite order" leads to the Euler method. This permits one to construct a special version of the walk on boundary algorithm.

Let $U = u_0 + u_1 + \ldots$ be a series, convergent or not, and let $s_0, s_1, \ldots$ be its partial sums. The generalized sum of the quantities $\{u_i\}$ is defined by some rule $\sum u$ which determines uniquely a real or a complex number U. The generalized summation is linear: $\sum(pu_n + qv_n)$ gives the generalized sum $pU + qV$. The generalized summation is regular if a series which converges to a sum A in the usual sense has the generalized sum coinciding with A.

Let us consider an example. Let $\{a_{nk}\}_{n,k=0}^\infty$ be an infinite matrix and suppose that a series $U = u_0 + u_1 + \ldots$ has the partial sums $s_0, s_1, \ldots$. Let $\sigma_n = \sum_{k=0}^n a_{nk} s_k$, then the limit $\sigma = \lim_{n\to\infty} \sigma_n$ (if it exists) gives an example of a generalized sum generated by the matrix $\{a_{nk}\}$. It is known [2.21] that this method is regular, i.e., $\lim_{n\to\infty} \sigma_n = \lim_{n\to\infty} s_n$ iff

$$\sum_{k=0}^{n} |a_{nk}| < M \,, \quad n = 0, 1, \ldots; \quad \lim_{n \to \infty} a_{nk} = 0 \,, \quad k = 0, 1, \ldots;$$

$$\lim_{n \to \infty} \sum_{k=0}^{n} a_{nk} = 1 \,.$$

We can define the Euler–Knopp generalized summation ["$E(r)$-method] as a special case when $a_{nk} = C_n^k r^k (1 - r)^{n-k}$, where r is a complex parameter.

We define the Borel generalized sum as

$$\lim_{x \to \omega} \sum_{n=0}^{\infty} s_n \varphi_n(x) \,, \tag{2.81}$$

where $\omega = \infty$, $\varphi_n(x) = \exp\{-x\} x^n / n!$.

The Poisson–Abel summation is defined by (2.81) where $\omega = 1$, $\varphi_n(x) = (1 - x) x^n$. In the Voronoj method $a_{nk} = p_k / P_n$ where $P_n = p_0 + \ldots + p_n$; $\{p_n\}$ is a positive sequence. A special case of this method is the Cesaro summation with $p_n = C_{n+k-1}^{k-1}$. Finally, the Euler sum of a series $\sum_{n=0}^{\infty} a_n$ is defined by

$$\sum_{p=0}^{\infty} 2^{-(p+1)} \left[a_0 + C_p^1 a_1 + C_p^2 a_2 + \ldots + C_p^p a_p \right] \,.$$

Questions concerning the regularity, the strength and other properties of the methods defined are studied in [2.21].

Suppose now that the initial element of the resolvent $R_\lambda f$ of an integral equation is given by

$$R_\lambda f = \sum_{k=0}^{\infty} c_k \lambda^k \,,$$

and let S be the Mittag–Leffler star of the function $R_\lambda f$, $\lambda_0 \in S$.

The question is how to calculate $R_{\lambda_0} f$ by a generalized summation method of the initial element $\sum_{k=0}^{\infty} c_k \lambda^k$. The answer has been given in [2.24]. Suppose that the series $\sum_{k=0}^{\infty} c_k z^k$ has a positive radius of convergence. If $\text{Re}\,\{r\} > 0$, then $z_0 \in S$ or z_0 is a pole of $R_\lambda f$. If $\text{Re}\,\{r\} > 0$ and $z_0 \in S$ then $R_{z_0} f$ is the "$E(r)$-sum" of the series $\sum_{k=0}^{\infty} c_k z_0^k$.

In the next section we present a Monte Carlo estimate based on $E(1/2)$-summation.

2.4.3 Transformation and Convergence Acceleration of Series. Euler Summation

Convergence of the series which arise in analytical continuation and generalized summation procedures is often slow. Consider some acceleration methods based on special transformations of series which generalize the method of [2.25]. We also give an example of an acceleration method which leads to a summation of a divergent series to a desired value of the resolvent. The acceleration methods are based on some a priori information about the coefficients of the series. The nature of this information can be very different, for example, it may be that

$$\lambda_1 = \lim_{n \to \infty} \frac{(K^{n+1} f, h)}{(K^n f, h)}$$

is known.

Suppose that it is desired to calculate

$$S_m = \sum_{k=m}^{\infty} c_k .$$
(2.82)

Assume first that the series converges. We introduce a sequence

$$\Delta_k^{(n)} = A(k) b_k^{(n)} - A(k+1) b_{k+1}^{(n)} ,$$
(2.83)

where $\{A(k) b_k^{(n)}\}_{k=m}^{\infty}$, $n = 1, 2, \ldots, p$, are arbitrary sequences such that

$$\lim_{k \to \infty} A(k) b_k^{(n)} = c^{(n)}$$
(2.84)

$$\lim_{k \to \infty} \frac{r_k^{(n-1)}}{\Delta_k^{(n)}} = \mu_n , \quad n = 1, 2, \ldots, p$$
(2.85)

where $r_k^{(0)} = c_k$; $r_k^{(n)} = r_k^{(n-1)} - \mu_n \Delta_k^{(n)}$. Then

$$\sum_{k=m}^{\infty} c_k = \sum_{n=1}^{p} \mu_n \left[A(m) b_m^{(n)} - c^{(n)} \right] + \sum_{k=m}^{\infty} r_k^{(p)} .$$
(2.86)

Note that the series in the right-hand side of (2.86) converges faster than the initial series (2.82) since

$$\lim_{k \to \infty} \frac{r_k^{(n)}}{r_k^{(n-1)}} = 0 .$$
(2.87)

Thus if information of the type (2.84, 85) is at hand, it is possible to use the transformation to accelerate the convergence of (2.82) as in (2.87).

Let us consider an example. Suppose that $\lim_{k \to \infty} [a_k / a_{k+1}]$ is known where $R_\lambda f = a_0 + a_1 \lambda + \ldots$. We put $A(k) = 1$, $b_k^{(n)} = \Delta^{n-1} a_k \lambda^k$ in (2.83). Then [at $p = 1$]:

$$\mu_1 = \lim_{k \to \infty} \frac{a_k \lambda^k}{\Delta_k^{(1)}} = \frac{1}{1 - \dfrac{\lambda}{\lambda_1}} ,$$

so that (2.86) takes the form

$$\sum_{k=m}^{\infty} a_k \lambda^k = \frac{a_m \lambda^m}{1 - \dfrac{\lambda}{\lambda_1}} + \sum_{k=m}^{\infty} \frac{\left[\dfrac{a_{k+1}}{a_k} - \lambda_1^{-1} \right]}{1 - \dfrac{\lambda}{\lambda_1}} a_k \lambda^{k+1} .$$

Let $\Delta^{k+1} a_m = \Delta^k a_m - \Delta^k a_{m+1}$. Suppose that

$$\lim_{k \to \infty} \frac{\Delta^n a_{k+1}}{\Delta^n a_k} = 1 , \quad n = 0, 1, \ldots, p .$$

Subsequent application of this procedure yields

$$\sum_{k=m}^{\infty} a_k \lambda^k = \sum_{n=0}^{p-1} (-1)^n \frac{\lambda^{m+n} \Delta^n a_m}{(1 - \lambda)^{n+1}} + (-1)^p \sum_{k=m}^{\infty} \frac{\lambda^{k+p} \Delta^p a_k}{(1 - \lambda)^p} . \tag{2.88}$$

In particular, taking $p \to \infty$ at $\lambda = -1$ (when the initial series diverges) we obtain from (2.88):

$$\sum_{k=m}^{\infty} (-1)^k a_k = (-1)^m \sum_{k=0}^{\infty} \frac{\Delta^k a_m}{2^{k+1}} . \tag{2.89}$$

The right-hand side of (2.89) coincides with the generalized $E(1/2)$-sum of the Euler method.

For illustration we compare the estimate of the walk on boundary algorithm (Chap. 3) based on the multiplication method

$$\xi_1 = 2 \sum_{k=0}^{N-1} (-1)^k g_k + (-1)^N g_N \tag{2.90}$$

and the Euler summation

$$\xi_2 = \sum_{k=0}^{N} \frac{\Delta^k g_0}{2^k} . \tag{2.91}$$

Note that (2.91) can be obtained (for $N = \infty$) by infinite applications of the multiplication method.

In (2.90, 91) g_k is the value of the boundary function in

$$\Delta u = 0 , \quad u|_{\partial G} = g$$

on the kth step of the walk on boundary process; N is the total number of steps. In calculations we used

$$G = \{ (x, y, z) : 0 \le x^2 + y^2 \le 4 , \quad -5 \le z \le 5 \} ,$$
$$g = 2x + y + z|_{\partial G} ,$$

and the solution was evaluated at the point $(1.7, 0, -1)$; the exact solution is equal to -2.4. In Table 2.4 the results of calculation of the quantities $u_i = M\xi_i$ and $\sigma = [D\xi_i/N]^{1/2} 10^2$, $i = 1, 2$ are shown.

Table 2.4. Comparison of (2.90) and (2.91)

Number of trials	N	u_1	σ_1	u_2	σ_2
10 000		2.26	9.48	2.26	6.49
20 000	6	2.26	6.68	2.32	4.55
40 000		2.30	4.70	2.33	3.20
80 000		2.33	3.34	2.34	2.27
10 000	8	2.44	10.90	2.42	7.34
20 000		2.35	7.63	2.36	5.17
40 000		2.32	5.38	2.33	3.65

As the results of the Table 2.4 show, the effectiveness of (2.91) is here approximately two times higher than that of (2.90). In conclusion, it should be noted that there also exist non-linear acceleration methods, for example, the Shanks method and others [2.26].

2.4.4 Padé Approximation of the Resolvent and Approximation by Continued Fractions

Up to now we have considered various methods of transformation of a series $U = \sum_{i=1}^{\infty} u_i$ based on linear transformations of the partial sums s_i. However, in the case when the resolvent $R_\lambda f$ is a meromorphic function of λ, it is natural to seek the approximation to $R_\lambda f$ in the form of a rational function. The well known rational approximation of a series is the Padé approximation. Practical calculations show that Padé approximations of the resolvent $R_\lambda f$ can accelerate the slowly converging Neumann series, and, moreover, extend the convergence domain.

A number of publications on the Padé approximations are available [2.27, 28]. A detailed bibliography and a survey of numerical aspects of Padé approximations are given in [2.27, 29].

Suppose that the initial element of the resolvent $R_\lambda f$ is given by

$$R_\lambda f = \sum_{i=0}^{\infty} c_i \lambda^i . \tag{2.92}$$

The Padé approximation of $R_\lambda f$ is defined as a rational function

$$\frac{U_{M,N}(\lambda)}{V_{M,N}(\lambda)} \equiv \left[\frac{N}{M}\right]_{R_\lambda f} = \frac{\alpha_{M,0} + \alpha_{M,1}\lambda + \ldots + \alpha_{M,N}\lambda^N}{\beta_{0,N} + \beta_{1,N}\lambda + \ldots + \beta_{M,N}\lambda^N} \tag{2.93}$$

such that the first $N + M$ coefficients of expansion of (2.93) in a power series coincide with the first coefficients of the power series (2.92). Consequently, the expansion of the function $R_\lambda f V_{M,N}(\lambda) - U_{M,N}$ must start with the term λ^{M+N+1}. Hence

$$c_0 \beta_{0,N} = \alpha_{M,0} \, ,$$

$$c_1 \beta_{0,N} + c_0 \beta_{1,N} = \alpha_{M,1} \, ,$$

$$\cdots \cdots \cdots \cdots$$

$$c_N \beta_{0,N} + c_{N-1} \beta_{1,N} + \ldots + c_{N-M} \beta_{M,N} = \alpha_{M,N} \, ,$$

$$c_{N+1} \beta_{0,N} + c_N \beta_{1,N} + \ldots + c_{N-M+1} \beta_{M,N} = 0 \, ,$$

$$\cdots \cdots \cdots \cdots$$

$$c_{N+1} \beta_{0,N} + c_{N+M-1} \beta_{1,N} + \ldots + c_N \beta_{M,N} = 0 \, ,$$

from which the coefficients α_{ij} and β_{ij} can be uniquely determined provided the known conditions are satisfied and $\beta_{0,N} = 1$. The Padé table is defined as the matrix $\{U_{M,N}(\lambda) / V_{M,N}(\lambda)\}$. The elements of this matrix can be considered as approximations to (2.92).

There exist many methods of numerical calculation of the Padé approximation. It is convenient to construct the approximation on the basis of its relation to the continued fractions: We assume that $\sum_{r=0}^{\infty} c_{m+r} \lambda^r$ is expanded in a continued fraction

$$\frac{c_m}{1-} \, \frac{q_1^{(m)} \lambda}{1-} \, \frac{e_1^{(m)} \lambda}{1-} \, \cdots \, \frac{q_r^{(m)} \lambda e_r^{(m)} \lambda}{1 - 1-} \, \cdots \, ,$$

then

$$\frac{U_{0,k}(\lambda)}{V_{0,k}(\lambda)} \, , \, \frac{U_{0,k+1}(\lambda)}{V_{0,k+1}(\lambda)} \, , \, \frac{U_{1,k+1}(\lambda)}{V_{1,k+1}(\lambda)} \, , \, \frac{U_{1,k+2}(\lambda)}{V_{1,k+2}(\lambda)} \, , \, \frac{U_{2,k+2}(\lambda)}{V_{2,k+2}(\lambda)} \, , \ldots .$$

are the convergents of the continued fraction

$$\sum_{i=0}^{k} c_i \lambda^i + \frac{c_{k+1} \lambda^{k+1}}{1-} \, \frac{q_1^{(k+1)} \lambda}{1-} \, \frac{e_1^{(k+1)} \lambda}{1-} \, \frac{q_2^{(k+1)} \lambda}{1-} \, \frac{e_2^{(k+1)} \lambda}{1-} \, \cdots .$$

For example, if $R_\lambda f$ belongs to the class of functions

$$R_\lambda f = \frac{s_0}{1+} \, \frac{g_1 \lambda}{1+} \, \frac{(1 - g_1) g_2 \lambda}{1+} \, \frac{(1 - g_2) g_3 \lambda}{1+} \, \cdots \, ,$$

$$|\arg (1 + \lambda)| < \pi \, , \quad s_0 > 0 \, , \quad 0 < g_n < 1 \, ,$$

then the coefficients in $R_\lambda f = s_0 - s_1 \lambda + s_2 \lambda^2 - s_3 \lambda^3 + \ldots$ [$s_i = (-1)^i c_i$] are calculated by the rhombus rule [2.19]:

$$g_0^{(m)} = 0 \, , \quad g_1^{(m)} = \frac{s_{m+1}}{s_m} \, ,$$

$$\left(1 - g_{2n+1}^{(m)}\right) \left(1 - g_{2n+2}^{(m)}\right) = \left(1 - g_{2n}^{(m+1)}\right) \left(1 - g_{2n+1}^{(m+1)}\right),$$

$$g_{2n}^{(m)} g_{2n+1}^{(m)} = g_{2n-1}^{(m+1)} g_{2n}^{(m+1)} ; \qquad g_n = g_n^{(0)} .$$

Consider a simple example. Representation of the number π in the form $\sum_{n=0}^{\infty} (-1)^n \frac{4}{2n+1}$ requires about 20×10^6 terms [2.27] to obtain the true value of π. If we apply the transformation $\begin{bmatrix} 1 \\ 1 \end{bmatrix}$ to the sum s_n [m times], i.e., if we turn to a new sequence

$$\Sigma_1 = \frac{s_{n-1} s_{n+1} - s_n^2}{s_{n-1} + s_{n+1} - 2 s_n} , \tag{2.94}$$

then, to achieve the same accuracy, it is necessary to take $n = 9$, $m = 5$.

As mentioned, a general class of non-linear transformations was introduced by *Shanks* [2.26]. However, this method generally requires calculation of a large number of determinants. Note also that the Shanks method, as well as the Padé approximation and methods based on continued fractions, all face the very complicated problem of investigation of the domain of convergence. The relation between the Padé approximations and the Fredholm series for the equation

$$\varphi = \lambda K \varphi + f \tag{2.95}$$

is studied in [2.27]. For a functional (h, φ) a Padé approximation $(N - 1)/N$ of the series

$$h(\lambda) = \left(h, \sum_{n=0}^{\infty} \lambda^n K^n f \right)$$

is constructed such that

$$(h, \varphi) \cong \sum_{i=1}^{N} \sum_{j=1}^{N} \frac{\omega_{i-1} V_{ij} \omega_{j-1}}{\det |U_{ij}|} , \tag{2.96}$$

where $\omega_i = (h, K^i f)$, $U_{ij} = \omega_{i+j-2} - \lambda \omega_{i+j-1}$; V_{ij} is a minor corresponding to U_{ij}. It is shown in [2.30] that in the case of compact operators

$$\lim_{N \to \infty} \left[\frac{N-1}{N} \right] = h(\lambda) .$$

Moreover, if some additional assumptions hold, the numerator and denominator converge, respectively, to the numerator and denominator of the well known Fredholm representation.

Note that it is possible to seek the solution to (2.95) directly in the form of a continued fraction [2.20]:

$$\varphi = \frac{\alpha_1}{1+} \frac{\alpha_2 \lambda}{1+} \frac{\alpha_3 \lambda}{1+} \cdots , \tag{2.97}$$

so that

$$\varphi = \frac{\alpha_1}{1+\varphi_1}, \quad \varphi_k = \frac{\lambda\alpha_{k+1}}{1+\varphi_{k+1}}, \quad k = 1,2,\dots,$$

and

$$\varphi - \left[\frac{\alpha_1}{1+}\,\frac{\alpha_2\lambda}{1+}\,\cdots\,\frac{\alpha_n\lambda}{1}\right] = 0(\lambda^n).$$

To find the coefficients $\{\alpha_i\}$, we substitute (2.97) into (2.95).
Thus we seek the solution in the form:

$$\varphi = \frac{\alpha_1}{1+}\,\frac{\alpha_2\lambda}{1+}\,\frac{\alpha_i\lambda}{1+\varphi_i}, \quad i = 2,\dots,n.$$

We denote the numerator and denominator of the rational function by U_n and V_n,

$$\frac{U_n}{V_n} = \frac{\alpha_1}{1+}\,\frac{\alpha_2\lambda}{1+}\,\cdots\,\frac{\alpha_n\lambda}{1}.$$

Then

$$\varphi = \frac{U_n + \varphi_n U_{n-1}}{V_n + \varphi_n V_{n-1}} = \frac{U_n}{V_n} - \frac{(-\lambda)^{n-1}\alpha_1\dots\alpha_n\varphi_n}{V_n(V_n + \varphi_n V_{n-1})}. \tag{2.98}$$

Let

$$G_n = \frac{\alpha_1\dots\alpha_n}{V_n(V_n + \varphi_n V_{n-1})}.$$

Then substituting (2.98) into (2.95) yields

$$\frac{U_n}{V_n} - (-\lambda)^{n-1}\varphi_n G_n - f - \lambda\int k(x,x')\frac{U_n}{V_n}dx'$$
$$- (-\lambda)^n \int k(x,x')\varphi_n G_n dx' = 0.$$

Combining the coefficients of the term λ^n and taking $\varphi_n = \lambda\alpha_{n+1}/(1+\varphi_{n+1})$ into account, we obtain

$$\alpha_{n+1} = \frac{(-1)^n}{\alpha_1\dots\alpha_n}\left[K^n f + v_{n,1}K^{n-1}f + \dots\right.$$
$$\left. + v_{n,[n/2]}K^{n-[n/2]}f\right], \tag{2.99}$$

where $v_{n,j}$ are determined from the expansion:

$$V_n = \sum_{j=0}^{[n/2]} v_{n,j}\lambda^j$$

obtained according to

$$V_{n+1} = V_n + \lambda \alpha_{n+1} V_{n-1} \ . \tag{2.100}$$

From (2.100) we have

$$v_{n+1,j} = v_{n,j} + v_{n-1,j-1} \alpha_{n+1} \ . \tag{2.101}$$

Thus, the coefficients in the expansion (2.97) can be obtained by the rucurrence relation (2.99, 101).

This method was applied to solve the problem mentioned in Sect. 2.4.3 (Table 2.4). Calculation of (2.97) was carried out by an inverse scheme starting with α_8; iterations $K^n f$ were calculated by the walk on boundary algorithm, the total number of trajectories taken was 4000, the statistical error was about 0.4%. An approximate solution of $u = 2.408$ was obtained in 12 min. (The exact solution is $u = 2.4$.) It should be noted that it is difficult to find the domain of convergence in this method.

2.4.5 Methods of Regularization of Analytical Continuation for Solving Integral Equations

The methods for solving integral equations described in previous sections are based on summation in a circle $S(0, R) = \{\lambda : |\lambda| < R\}$ of a series:

$$R_\lambda f = \sum_{n=0}^{\infty} a_n \lambda^n \ ,$$

where a_i are coefficients determined through $K^n f$. The iterations $K^n f$ are calculated by the Monte Carlo method, therefore, there arise some statistical errors. Suppose that $\tilde{a}_n$ is the calculated approximation to a_n; let $\delta_n = a_n - \tilde{a}_n$. We denote by

$$R^{(N)}(\lambda) = \sum_{n=0}^{N} \tilde{a}_n \lambda^n$$

the approximation to $R_\lambda f$. Then

$$R_{\lambda_*} f - R^{(N)}(\lambda_*) = \sum_{n=0}^{N} \delta_n \lambda_*^n + \sum_{n=N+1}^{N} a_n \lambda_*^n \ .$$

Two situations may arise:

1) The point λ_* lies inside of the circe of convergence of the series $\sum_{n=0}^{\infty} \delta_n \lambda^n$; in this case, N can be chosen to be sufficiently large.
2) λ_* does not lie in this circle; in this case N must be chosen according to the information about the errors δ_n.

Suppose that

$$\sup_{|\lambda|=R} |R_\lambda f| = M_0 < \infty \ .$$

Consider three cases:

1) $|\delta_n| \le \beta_n$, $n = 1, 2, \ldots$.
2) $\{\delta_n\}_{n=1}^{\infty}$ are independent random numbers such that

$$M\delta_n = 0, \quad D\delta_n = \sigma_n^2 < \infty.$$

3) $\{\delta_n\}$ are dependent random numbers such that

$$M\delta_n = 0, \quad M\delta_n\delta_m = \varrho_{nm}.$$

Consider the question of how to choose the optimal value of N, say N_0. By the Cauchy inequality $|a_n| \le M_0/R^n$, therefore

$$|R_{\lambda_*}f - R^{(N)}(\lambda_*)| = \sum_{n=0}^{N} \beta_n|\lambda_*|^n + \sum_{n=N+1}^{N} M_0 \left[\frac{|\lambda_*|}{R}\right]^n$$

$$= \sum_{n=0}^{N} \beta_n|\lambda_*|^n + M_0 \left[\frac{|\lambda_*|}{R}\right]^{N+1} \frac{R}{R - |\lambda_*|}.$$

Hence, the optimal value N_0 can be obtained in case 1 from the condition

$$M_0 \left(\frac{|\lambda_*|}{R}\right)^{N_0+1} \frac{R}{R - |\lambda_*|} < \sum_{n=0}^{N_0-1} \beta_n|\lambda_*|^n < M_0 \left(\frac{|\lambda_*|}{R}\right)^{N_0} \frac{R}{R - |\lambda_*|}.$$

In case 2, it is natural to minimize the mean squared error

$$M\left|R_{\lambda_*}f - R^{(N_0)}(\lambda_*)\right|^2 = \min_{N} M\left\{\left|\sum_{n=0}^{N} \delta_n\lambda_*^n\right|^2\right.$$

$$+ M_0 \left(\frac{|\lambda_*|}{R}\right)^{2(N+1)} \frac{R^2}{(R - |\lambda_*|)^2}\right\}$$

$$= \min_{N}\left\{\sum_{n=0}^{N} \sigma_n^2\lambda_*^{2n} + M_0^2 \left(\frac{|\lambda_*|}{R}\right)^{2(N+1)} \frac{R^2}{R - |\lambda_*|^2}\right\}.$$

By analogy, in case 3 N_0 is chosen from the condition

$$\min_{N} M\left\{\left|\sum_{n=0}^{N} \delta_n\lambda_*^n\right|^2 + M_0 \left(\frac{|\lambda_*|}{R}\right)^{2(N+1)} \frac{R^2}{(R - |\lambda_*|)^2}\right\},$$

i.e., from

$$\left\{\sum_{n,m=1}^{N} \varrho_{nm}\lambda_*^n\lambda_*^m + M_0 \left(\frac{|\lambda_*|}{R}\right)^{2(N+1)} \frac{R^2}{(R - |\lambda_*|)^2}\right\} \to \min.$$

In case 1, when $|a_n - \tilde{a}_n| \leq \delta$ we obtain

$$\left| R_{\lambda_*} f - R_0^{(N)}(\lambda_*) \right| \geq \frac{\delta^{1 - \ln[|\lambda_*|]/\ln(R)}}{|\lambda_*|} M_0 \left(\frac{R|\lambda|}{R - |\lambda_*|} \right)^{\ln[|\lambda_*|]/\ln(R)} , \qquad (2.102)$$

i.e., the best estimation has the order $\delta^{1 - (\ln[|\lambda_*|]/\ln(R))}$ at $N_0 = \ln[M_0|\lambda_*|/(\delta(R - |\lambda_*|))]/\ln(R)$.

When the solution of the Dirichlet problem for the Poisson equation is constructed in the form of a simple layer potential, the density satisfies the known boundary integral equation of the first kind. Therefore, it is interesting to construct the Monte Carlo methods for solving integral equations of the first kind. This problem is ill-posed, so we use some known regularization methods [2.15, 30].

For a compact operator K, let us consider the equation

$$K\varphi = f . \qquad (2.103)$$

We use the iterative process proposed in [2.30] for the case $K = K^*$:

$$\varphi_n = h \sum_{k=1}^{n} (I - hK)^{k-1} f \xrightarrow[n \to \infty]{} \varphi , \quad 0 < h < 2\|K\|^{-1} , \qquad (2.104)$$

where h is the regularization parameter.

The unbiased estimate of (2.104) has the form

$$\xi(n) = - \sum_{i=1}^{n} C_n^i (-h)^i Q_{i-1} f(y_{i-1}) , \quad M\xi(n) = \varphi_n ,$$

where $\{y_i\}$ is a Markov chain of points in G; $y_0 = x$, $p(y_{i-1}, y_i)$ is an appropriate transition density, $Q_0 = 1$, $Q_i = Q_{i-1} k(y_{i-1}, y_i)/p(y_{i-1}, y_i)$.

There exists another iterative process

$$\varphi_n(x) = (I + hK)^{-1} \varphi_{n-1} - h(I + hK)^{-1} f \xrightarrow[n \to \infty]{} \varphi(x) .$$

The corresponding random estimate can be written as

$$\xi(n) = h \left[n f(x) + \sum_{k=1}^{n} C_{n+1}^{k+1} (-h)^k Q_k^* f(y_k) \right] , \qquad (2.105)$$

$$M\xi(n) = \varphi_{n-1} , \quad \varphi_0 = h[I + hK]^{-1} f ,$$

where $Q_0^* = 1$, $Q_k^* = Q_{k-1}^* r(y_{k-1}, y_k)/p(y_{k-1}, y_k)$; $r(x, y)$ is the kernel of the resolvent operator defined by the relation $(I + hK)^{-1} = I - hR$. Applying the double randomization principle it is possible to utilize in (2.105) unbiased estimates ζ_k which can be constructed on the basis of solving the known equation

$$r(x, y) = -hKr(x, y) + k(x, y) , \quad x \in G .$$

For $0 < h < \|K\|^{-1}$ we have

$$\zeta_k = \sum_{i=0}^{n} Q_i^{(k)} k\left(y_i^{(k)}, y\right) \,, \quad Q_0^{(k)} = 1 \,,$$

$$Q_i^{(k)} = Q_{i-1}^{(k)} \left[-h \frac{k\left(y_{i-1}^{(k)}, y_i^{(k)}\right)}{l\left(y_{i-1}^{(k)}, y_i^{(k)}\right)}\right] \,,$$

where $y_i^{(k)}$ is a terminating finite Markov chain with the transition density $l(x,y)$, $y_0^{(k)} = y_{k-1}$; N is the random number of iterations.

3. Monte Carlo Algorithms for Solving Boundary Value Problems of the Potential Theory

Numerical methods for solving two-dimensional boundary value problems are well developed. However, multidimensional problems are difficult to be solved by all numerical methods. Therefore we construct here the walk on spheres algorithms and the walk on boundary algorithms for dimensions $m \geq 3$.

3.1 The Walk on Boundary Algorithms for Solving Interior and Exterior Boundary Value Problems of the Potential Theory

3.1.1 Boundary Integral Equations

Let G be a bounded simple connected domain of the Euclidian space $\mathbf{R}^m$, $m \geq 3$, and let ∂G be the boundary of G which is assumed to be piece-wise smooth after Lyapunov (Lyapunov-smooth). This means that

1) there exists a vector normal to ∂G at each point and $\theta \leq c|x-y|^\lambda$, $\lambda \in (0,1)$ where θ is the angle between the normal vectors at the points x and y, λ and c are some constants, and

2) there exists a number $d > 0$ such that the lines which are parallel to the normal vector at an arbitrary boundary point x have one intersection with the part of the surface ∂G lying inside the sphere $S(x,d)$.

Let $\bar{G} \cup \partial G$ and $G_1 = \mathbf{R}^m \backslash \bar{G}$. Assume that the area of ∂G is finite. Let

$$Lu = \sum_{i,j=1}^{} \frac{\partial}{\partial x_i} \left(a_{ij}(x) \frac{\partial}{\partial x_i} u \right) \tag{3.1}$$

be a self-adjoint elliptic operator; u is a twice continuous differentiable function in $\mathbf{R}^m$.

There exists a fundamental solution for L if $a_{ij}(x) \in C^3(\mathbf{R}^m)$; we denote it by $\mathcal{E}(x,y)$. Then [3.2]

$$\mathcal{E}(x,y) = -\frac{1}{(m-2)\sigma_m}[\varrho(x,y)]^{2-m}, \quad m \geq 3, \tag{3.2}$$

where $\sigma_m = 2\pi^{m/2}/\Gamma(m/2)$ is the area of a unit sphere in $\mathbf{R}^m$ and ϱ is a distance function such that

$$m|x - y| \le \varrho(x, y) \le M|x - y|$$

for arbitrary $x, y \in \mathbf{R}^m$; m, M being some positive constants. If the coefficients in (3.1) are constants, then [3.2, 3]

$$\varrho(x, y) = [\det\{a_{ij}\}^{(m-2)/2}\Big\{\sum_{i,j=1}^m a_{ij}^*(x_i - y_i)(x_j - y_j)\Big\}^{1/2},$$

where $\{a_{ij}^*\} = \{a_{ij}\}^{-1}$. In the case of the Laplace equation $\varrho(X, Y) = |X - Y|$.

We introduce the simple layer potential for (3.1):

$$V(x) = \int_\Gamma \frac{2}{(m-2)\sigma_m}[\varrho(x, y)]^{2-m}\mu(y)d\sigma(y) \tag{3.3}$$

and the double layer potential

$$W(x) = \int_\Gamma \frac{2}{(m-2)\sigma_m}\frac{\partial}{\partial n(y)}[\varrho(x, y)]^{2-m}\mu(y)d\sigma(y) , \tag{3.4}$$

where $V(X)$, $W(x) \in C^2(\mathbf{R}^m\backslash\partial G)$, μ is a bounded piece-wise continuous function, $d\sigma$ is a surface element, i.e. σ is a finite measure defined on $\Gamma = \partial G\backslash\Gamma_0$; here Γ_0 is a set of zero surface measure. Further,

$$\frac{\partial}{\partial n(y)} = \sum_{i,j=1}^m n_i(y)a_{ij}(y)\frac{\partial}{\partial y_i} \tag{3.5}$$

is the derivative (not normalized) with respect to the inner conormal, $n_i(y)$ are the components of the inner normal at a point $y \in \Gamma$.

Suppose that the fundamental solution is normalized so that for $\mu = 1$

$$W(x) = \begin{cases} 2, & x \in G \\ 1, & x \in \Gamma \\ 0, & x \in G_1 \end{cases} \tag{3.6}$$

(Sternberg normalization [3.2]). Note that the double layer potential W multiplied by $\sigma_{m/2}$ corresponds to the solid view angle of Γ from the point x.

Consider an inner Dirichlet problem in the classical formulation

$$Lu(x) = 0 , \quad x \in G , \quad u(y) = g(y) , \quad y \in \partial G$$

where ∂G is a smooth surface, and g is a continuous function. We seek the solution of this problem in the form of a double layer potential. Then the unknown density μ satisfies the boundary integral equation [3.1]

$$\frac{\sigma_m}{2}\mu(y) = -\int_{\partial G} \frac{\partial}{(m-2)\partial n(y')}[\varrho(y, y')]^{2-m}\mu(y')d\sigma(y') + g(y) . \tag{3.7}$$

In the case of a piece-wise smooth boundary ∂G, equation (3.7) remains true at all boundary points apart from the vertex points where the coefficient in the left-hand side of (3.7) must be changed. Note that the value of the double layer potential W inside the domain does not change if the integrals in (3.7) and (3.4) are taken over $\Gamma = \partial \backslash \Gamma_0$ since the measure of the vertex points of Γ_0 is equal to zero. The boundary conditions are considered only at $y \in \Gamma$. To satisfy the boundary conditions at Γ_0 it is sufficient to define the function μ as follows. Let $y_0 \in \Gamma_0$, and let

$$\alpha(y_0) = \int_\Gamma \frac{\partial}{(m-2)\partial n(y)} [\varrho(y_0, y)]^{2-m} d\sigma(y)$$

be the angle of view of Γ from y_0. Then we put

$$\mu(y_0) = \frac{1}{S_\alpha} \left\{ - \int_\Gamma \frac{1}{(m-2)} \mu(y) \frac{\partial}{\partial n(y)} [\varrho(y_0, y)]^{2-m} d\sigma(y) \right\} + g(y_0)$$

where $S_\alpha = \sigma_m - \alpha(y_0)$.

If the function $g(y)$ is piece-wise continuous then we eliminate the points of discontinuity of g. This procedure does not change the function $u(x)$ at $x \in G$ [3.4].

Thus we shall consider in this chapter the following formulation of boundary value problems. It is necessary to find the solution of the equation $Lu(x) = 0$ at the points $x \in G$ (or $x \in G_1$ in the case of exterior problems) such that $Bu(y) = g(y)$, $y \in \Gamma = \partial G \backslash \Gamma_0$ where Γ_0 is a set of zero measure including the vertex points of ∂G and the points of discontinuity of the piece-wise continuous bounded function g. The boundary operator B defines three main boundary value problems:

I) $B = I -$ identity operator
II) $B = \partial/\partial n$
III) $B = \partial/\partial n - H(y)$

Existence and uniqueness of the solutions in the form of simple and double layer potentials to the problems posed follow from the Fredholm theorems for the integral equations for μ. Applicability of these theorems follows from the fact that the kernels of these equations $k(y, y')$ have weak singularities as $y \to y'$ provided that the parts of ∂G intersecting under non-zero angles are Lyapunov surfaces [3.5].

To handle non-homogeneous equations $Lu = f$ we turn from the original problem to homogeneous equations using the superposition principle or the R-function method [3.6].

3.1.2 Interior Dirichlet and Exterior Neumann Problems

It is convenient to consider the interior Dirichlet problem

$$Lu(x) = 0 , \quad x \in G , \quad u(y) = g(y) , \quad y \in \Gamma \tag{3.8}$$

and the exterior Neumann problem

$$Lu(x) = 0 , \quad x \in G , \quad \frac{\partial u}{\partial n}(y) = g^*(y) , \quad y \in \Gamma \tag{3.9}$$

simultaneously.

Solutions to these problems will be constructed in the form of double and simple layer potentials with unknown densities. Note that if we seek the solution of the problem (3.8) in the form of a double layer potential than we come to an integral equation of the second kind. If, however, we seek this solution in the form of a simple layer potential than we come to an integral equation of the first kind. Here we consider only the boundary integral equations of the second kind.

Thus we seek the solution to (3.8) in the form of a double layer potential with an unknown density μ, and the solution of (3.9) in the form of a simple layer potential with a density μ^*. Then the densities μ and μ^* satisfy the following adjoint integral equations [3.1].

$$\mu(y) = \lambda \int_\Gamma k(y, y')\mu(y')d\sigma(y') + \lambda g(y) , \quad \lambda = 1 , \tag{3.10}$$

$$\mu^*(y) = \lambda \int_\Gamma k^*(y, y')\mu^*(y')d\sigma(y') + \lambda g^*(y) , \quad \lambda = 1 , \tag{3.11}$$

where

$$k(y, y') = \frac{2}{(m-2)\sigma_m} \frac{\partial}{\partial n(y')}[\varrho(y, y')]^{2-m} , \quad k^*(y, y') = k(y', y)$$

and g^* is the value of normal derivative $\partial u / \partial n$ on the boundary.

Spectral properties of (3.10, 11) can be derived as in the case of the Laplace equation [3.1, 3]. Below we use the following properties:

1) all the poles of (3.10, 11) are real
2) all the characteristic numbers are simple
3) there is no pole between -1 and $+1$
4) the number $\lambda = -1$ is an eigen-value of (3.10, 11).

Thus, $\varrho(K) = \varrho(K^*) = 1$ where K is the integral operator with the kernel $k(y, y')$. Therefore, the Neumann series for (3.10, 11) diverges, however, it is possible to use the methods described in Chap. 2.

Note that the double layer potential is a linear functional of the solution to the integral equation (3.10), $u = (\mu, h)$, where

$$h(y) = \frac{2}{(m-2)\sigma_m} \frac{\partial}{\partial n(y)}[\varrho(x, y)]^{2-m} .$$

Therefore, to construct a Monte Carlo estimate for $u = (\mu, h)$, it is sufficient, in the context of Chap. 2, to construct estimates for $(K^i g, h)$, $i = 1, 2, \ldots$, since the power series in η converges in norm; consequently, it is possible to integrate the series termwise.

Since $\lambda_1 = -1$, the transformation $\lambda = \psi(\eta)$ can be chosen as follows:

1) $\lambda = \psi(\eta) = \eta/(1 - \eta)$, $\eta_* = 1/2$, or
2) $\lambda = \psi(\eta) = 4\eta/(1 - \eta)^2$, $\eta_* = 3 - 2(2)^{1/2}$.

These transformations were studied in Chap. 2. Let us derive some real statistical estimates for $u(x)$. Let $\{y_k\}$ be the isotropic walking on boundary process starting at x. Then, to calculate $(h, K^k g)$, it is possible to use non-biased direct and adjoint estimates, respectively,

$$\xi_k = Q_k h(y_k) ; \quad \xi_k^* = Q_k^* g(y_k)$$

so that

$$M\xi_k = M\xi_k^* = (h, K^k g).$$

Consequently, the estimates

$$\xi = \sum_{k=0}^{n} a_k Q_k h(y_k) , \quad \xi^* = \sum_{k=0}^{n} a_k Q_k g(y_k) \tag{3.12}$$

are δ-biased for $u(x)$; the coefficients a_k are obtained from the transformation used. For example, if $\psi = 2\eta/(1 - \eta)$, then

$$a_k = \sum_{i=k}^{n} 2^k 3^{-i} C_{i-1}^{k-1} ,$$

i.e., $a_0 = 1$, $a_1 = 80/81$, $a_2 = 8/9$, $a_3 = 16/27, \ldots$. The weights Q_k, Q_k^* are calculated according to the formula presented in Sect. 1.1.1. The error δ in these δ-biased estimates can be found from

$$\mu = \sum_{k=0}^{n} a_k K^k g + \varepsilon(n) , \quad |\varepsilon(n)| \le Cq^n , \quad 0 > q > 1 ,$$

$$u = \sum_{k=0}^{n} a_k (h, K^k g) + \delta(n) .$$

For simplicity, let us present Monte Carlo estimates for (3.8,9) for the Laplace equation in the case when G is convex (generalization to non-convex domains can be carried out as described in Chap. 1):

$$\xi^* = \sum_{i=0}^{n} 2a_i (-1)^i g(y_i) ,$$

$$\xi = \sum_{i=0}^{n} 2a_i (-1)^i \frac{g^*(y_0)}{(m - 2)\sigma_m p(y_0)} |x - y_i|^{2-m} . \tag{3.13}$$

In the general case of non-convex domains when direct and adjoint estimates (3.12) are used for calculating solutions to the first interior and the second exterior problems, respectively, we have

$$Q_k = Q_0 T_k (-1)^k , \quad Q_k^* = T_k (-1)^k ,$$

$$T_k = \prod_{1=0}^{k-1} q(y_{i-1}) \operatorname{sign} [(n(y_i), y_{i-1} - y_i)] ,$$

$$Q_0 = \frac{g(y_0)}{p_0(y_0)} , \quad Q_0^* = \frac{h(y_0)}{p_0(y_0)} , .$$

and $T_k = 1$ for convex domains (Sect. 1.2.2). Taking into account that $p_0 = h(y)/2$ (for convex G) and assuming (in the second boundary value problem) that y_0 is uniformly distributed on Γ we obtain

$$\xi^* = \sum_{i=0}^{n} 2a_i(-1)^i q(x, y_0) \operatorname{sign} (n(y_0), x - y_0) T_i g(y_i)$$

$$\xi = \sum_{i=0}^{n} a_i(-1)^i \sigma(\Gamma) g(y_0) T_i h^*(y_i)$$

(3.14)

where

$$h^*(y) = \frac{2|x - y|^{2-m}}{(m-2)\sigma_m} ,$$

$q(y_{i-1}, y_i) \geq 1$ is the number of intersections of the line going through y_i, y_{i-1}, with the surface Γ. Note that here it is possible to apply the multiplication method to construct the resolvent for (3.10, 11) since λ_1 is exactly known. Indeed,

$$R_\lambda = (\lambda + 1)^{-1} R^1(\lambda)$$

(3.15)

where R^1 is analytic in the disk $|\lambda| < |\lambda_2|$. In our case $|\lambda_2| > 1$, so we have $R_1 = R^1(1)/2$. From this we obtain that in (3.14) the coefficients $a_k = 1$ for $k = 0, 1, \ldots, n - 1$, and $a_n = 1/2$ can be used. Note that in the case $\lambda = \psi(\eta) = 4\eta/(1-\eta)^2$ $a_0 = 1$, $a_1 = 0.6461$, $a_2 = 0.4207$, $a_3 = 0.2299$, $a_4 = 0.0701$. In [3.6], (3.14) was studied to obtain numerically optimal choice of the coefficients a_k.

Consider now again a situation where the Neumann series for $\varphi = K\varphi + f$ diverges, and we will present another approach based on a transformation of the integral equation. More exactly, we seek an integral operator K_1, a constant c_0 and a function f_1 such that $R_1 f + f(1 - c_0)$ can be represented in the form of a Neumann series for the equation

$$\varphi_1 = K_1 \varphi_1 + f_1 .$$

If this is true, then the solution of the original integral equation is represented as

$$\varphi = c_0 f + \sum_{i=0}^{\infty} K_1^i f_1 .$$

On this basis the functional $u = (\varphi, h)$ can be calculated using the estimates

$$\xi = c_0 \frac{f(y_0)h(y_0)}{p_0(y_0)} + \sum_{i=1}^{N} Q_i h(y_i)$$

$$\xi^* = c_0 \frac{f(y_0)h(y_0)}{p_0(y_0)} + \sum_{i=1}^{N} Q_i^* f_1(y_i)$$

(3.16)

where in the representations of Q_i, Q_i^* the kernel k is replaced by k_1, and f by f_1. For example, the multiplication in (3.15) corresponds to the case $c_0 = 1/2$, $k_1 = k$, $f = (Kf + f)/2$. Therefore, to obtain (3.16) in this case, the Markov chain is constructed as previously; however, to calculate $f_1(y_i)$, a randomization can be used. For example, in the walking on boundary algorithm a point Y_i must be sampled according to a density $p(y_{i-1}, y_i)$ independently of y_{i+1}. Then the estimate (3.16), to solve (3.8) takes the form

$$\xi^* = \frac{1}{2} Q_0^* g(y_0) + \frac{1}{2} \sum_{i=0}^{n-1} Q_i^* \left[g(y_i) + \frac{k(y_i, Y_i)}{p(y_i, Y_i)} g(Y_i) \right]$$

and Q_i^* are defined as described in Sect. 1.1.1.

In the case of (3.9) a randomization is needed only to calculate

$$f_1^*(y_0) = [K^* g^*(y_0) + g^*(y_0)]/2 .$$

In the direct estimate for (μ, h), $c_0 = 1/2$, $N = n - 1$, and

$$Q_i = Q_{i-1} \frac{k^*(y_i, y_{i-1})}{p(y_{i-1}, y_i)} , \quad Q_0 = \left\{ \frac{k^*(y_0, Y_0)g^*(Y_0)}{p^*(Y_0)} \right\} \Big/ (2p_0(y_0)) ,$$

where p^* is an appropriate density.

Below, the transformation $\lambda = \eta/(1 - \eta)$ corresponds to $c_0 = 1$, $f_1 = Kf/2$, $K_1 = (K + I)/2$. The Markov chain is constructed in accordance with K_1: y_0 is chosen with a density $p_0(y)$; $y_i = y_{i-1}$ with a probability of $1/2$ and y_i also with probability $1/2$, is simulated as previously, i.e., y_i is the next (after y_{i-1}) point in the walk on boundary process; to calculate $f_1(y_i)$ a randomization is used. Thus the corresponding estimate for (3.8) takes the form

$$\xi^* = \frac{g(y_0)h(y_0)}{p_0(y_0)} + \frac{1}{2} \sum_{i=0}^{n-1} \frac{h(y_0)}{p_0(y_0)} \beta_i \frac{k(y_i, Y_i)}{p(y_i, Y_i)} g(Y_i)$$

where $\beta_0 = 1$, $(\beta_i) = \prod_{j=1}^{i} b(y_{j-1}, y_j)$, and $b(y_{j-1}, y_j) = 1$, or $b(y_{j-1}, y_j) = k(y_{j-1}, y_j)/p(y_{j-1}, y_j)$, both with probability $1/2$. The corresponding estimate for (3.9) has the form

$$\xi = \frac{g^*(y_0)h^*(y_0)}{p_0(y_0)} + \frac{1}{2} \sum_{i=0}^{n-1} Q_i \beta_i h^*(y_i)$$

$$Q_0 = \frac{k^*(y_0, Y_0)g^*(Y_0)}{p^*(Y_0)p_0(y_0)} ; \quad Q_i = Q_{i-1} \frac{k^*(y_i, y_{i-1})}{p(y_{i-1}, y_i)} .$$

Finally, the transformation $\lambda = 2\eta/(1-\eta)$ corresponds to $c_0 = 1$, $K_1 = (2K+I)/3$, $f_1 = 2Kg/3$. The initial point of the Markov chain is sampled according to $p_0(y_0)$, $y_i = y_{i-1}$ with probability $1/3$, and y_i is simulated according to $p(y_{i-1}, y_i)$ with probability $2/3$. The estimate for (3.8) has the form

$$\xi^* = \frac{g(y_0)h(y_0)}{p_0(y_0)} + \frac{2}{3} \sum_{i=0}^{n-1} \frac{h(y_0)}{p_0(y_0)} \beta_i \frac{k(y_i, Y_i)}{p(y_i, Y_i)} g(Y_i)$$

where $b(y_{j-1}, y_j) = 1$ with probability $1/3$, and with probability $2/3$

$$b(y_{j-1}, y_j) = \frac{k(y_{j-1}, y_j)}{p(y_{j-1}, y_j)} \; ;$$

$$\beta_0 = 1 \, , \quad \beta_i = \prod_{j=1}^{i} b(y_{j-1}, y_j) \, .$$

Note that the general form of such iterative methods is generated by the transformations $\lambda = \eta(\eta) = \alpha\eta/(1 - \beta\eta)$ studied in Chaps. 1, 2.

An important question is how to calculate derivatives simultaneously with the solutions of the boundary value problems (3.8, 9) at desired points of G. At inner points, the estimates for the derivatives can be obtained very simply: the function h in all the estimates presented must be changed with the corresponding derivative. This procedure is allowed since the derivatives of the potentials can be calculated by direct differentation under the integral sign. For example, let us present concrete estimates for the Laplace equation in the case of a convex domain G.

Suppose that it is desired to calculate the derivative $\partial u/\partial x_1$ where u is the solution of the interior Dirichlet problem. This procedure leads to

$$\xi^* = \sum_{i=0}^{n} \left\{ \frac{n_1(y_0)}{(n(y_0), x - y_0)} - m \frac{x_1 - y_{01}}{|x - y_0|^2} \right\} a_i (-1)^i g(y_i) \, . \tag{3.17}$$

Here $n(y_0) = (n_1(y_0), \ldots, n_m(y_0))$ is the inner normal to Γ at the point $y_0 = (y_{01}, \ldots, y_{0m})$. In the case of the exterior Neumann problem the estimate for $\partial u/\partial x_1$ is

$$\xi = \sum_{i=0}^{n} \frac{g^*(y_0)}{p_0(y_0)} a_i \frac{2}{\sigma_m} (y_{i1} - x_i)|x - y_i|^{-m} \, . \tag{3.18}$$

Note that the estimates (3.17, 18) fail as the point x approaches the boundary ∂G since the projection of the gradient of the potential V on the direction n has a discontinuity on the boundary, and the derivative with respect to the tangent direction on the boundary is represented in the form of a singular integral. However, if we seek the solution of

$$\Delta u(x) = 0 \, , \quad x \in G \, , \quad u(y) = g(y) \, , \quad y \in \partial G \tag{3.19}$$

in the form of a simple layer potential then we come to an integral equation of the first kind:

$$\int_{\partial G} k(y, y')\mu(y')d\sigma(y') = g(y) , \quad y \in \partial G .$$ (3.20)

Using the representation

$$\frac{\partial u}{\partial x_i}(x) = \frac{\partial u}{\partial n}(x)n_i + \sum_{J=1}^{2} \frac{\partial u}{\partial \tau_j}(x)\tau_j$$ (3.21)

(where $n(y) = (n_1, n_2, n_3)$ is the normal to ∂G at a point $y \in \partial G$ nearest to the point x, $\partial G/\partial n(x) = (\text{grad}\, u(x), n)$, and $\tau_j(y)$ are the orthonormalized basis vectors lying in the tangent plane to ∂G), along with the estimates in Sect. 2.4.5, it is possible to construct random estimates for $\partial u/\partial x_i$ [3.7]. Indeed, let us take the follwing estimate (Sect. 2.4.2) for the solution of (3.20):

$$\xi(n) = - \sum_{i=1}^{n} C_n^i (-\kappa)^i Q_{i-1} g(y_{i-1}) ,$$

$$M\xi(n) = \mu_n , \quad \mu_n \xrightarrow[n \to \infty]{} \mu(y) , \quad 0 < \kappa < 2\|K\|^{-1} ,$$

or

$$\xi^*(n) = \kappa \left[ng(y_0] + \sum_{k=1}^{n} C_{n+1}^{i+1} (-\kappa)^k Q_k^* g(y_k) \right]$$

(Sect. 2.4.5), and consider the terms of (3.21). The integrals

$$\frac{\partial u}{\partial \tau}(x) = \frac{1}{2\pi} \int_{\partial G} \frac{(\tau, y - x)}{|y - x|^3} \mu(y)d\sigma(y)$$ (3.22)

can be estimated [because (3.22) is singular at $x \in \partial G$] by using two random points which are chosen symmetric with respect to $y \in \partial G$, nearest to x, in the plane including the normal $n(y)$ (see Chap. 4 for details concerning estimates for singular integrals).

Now assume thate

$$\frac{\partial u}{\partial n}(x) = v(y) + f(x, y)$$ (3.23)

where

$$v(y) = \lim_{x \to y} \frac{\partial u}{\partial n}(x) ,$$

f is unknown. Note that if the Neumann problem is solved, then $v = g$; in the case of the Dirichlet problem v can be found from the well-known formula

$$v(y) = \pm\mu(y) + \frac{1}{2\pi} \int_{\partial G} \frac{(n(y), y' - y)}{|y - y'|^3} \mu(y')d\sigma(y') ,$$

where y is a point of continuity of the function μ; the plus sign corresponds to the exterior problem while the minus sign is taken in the case of the interior problem. From (3.23) we obtain f :

$$f(x,y) = \frac{1}{2\pi} \int_{\partial G} \left(n(y), \frac{y-x}{|y-x|^3} - \frac{y-y'}{|y-y'|^3} \right) \mu(y') d\sigma(y') \mp \mu(y) .$$

3.1.3 Interior Neumann, Exterior Dirichlet, and the Third Boundary Value Problems

Consider the inner Neumann problem

$$Lu(x) = 0 , \quad \frac{\partial u}{\partial n}(y) = g(y) \quad y \in \Gamma . \tag{3.24}$$

Suppose that this problem is uniquely solvable, in particular, it is assumed that the following condition holds:

$$\int_\Gamma g(y) d\sigma(y) = 0 . \tag{3.25}$$

Let us also consider the exterior Dirichlet problem

$$Lu(x) = 0 , \quad x \in G_1 , \quad u(y) = g(y) , \quad y \in \Gamma . \tag{3.26}$$

We preserve all assumptions about the regularity of the boundary ∂G.

Now we seek the solution to (3.24) in the form of a simple layer potential (3.3) with unknown density μ^*, and the solution to (3.26) in the form of a double layer potential (3.4) with unknown density μ. Then the equations describing μ and μ^* have also the form of (3.10, 11), respectively, with the only difference that $\lambda = -1$. We write these equations in the form

$$\mu = \lambda K\mu - g \tag{3.27}$$

$$\mu^* = \lambda K^*\mu - g^* , \tag{3.28}$$

where $\lambda = -1$. Thus, $\lambda = -1$ is an eigen-value having minimal modulus, and the constant is the unique eigen-function of the operator K^* corresponding to $\lambda_1 = -1$. Hence by the Fredholm theorem, taking into account (3.25), we obtain that (3.28) is uniquely solvable and its solution is represented in the form of a uniform and absolute convergent Neumann series.

When estimates for calculating $u = (\mu^*, h)$, where

$$h(y) = \frac{2}{(m-2)\sigma_m} [\varrho(x,y)]^{2-m}$$

are constructed, the condition (3.25) can be taken into account using the follwing representation

100

$$u(x) = (\mu^*, h) = \sum_{i=0}^{\infty} (-1)^{i+1} (K^{*i} g, h)$$

$$= \sum_{i=0}^{\infty} (-1)^{i+1} \left\{ (K^{*i} g_+, h) - (K^{*i} g_-, h) \right\} \qquad (3.29)$$

where $g_+ > 0$ is the positive, and $g_- > 0$ is the negative part of the function g. We introduce the notation

$$\Gamma_+ = \{y \in \Gamma : g(y) \geq 0\}, \quad \Gamma_- = \{y \in \Gamma : g(y) \leq 0\}.$$

Let $\{y_i^1\}$, $\{y_i^2\}$ be two independent walk on boundary processes such that in the first process the initial density p_0^+ is defined on Γ_+ while in the second one the initial density p_0^- is defined on Γ_-; the points y_i^1, $i = 1, 2, \dots$ and y_i^2, $i = 1, 2, \dots$ are sampled as in the standard walk on boundary process. Then the estimate for $u = (\mu^*, h)$ has the form

$$\xi(x) = \sum_{i=0}^{n} \frac{2(-1)^{i+1}}{(m-2)\sigma_m} \left[Q_i^+ [\varrho(x, y_i^1)]^{2-m} - Q_i^- [\varrho(x, y_i^2)]^{2-m} \right] \qquad (3.30)$$

where the weights Q_i^+ and Q_i^- are determined through g^+, $p_0 = p_0^+$ and g^-, $p_0 = p_0^-$, respectively, using standard formulas (Sect. 1.1.1) for direct estimates. For example, if $L = \Delta$ and G is convex, and y_0^1, y_0^2 are chosen according to the density c^{-1}, where

$$c = \int_{\Gamma} |g(y)| d\sigma(y),$$

then

$$\xi(x) = \sum_{i=0}^{n} \frac{c(-1)^{i+1}}{(m-2)\sigma_m} \left[|x - y_i^1|^{2-m} - |x - y_i^2|^{2-m} \right]. \qquad (3.31)$$

Estimates for the derivatives at inner points can be obtained by differentiation of (3.30, 31), since as mentioned above, the derivatives of the potentials V and W can be evaluated by differentiation under the integral sign. It is known that the solution of the problem (3.24) is unique to within an additive constant [3.5]. Usually, the following information is given: $u(x_0) = u_0$ where x_0 is a prescribed point in G. Then the desired estimate for $u(x)$ has the form $\eta(x) = \xi(x) + u_0 - \xi(x_0)$ where $\xi(x)$ is the estimate presented above.

Consider now the exterior Dirichlet problem (3.26). Direct utilization of W leads to asymptotics $W = 0(|x|^{1-m})$ as $x \to \infty$. It is often desired in physical problems to have asymptotics $u = 0(|x|^{2-m})$ as $x \to \infty$. To construct the algorithm with the desired asymptotics we use the following approach [3.5, 7]. We seek the solution $u(x)$ of (3.26) in the form

$$u(x) = W(x) + \alpha |x|^{2-m}$$

(it is assumed that G includes the origin of the coordinate system). Then (3.27) becomes

$$\mu = \lambda K\mu + \alpha|y|^{2-m} - g(y) , \quad y \in \Gamma \tag{3.32}$$

hence, $\lambda = -1$ is also a pole of the resolvent of (3.32) having minimal modulus.

We denote by μ_0 the eigen-function corresponding to the eigen-value $\lambda = -1 : \mu_0 = K^*\mu_0$. Then by the Fredholm theorem, the Neumann series for (3.32) converges absolutely at $\lambda = 1$ provided that

$$\int_\Gamma \{(\alpha|y|^{2-m} - g(y))\}\mu_0(y)d\sigma(y) = 0 . \tag{3.33}$$

From (3.33) we obtain

$$\alpha = \left[\int_\Gamma g(y)\mu_0(y)d\sigma(y)\right]\left[\int_\Gamma |y|^{2-m}\mu_0(y)d\sigma(y)\right]^{-1} .$$

The integrals in this expression can be calculated by estimates of Sect. 2.3.1. Another estimate [3.7] is based on the relation

$$\mu_0(y) = \lim_{n\to\infty} (K^*)^n l(y)$$

(uniform with respect to y), where $l(y) \in C(\Gamma)$ is an arbitrary function. Thus

$$\alpha = \lim_{n\to\infty} \left[\frac{M\xi_1^{(n)}}{M\xi_2^{(n)}}\right] ,$$

where

$$\xi_1^{(n)} = Q_n g(y_n) ; \quad \xi_2^{(n)} = Q_n|y_n|^{2-m}$$

and Q_n are weights calculated by standard formulas (Sect. 1.1.1). It remains to calculate the term $W = (h, \mu)$. To this end we apply the scheme (3.29) using two independent walk on boundary processes $\{y_i^1\}, \{y_i^2\}$:

$$W = (h, \mu) = \left(h, \sum_{i=0}^{\infty}(-K)^i[\alpha|y|^{2-m} - g(y)]\right)$$

$$= \sum_{i=0}^{\infty}\{((-K^*)^i h_+, \alpha|y|^{2-m} - g(y)) - ((-K^*)^i h_-, \alpha|y|^{2-m} - g(y))\}$$

where h_+ is the positive and h_- is the negative part of the function h. To complete the proof, the steps used in the construction of (3.30,31) must be repeated.

Note that μ_0 is the solution of the Robin problem. Special estimates based on the ergodic property of the walk on boundary process can be constructed [3.7].

Consider now a third boundary value problem, the interior ($x \in G$) and the exterior one ($x \in G_1$):

$$Lu(x) = 0 , \quad x \in G(x \in G_1)$$

$$\frac{\partial u}{\partial n}(y) - H(y)u(y) = g(y) , \quad y \in \Gamma \tag{3.34}$$

where $H(y)$ is a continuous function, $H \not\equiv 0$. We seek the solution to (3.34) in the form of a simple layer potential V. Then [3.5]

$$\mu(y) = \lambda \int_\Gamma \frac{1}{(m-2)\sigma_m} \left[\frac{\partial}{\partial n(y)} [\varrho(y', y)]^{2-m} \right.$$

$$\left. - H(y)[\varrho(y', y)]^{2-m} \right] \mu(y') d\sigma(y') + f(y) \tag{3.35}$$

where $\lambda = 1$, $f = -g$ in the case of the interior problem, and $\lambda = -1$, $f = g$ for the exterior problem.

To apply the estimates of Chap. 2 to solve (3.35), it is necessary to have spectral properties of this integral equation. Using properties of the potential V and taking into account that (3.35) has a weak singularity we obtain [3.5, 7] that the set of characteristic values is real, discrete and does not have finite concentration points. Hence, there exists a unique solution to the interior problem if $H \geq 0$, and the poles of (3.35) lie on the ray $\lambda \leq a$ or on the ray $\lambda \geq 1 + b$, $a, b > 0$.

In the case of the exterior problem a solution in the form of the potential V exists, is unique, and the poles of (3.35) lie on $\{\lambda \geq a\}$ or on the ray $\{\lambda \leq -1 - b\}$. The characteristic numbers having minimal modulus $-1 - b$, $-a$, a or $1 + b$ can be calculated by one of the algorithms described in Sect. 2.3.

If information about the numbers a, b is available, it is possible to apply one of the methods of Chap. 2 to construct the resolvent of (3.35). In particular, if the numbers $a < 1$ and $b > 0$ are known, then in the case of the interior problem the following transformation can be used

$$\lambda = 4a\eta \left[(1 - \eta)^2 + \frac{a(1+\eta)^2}{1+b} \right] ,$$

or the transformation (2.16), or

$$\lambda = 4(1 + b)\eta / [(1 - \eta)^2 + (1 + b)\frac{(1+\eta)^2}{a}] .$$

In the case of the exterior problem the transformation (2.16) with the minus sign can be used.

Calculations show that the transformation (2.16.) is the best one, and when $b > 0.5$, the transformation $\lambda = a\eta/(1 - \eta)$ works well.

If $a > 1$, the Neumann series for (3.35) converges, therefore the estimate can be constructed by standard formulas for the walk on boundary process by introducing some positive terminating probability (Sect. 1.2.2). However, if b is small [$b \leq 1$], the Neumann series canverges very slowly, and it is then necessary to accelerate the convergence by the methods described in Sect. 2.4.

The simplest way is to use the transformation $\lambda = (\psi(\eta)) = b\eta/(\eta - 1)$. Iterations of the integral operator in (3.35) must be calculated by the direct estimate (Sect. 1.1.1). We present the estimate for the case $A = \Delta$ (G is convex, the value of a is known)

$$\xi = \sum_{i=0}^{n} a_i Q_i |x - y_i|^{2-m}$$

where $Q_0 = f(y_0)/p_0(y_0)$, and

$$Q_i = \lambda Q_{i-1} \left\{ 1 - \frac{H(y_i)|y_i - y_{i-1}|^2}{(m-2)(n(y_i), y_{i-1} - y_i)} \right\},$$

p_0 is an appropriate density of the initial point of the process $\{y_i\}$; the coefficients a_i are generated by the transformation $\psi(\eta)$.

If we construct the solution of interior (or exterior) Dirichlet problem in the form of a sum of simple and double layer potentials with densities $-H\mu$ and μ, respectively, then we come to an integral equation, adjoint to (3.35), for the exterior (or interior one, respectively) third boundary value problem. Such estimates for the Laplace equation are given in [3.7].

3.1.4 Dirichlet and Neumann Problems for the Helmholtz Equation

Formally, the above described algorithms could be easily generalized on the case of Helmholtz equation. However, analysis of the spectral properties of the corresponding boundary integral equations is more complicated.

Consider boundary value problems (interior and exterior) in a bounded domain $G \subset \mathbf{R}^3$ with a smooth boundary ∂G (using the notation of Sects. 3.1.1–3):

$$\Delta u + k^2 u = 0, \quad x \in G[x \in G_1], \quad Bu(y) = g(y), \quad y \in \partial G \qquad (3.36)$$

where $B = I$ in the case of Dirichlet problem, and $B = \partial/\partial n(y)$ for the Neumann problem.

We assume that these problems are uniquely solvable in certain classes of functions. In particular, the Sommerfeld radiation condition must hold in the case of exterior problems:

$$u(x) = 0(|x|^{-1}), \quad \frac{\partial u(x)}{\partial |x|} - i\, k u(x) = 0(|x|^{-1}) \quad \text{as} \quad |x| \to \infty .$$

Let

$$V_1 = \int_\Gamma \frac{\exp\{ik|x - y|\}}{|x - y|} \mu(y) d\sigma(y) ,$$

$$V_2 = \int_\Gamma \frac{\partial \exp\{ik|x - y|\}}{\partial n_y |x - y|} \nu(y) d\sigma(y)$$

be simple and double layer potentials with densities μ and ν, respectively, where n_y is the exterior normal vector at $y \in \partial G$. If we seek the solution of Dirichlet problem (the interior and exterior one with the radiation condition) in the form of a double layer potential V_2 with unknown density $\nu \in C(\Gamma)$, then we come, as in the case of Laplace equation, to the integral equation [3.1]

$$\nu(y) = \lambda \int_\Gamma k(y, y')\nu(y')d\sigma(y') + f(y) , \quad y \in \Gamma \tag{3.37}$$

where $\lambda = 1$, $f = -g/2\pi$ for the interior Dirichlet problem, and $\lambda = -1$, $f = g/2\pi$ for the exterior Dirichlet problem, and

$$k(y, y') = \frac{1}{2\pi} \frac{\partial}{\partial n_y} \frac{e^{ik|y-y'|}}{|y - y'|} = [1 - ik|y - y'|]\frac{\cos(\varphi_{yy'})}{|y - y'|^2}e^{ik|y-y'|} . \tag{3.38}$$

If we seek the solution of Neumann problem (interior and exterior with the radiation condition) in the form of a simple layer potential with unknown density $\mu \in C(\Gamma)$, then we come to equation adjoint to (3.37):

$$\mu(y) = \lambda \int_\Gamma k^*(y, y')\mu(y')d\sigma(y') + f_1(y) , y \in \Gamma \tag{3.39}$$

where $\lambda = -1$, $f_1 = g/2\pi$ for the interior, and $\lambda = 1$, $f = -g/2\pi$ for the exterior Neumann problem.

The case $k = i\kappa$, $\kappa \geq 0$ is not difficult, and all estimates of Sects. 3.1.2, 3 are applicable here; it is only necessary to introduce a factor

$$[1 - ik|y_{i-1} - y_i|] \exp\{ik|y_{i-1} - y_i|\} .$$

Now se suppose that $\lambda = k^2 \neq \mu_i$ where $\mu_i[i = 1, 2, \ldots]$ are the eigen-values of the interior Dirichlet (or Neumann) problem for the Laplace equation. We assume also that some information about the characteristic numbers of (3.37, 39) is known, for example,

$$\lambda_1 < \lambda_2 < \ldots < \lambda_n < \ldots . \tag{3.40}$$

To solve the boundary value problems for $k^2 \neq \mu_i$ the following approaches can be used:

I) We can choose the appropriate transformation $\lambda = \psi(\eta)$ in (3.37, 39) using the information of (3.40); next use the estimates of Sect. 2.1.2.
II) If it is only known that $\{\lambda, \ldots, \lambda_m\} \subset \mathbf{R}$, then we can use the mapping of the half plane $\text{Im}\{\lambda\} \geq 0$ on the disk $|\eta| \leq 1$; next we can apply one of the generalized summation methods (such an example is given in Sect. 2.4.1)
III) If approximate values of first characteristic numbers $\lambda_1, \lambda_2, \ldots$ are known, then we can use estimates based on the Padé approximations (Sect. 2.4.4), which work quite well.
IV) The analytical continuation can be applied directly to the equation $\Delta u + k^2 u = 0$, since it is possible to calculate the initial element of the analytic function

$u(x, k)$ by the walk on boundary algorithm. However, it is then necessary to calculate iterations of the differential operator $(\Delta - \lambda_0)$ (Chap. 4).

Note that the desired information about characteristic numbers can be obtained by the estimates of Sect. 2.3.

Now we give some examples. In the cube $0 \leq x_1, x_2, x_3 \leq \pi$ we want to solve the following boundary value problem

$$\Delta u + \lambda u = 0 ,$$

$$u|_\Gamma = \begin{cases} \cos\{(\lambda/3)^{1/2}[x_1 + x_2 + x_3]\} & \lambda \geq 0 , \\ \exp\{(\lambda/3)^{1/2}[x_1 + x_2 + x_3]\} , & \lambda \leq 0 . \end{cases}$$

We seek the solution in the form

$$u(x, \lambda) = \sum_{n=0}^{\infty} u_n(x)(\lambda - \lambda_0)^n .$$

To continue the solution $u(x, \lambda)$ to the domain $\lambda > \lambda_1$ we first calculate λ_1 on the basis of the relation $\lambda_1 \approx u_n/u_{n+1}$. The coefficients u_n are calculated by the Cauchy formula

$$u_n(x) = \int_{C_A} \frac{u(x, \lambda)d\lambda}{(\lambda - \lambda_0)^{n+1}} , \tag{3.41}$$

where C_A is a closed curve lying in the half plane $\text{Re}\{\lambda\} < \lambda_1$ and including the point λ_0. For example, for C_A we choose a circle of radius r centered at A. Then (3.41) takes the form

$$u_n = \frac{r}{2\pi} \int_0^{2\pi} \frac{u(A + r\exp\{i\varphi\})\exp\{i\varphi\}}{(A - \lambda_0 + r\exp\{i\varphi\})^{n+1}} d\varphi . \tag{3.42}$$

Using the dependent sampling technique we can calculate u_n, $n = 0, 1, \ldots$, on the basis of (3.42) where the solution u is calculated simultaneously at $\{\lambda^k\}$ on one trajectory of the walk on boundary process, λ^k being some quadrature nodes.

In calculations we used Simpson's rule with 24 nodes. The solution at these nodes was calculated by estimates (3.14) for (3.37), four numbers of terms in the walk on boundary algorithm were taken. In Table 3.1 we show the numerical results for λ_1 where $A = (-1.47, 0)$; $r = 0.8$, n is the number of iterations, N is the number of the trajectories of the walk on boundary process, the relative statistical error is on the order of 1%. Using $\lambda = \psi(\eta)$ from Sect. 2.1.2 and the algorithm of Sect. 2.3.1, the number λ_2 and $u(x, \lambda)$ for $\lambda \in (\lambda_1, \lambda_2)$ were calculated. This experiment was carried out in 61 min of BESM-6 computer time, the approximate values lie within 1% of the correct solution. Note that this method failed when the solution $u(x, \lambda)$ for $\lambda > \lambda_2$ was calculated.

In the follwing experiment we compare methods II and III with the method based on the transformation (2.23). In the square $0 \leq x, y \leq \pi$ the problem

$$\Delta u + k^2 u = 0 , \quad u(x, y)|_\Gamma = \cos 2^{-1/2}(kx + ky)|_\Gamma$$

Table 3.1. Approximation for λ_1

N	$n = 1$	$n = 2$	$n = 3$	$n = 4$
500	0.91	2.13	2.69	2.94
1000	0.89	2.11	2.71	2.99

was solved. In the region $\lambda_1 < k^2 < \lambda_2$ all the methods give satisfactory results (λ_1 and λ_2 were calculated by the estimate of Sect. 2.3). At $k^2 = 2.42$, $x = 0.2\pi$, $y = 0.4\pi$, method II (the Lindelef summation, $\sigma = 0.01$) gives the result 0.4604, method III (estimate of Sect. 2.4.4) gives the value 0.4628, the transformation (2.33) leads to 0.4697, the exact value is 0.4818. In the region $\lambda_2 < k^2 < \lambda_3$ satisfactory results were obtained by methods II, III. In the region $\lambda_3 < k^2 < \lambda_4$ only method II worked. In the region $\lambda_4 < k^2 < \lambda_5$ the error of method II was about 20%. Thus the methods described enable one to calculate only a few harmonics of the solution to Helmhotz equation.

3.1.5 The Variance, the Error and the Cost of the Walk on Boundary Algorithms

The computational cost of the walk on boundary algorithms can be estimated on the basis of results of Sect. 2.1.2. Consider an estimate written in the general form

$$\zeta_\varepsilon^{(m)} = \sum_{k=1}^{m} \zeta(k) l_k^{(m)} , \tag{3.43}$$

where

$$l_k^{(m)} = \sum_{n=k}^{m} d_k^{(n)} \eta_*^n , \tag{3.44}$$

and $\zeta(k)$ is one of the non-biased estimates of $K^k f$, K is the boundary integral operator of (3.19).

Theorem 3.1. The variance of the walk on boundary algorithm based on the transformation $\lambda = \psi(\eta) = 4\eta/(1 - \eta)^2$ has the order $0(m)$ for convex domains. The computational cost of this method necessary to achieve an error ε is given by

$$T_\varepsilon = 0(|\ln(\varepsilon)|^3/\varepsilon^2) \quad \text{as} \quad \varepsilon \to 0 .$$

Proof. Note that m is chosen from the condition that the rest of the series

$$\sum_{n=1}^{\infty} b_n \eta_*^n , \quad b_n = \sum_{k=1}^{n} d_k^{(n)} K^k f \tag{3.45}$$

is of order $0(\varepsilon)$. Hence, $m = 0(|\ln(\varepsilon)|)$, since (3.45) converges as a geometric progression [3.8]. Note that $|l_k^{(m)}| \leq 1$ for $\lambda = 4\eta/(1-\eta)^2$ at $\lambda_* = 1$ (Chap. 2), hence

$$D\zeta_\varepsilon^{(m)} \leq m \sum_{k=1}^{m.} D\zeta(k) . \tag{3.46}$$

For convex domains $\|K\|_1 = \|K^*\|_1 = 1$, therefore $D\zeta(k) \leq \sigma^2$. Consequently, we obtain from (3.46) $D\zeta_\varepsilon^{(m)} = 0(m^2)$. Substituting $m = 0(|\ln(\varepsilon)|)$ yields

$$T_\varepsilon = 0(|\ln(\varepsilon)|^3/\varepsilon^2) . \quad \square \tag{3.47}$$

Remark. The total error, when an estimate of the type $\zeta^{(m)}$ is used, has the form $\varepsilon(m) = \varepsilon_1(m) + \varepsilon_2(m)$ where $\varepsilon_1(m)$ is a deterministic bias and $\varepsilon_2(m)$ is the statistical error of the estimate (3.43). To illustrate, we show in Fig. 3.1 three variants of transformations for values of η_* that are small, intermediate and close to 1. The curves show that the optimal choice of $m = m_0$ depends on the rate of convergence of the series and on the rate of divergence of the variance. If for some two estimates the variance curves are similar (curves 1 and 2, for instance) but the corresponding rates of convergence significantly differ, then the optimal value of m corresponds to the point of intersection of curves 1. If the curves of the rate of convergence are similar but the variance divergence curves differ (cases 2, 3) then the optimal value corresponds to the point of intersection of curves 3.

Consider the estimate for the solution to the Dirichlet problem for the Laplace equation in a convex bounded domain G, based on the representation (3.15):

$$\xi(n) = \sum_{i=0}^{n-1} 2g(y_i)(-1)^i + (-1)^n g(y_n) .$$

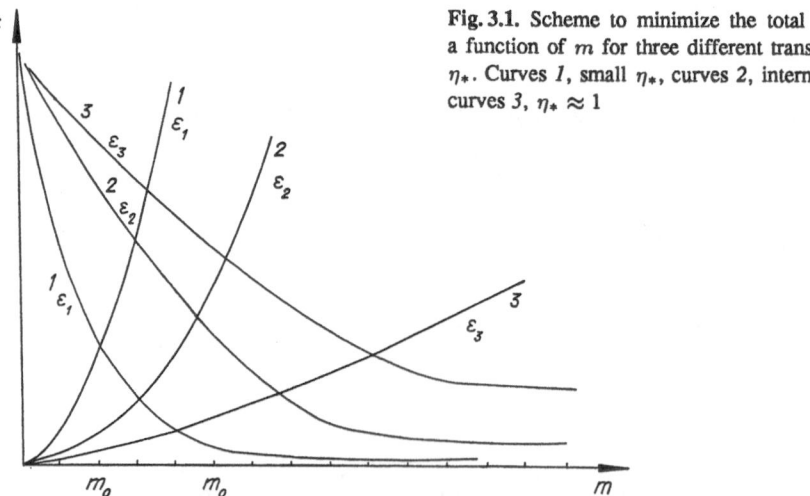

Fig. 3.1. Scheme to minimize the total error ε as a function of m for three different transformations η_*. Curves *1*, small η_*, curves 2, intermediate η_*, curves 3, $\eta_* \approx 1$

Let

$$s = \inf_{y \in \partial G} g(y) \geq 0 , \quad S = \sup_{y \in \partial G} g(y) , \quad \underset{\partial G}{\mathrm{osc}}\, g(y) = S - s .$$

We put $\xi(0) = g(y_0)$ and rewrite this estimate in the form

$$\xi(n) = \sum_{i=0}^{n-1} (-1)^i [g(y_i) - g(y_{i+1})] + g(y_0) , \quad n \geq 1 . \tag{3.48}$$

From this we obtain

$$M\xi^2(n) = Mg^2(y) + 2\sum_{i=0}^{n-1} (-1)^i M\{\xi(i)[g(y_i) - g(y_{i+1})]\}$$

$$+ \sum_{i=0}^{n-1} M[g(y_i) - g(y_{i+1})]^2 . \tag{3.49}$$

Using the fact that $k(y_{i-1}, y_i) = -p(y_{i-1}, y_i)$ and

$$M f_1(y_j) f_2(y_i) = (-1)^i (h, K^j(f_1 K^{i-j} f_2)) , \quad i \geq j ,$$

(p is the transition density of the walk on boundary process) we get from (3.49):

$$M\xi^2(n) = \left(h, g^2 + 2\sum_{i=0}^{n-1} \left(2\sum_{j=0}^{i-1} K^j g K^{i-j}(g + Kg) \right) \right.$$

$$\left. + (-1)^i K^i(g(g + Kg)) \right) + \sum_{i=0}^{n-1} (-1)^i K^i(g(g + Kg))$$

$$- K^i(g^2 + Kg^2) . \tag{3.50}$$

Here K is the integral operator of (3.10). Since $\|K\| = 1$ we obtain from (3.50)

$$D\xi(n) \leq c(\underset{\partial G}{\mathrm{osc}}\, g)^2 n^2 .$$

Hence, in this case

$$T_\varepsilon = 0(|\ln(\varepsilon)|^3 / \varepsilon^2) .$$

Note that the condition $D\zeta(k) = D(k) < \infty$ holds for the boundary integral equations, since the kernel

$$k(y, y') = \frac{2}{(m-2)\sigma_m} \frac{\partial}{\partial n_{y'}} [\varrho(y, y')]^{2-m}$$

and the transition density

$$p(y', y) = \frac{2}{\sigma_m} |\cos(n(y), y' - y)| \, |y - y'|^{1-m}$$

have the same singularity order, namely, $O(|y'-y|^{1-m+\lambda})$ as $|y-y'| \to 0$ where λ is the Lyapunov parameter [3.5]. In the case of convex domains, the density $p(y,y')$ is admissible with respect to $k(y',y)$ and $k(y,y')$. However, in the general case $p(y,y')$ is admissible only with respect to $k(y,y')$; accordingly direct and adjoint estimates are constructed.

Note that if standard Monte Carlo estimates are used which are based on the Neumann series, for example, estimates (3.30, 31), then the variance is studied on the basis of the relation [3.7]:

$$M\xi^2 = (c, h(2\mu^* - h)) \tag{3.51}$$

where μ^* is the Neumann series for the adjoint equation (3.28) with the right-hand side h. As (3.51) shows, $M\xi^2$ is infinite, since $\mu^* = \infty$, and the Neumann series χ for the integral equation with the kernel $k^2(y,y')/p(y,y')$ diverges.

3.2 Walk Inside the Domain Algorithms

We turn now to algorithms of approach II based on the local integral equations with a generalized kernel. Although algorithms of this type are well known [3.9, 10] and often used in calculations undergoing further development [3.11–15], there exists a series of unsolved problems in this field. In particular, it is difficult to derive the corresponding adjoint equation and the Fredholm theorems.

The next problem is to generalize the walk on spheres algorithms to obtain the solution at a number of points simultaneously. It is also interesting to obtain algorithms which permit one to calculate derivatives at arbitrary points of the domain. There exist also a number of optimization problems, for example, finding the optimal choice of the radius in the walk on spheres algorithms, the optimal choice of the transition density, etc. An important question is the effective implementation of the walk inside the domain algorithms. All the questions mentioned are considered in this section and in Chaps. 4 and 5 .

3.2.1 General Scheme for Constructing Monte Carlo Estimates on the Walk Inside the Domain Processes

Consider a Dirichlet problem in a domain G which is a union of interesting subdomains G_i:

$$Lu(x) = 0 , \quad x \in G \subset \mathbf{R}^m , \quad u(y) = g(y) , \quad y \in \partial G \quad \text{(or } y \in \Gamma) \tag{3.52}$$

where L is a differential operator for which an ε-biased estimate on the walk on spheres process described in Sects. 1.2.1, 2 can be constructed.

It is assumed that the set of subdomains is devided into classes such that each class involves subdomains G_i for which a fixed type of random estimate for solving the problem (3.52)

$$\xi_x = Q(x, \{x_k\}, \{y_k\}, G_i)\Phi(g(\{y_k\}))$$ (3.53)

is used. Here $\{x_k\} \subset G$, $\{y_k\} \subset \partial G$ are the random points of the walk on spheres or the walk on boundary processes; the random weight Q depends only on the points x, $\{x_k\}$, $\{y_k\}$, and subdomains G_i; the functional Φ is defined by the type of random walk used. For example, $\Phi(g) = g$ for the walk on spheres process, and $\Phi(g) = \sum_{k=1}^{n} a_k[g(y_k)]$ in the case of the walk on boundary process. In Fig. 3.2 an example of a domain $G \subset \mathbb{R}^3$ is shown. Here a non-bounded domain $G = \cup G_i$ is represented as a union of regions bounded by intersecting dihedral angles G_i. In general, G_i can be chosen arbitrarily, it is only desired that the random estimate (3.53) needed to solve the boundary value problem in G_i is effective. For example, in the case of the walk on boundary process it is natural to use convex subdomains G_i. The idea of applying the scheme of iteration on subdomains was suggested by this author in connection with the necessity of constructing an effective generalization of the walk on boundary algorithm to solve boundary value problems in non-convex domains.

We assume that the number of subdomains G_i is finite, $i = 1, \ldots, n$. Suppose also that the boundary ∂G_i consists of two parts, one part of ∂G_i belongs to ∂G, ∂G_i^r, and the rest to ∂G_i^f. The part ∂G_i^f is called a fictitious boundary.

We introduce also the following notations. Let

$$d_i(x) = \inf\{|y - x|, \quad y \in \partial G_i\}$$

be the distance from the point x to ∂G_i, and let

$$\partial G_{i\varepsilon} = \{x \in G_i : d_i(x) < \varepsilon\}$$

be the ε-neighborhood of ∂G_i. We decompose it analogously as we decomposed ∂G:

$$\partial G_{i\varepsilon} = \partial G_{i\varepsilon}^r \cup \partial G_{i\varepsilon}^f .$$

We assume that there exists a subdomain G_i such that $\partial G_{i\varepsilon}^f \subset G_j$ for all G_i, and the set $v_{ij}(\varepsilon) = \partial G_{j\varepsilon}^f \cap \partial G_{i\varepsilon}^r$ is not empty (maximum for one value $j \neq i$). It is supposed that in this set it is possible to approximate the solution of (3.52) to within $O(\omega(\varepsilon))$ where ω is the continuity modulus of the function u in ∂G_ε.

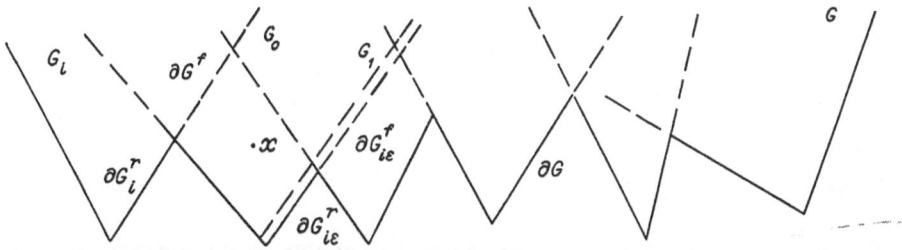

Fig. 3.2. Subdomains G_i for walk on subdomains algorithm

The random walk is constructed as follows (for simplicity, using the walk on spheres process). Assume that it is desired to calculate the solution $u(x)$. Then there exists an i such that $x \in G_i$: assume that $i = 0$; thus $x \in G_0$. The transition from the state n to the state $n + 1$ is carried out by

$$x(n + 1) = x(n) + d_i(x(n))\omega_n , \quad n = 0, 1, \ldots, N_1 - 1 ,$$

where $\{\omega_n\}$ is a sequence of independent isotropic random unit vectors, N_1 is a random number of the chain transitions before the break in $\partial G_{i\varepsilon}$. If $x(N_1) \in \partial G_{i\varepsilon}^{\tau}$ or if $x(N_1) \in \partial G_{i\varepsilon}^{f} \cap \partial G_{j\varepsilon}$ for some $j \neq i$ then the corresponding contribution to the estimate is stored. The process is continued if

$$x(N_1) \in A_{ij} \equiv \{\partial G_{i\varepsilon}^{f} \cap (G_j \backslash \partial G_{j\varepsilon}) , \quad j \neq i\} ,$$

where d_i are changed with d_j. The process is stopped if $x(N_1 + N_2) \in \partial G_{j\varepsilon}$, and it is continued if the event A_{jk} occured for some $k = j$, etc.

Thus the total number of steps of the random walk before its break in ∂G_ε is equal to $N = N_1 + \ldots + N_n$. Simple arguments show that the well-known logarithmic estimation for the mean number of steps of the walk on spheres process is conserved here if

$$MN_i = c_{1i}|\ln(\varepsilon)| + c_{2i}(\varepsilon)$$

holds for all subdomains, where

$$c_{2i}(\varepsilon) = 0(|\ln(\varepsilon)|) \quad \text{as} \quad \varepsilon \to 0 , \quad i = 1, \ldots, n .$$

Indeed, let $c_1 = \max\limits_{i} c_{1i}$, $c_2 = \max\limits_{i} |c_{2i}(\varepsilon)|$. Then

$$MN = MM(N|k) = \sum_{j=1}^{\infty} \left\{ \sum_{i=1}^{j} MN_i \right\} P(k = j)$$

$$= \sum_{i=1}^{\infty} MN_i P(k \geq i) \leq [c_1|\ln(\varepsilon)| + c_2(\varepsilon)] \sum_{i=1}^{\infty} P(\leq i) .$$

By the definition of the random walk, there exists $p > 0$ for arbitrary i, j, k, such that

$$P(A_{jk}|A_{ij}) < 1 - p .$$

Consequently,

$$P(k \geq i) \leq (1 - p)^{i-1}$$

and hence

$$MN \leq [c_1|\ln(\varepsilon)| + c_2(\varepsilon)]p^{-1} .$$

To complete the description of the process of iterations on subdomains, it is necessary to estimate the variance, the error and the computational cost of the algorithm.

Suppose that

$$G_i \cap \bigcup_{k \neq i} G_k \neq \emptyset$$

for all $i = 1, \ldots, n$. Then

$$\partial G = \bigcup_{i=1}^{n} \partial G_{ii}, \quad \text{and} \quad \partial G_i = \Gamma_{i1} \cup \Gamma_{i2}$$

where $\partial G_{ii} = \Gamma_{i1}$ is a part of G_i lying in ∂G, and

$$\Gamma_{i2} = \bigcup_{i=1}^{n} \partial G_{ik}, \quad \partial G_{ik} = \partial G_i \cap G_k$$

is a part of ∂G_i lying in G_k.

Assuming that in each subdomain the estimate (3.53) is used, we present the estimate for solving the problem (3.52) at the point x. Suppose that x lies in G_0. Next we construct a Markov chain of points $x_1^0, \ldots, x_{N_0}^0$ lying in $\bar{G}_0$, and $y_0^0 \in \partial G_0$. Then either $y_0^0 \in \Gamma_{01}$ or $y_0^0 \in \Gamma_{02}$. In the first case, the chain is terminated. If $y_0^0 \in \Gamma_{02}$, we proceed to simulate the Markov chain in a subdomain $\bar{G}_1$ where the point y_0^0 appears. The process is stopped when a random point, say y_0^κ, hits $\Gamma_{\kappa 1}$. The corresponding random estimate has the form

$$\zeta_x = \left(\prod_{k=0}^{\kappa} Q(y_0^{k-1}; \{y_0^k\}, x_0^k, \ldots, x_N^k, G_k) \right) g(y_0^\kappa) \tag{3.54}$$

where we put $y_0^{-1} \equiv x$.

Theorem 3.2. Assume that an unbiased estimate of the form (3.53) is chosen for each subdomain G_i. Then (3.54) is an unbiased estimate for solving (3.52) in the domain $G = \bigcup_{i=0}^{n} G_i$.

The proof follows from the Schwarz lemma [3.3, 16] for the case when the boundary of the intersecting domains G_1 and G_2 consists of a finite number of surfaces having a continuously varying normal vector; in addition, it is assumed that the angle of intersection of ∂G_1 and ∂G_2 does not vanish. In this case the statement of the Schwarz lemma can be formulated as follows. Let f be a function such that $Lf = 0$ in G_1, f is continuous in $\bar{G}_1$, $f = 0$ on Γ_{11} and $|f| \leq 1$ on Γ_{12}. Then there exists a constant $q \in (0, 1)$ depending on G_1, G_2 such that $|f| \leq q$ on Γ_{22}. Here

$$\Gamma_{12} = \partial G_1 \cap \partial G_2; \quad \Gamma_{22} = \partial G_2 \cap G_1;$$
$$\Gamma_{11} = \partial G_1 \backslash \Gamma_{12}; \quad \Gamma_{21} = \partial G_2 \backslash \Gamma_{22}.$$

Thus we suppose that the angles of intersection of the surfaces ∂G_i do not vanish. We define in G_i a function $v_0(x)$ as the solution of (3.52) in G_i satisfying the boundary conditions

$$v_0(y) = \begin{cases} 0 & \text{, if } y \in \partial G_{ik}, \ k \neq i, \\ g(y) & \text{, if } y \in \partial G_{ii} \end{cases} \tag{3.55}$$

We define also in each subdomain G_i a function $v_1(x)$ as the solution of the problem

$$Lv_1(x) = 0 \ , x \in G_i \ ,$$

$$v_1(y) = \begin{cases} 0 & \text{, if } y \in \partial G_{ii}, \\ v_0(y) & \text{, if } y \in \partial G_{ik}, \ k \neq i \ . \end{cases}$$

The functions $v_j(x)$, $j \geq 2$ are defined analogously. Obviously,

$$v_j(x) = M\{\zeta_x \ ; \kappa = j\} = M\{\zeta_x | \kappa = j\} P(\kappa = j) \ .$$

In each subdomain G_i the function

$$u_a(x) = \sum_{j=0}^{a} v_j(x)$$

is defined. It follows from (3.55) that the function $v = u(x) - v_0(x)$ satisfies the conditions $Lv(x) = 0$ in G_i,

$$v(y) = \begin{cases} 0 & \text{, if } y \in \partial G_{ii} \ , \\ u(y) & \text{, if } y \in \partial G_{ik}, \ k \neq i \end{cases}$$

where $u(x)$ solves (3.52) in G.

We put $q = \max_{i,k} q_{ik}$, where $q_{ik} < 1$ is the constant for a pair of subdomains G_i, G_k of the Schwarz lemma. Let $S = \sup_{y \in \partial G} |g(y)|$. Then

$$|u(x) - v_0(x)| \leq q \sup_{x \in G} |u(x)| \leq qS$$

for an arbitrary surface ∂G_{ik}, $k \neq i$, for all i. Applying recursively the Schwarz lemma to the functions $u(x) - u_a(x)$, $a \geq 1$, we obtain a set of inequalities $|u(x) - u_a(x) \leq q^a S$, hence

$$u(x) = \lim_{a \to \infty} u_a(x) = \lim_{a \to \infty} M\{\zeta_x \ ; \kappa \leq a\} = M\zeta_x \ ,$$

which proves the Theorem. □

Suppose now that (3.53) is a biased estimate. Then, using a property of the Schwarz's sequence u_{2a-2}, the following result can be obtained.

Theorem 3.3. Suppose that $\sigma(x)$ is a bias of the estimate (3.53) for solving (3.52) in G_i:

$$M\xi_x = u(x) + \sigma(x), \quad |\sigma(x)| \leq \varepsilon \sup_{x \in G_i} |u(x)|.$$

Then the bias of the random estimate (3.54) for solving the problem (3.52) in G has the following estimation

$$\cdot\ |M\zeta_x - u(x)| \leq \frac{\varepsilon}{1 - (\varepsilon + q)} \sup_{x \in G} |u(x)| \tag{3.56}$$

where $q = \max_{i,k} q_{ik}$. The variance of ζ_x is finite.

Proof. Let

$$g_0(y) = \begin{cases} 0 & , \text{ if } y \in \Gamma_{02}, \\ g(y) & , \text{ if } y \in \Gamma_{01} \end{cases}$$

and let

$$\xi_x^0 = Q(x;\ y_0^0,\ x_0^0,\ \ldots,\ x_{N_0}^0)g_0(y_0^0).$$

Then, obviously,

$$M\{\zeta_x\ ;\kappa = 0\} = M\xi_x^0,$$

i.e., ξ_x^0 is an estimate of type (3.53) for $v_0(x)$, $x \in G_0$. Therefore,

$$M\{\zeta_x\ ;\kappa = 0\} = v_0 + \sigma(x).$$

where

$$|\sigma_0(x)| \leq \varepsilon \sup_{x \in G} |g(x)|.$$

Hence,

$$\begin{aligned}
M\{\zeta_x;\kappa = 1\} &= M_{\{y_0^0 \in \Gamma_{02}\}}\big[Q(x; y_0^0, x_0^0, \ldots, x_{N_0}^0)M_{\{y_0' \in \Gamma_{11}\}} \\
&\quad \times Q(y_0^0; x_0^1, \ldots x_N^1)g(y_0^1)|y_0^0\big] = M_{\{y_0^0 \in \Gamma_{02}\}} \\
&\quad \times \big[Q(x; y_0^0, x_0^0, \ldots, x_{N_0}^0)(v_0(y_0^0) + \sigma_0(y_0^0))\big].
\end{aligned}$$

From this,

$$\begin{aligned}
M\{\zeta_x;\kappa \leq 1\} &= M\big[Q(x; y_0^0, x_0^0, \ldots, g_1(y_0^0)\big] \\
&\quad + M_{\{y_0^0 \in \Gamma_{02}\}}\big[Q(x; y_0^0, x_0^0, \ldots, x_{N_0})\sigma_0(y_0^0)\big],
\end{aligned}$$

where

$$g_1(y) = \begin{cases} v_0(y), & \text{if } y \in \Gamma_{02} \\ g(y), & \text{if } y \in \Gamma_{01} \end{cases},$$

and the first summand is equal to $u_1(x) + \sigma_0(x)$. Induction in Schwarz's scheme yields

$$M\{\zeta_x, \kappa \leq a\} = u_a(x) + \delta_0(x) + \sum_{k=0}^{a-1} M_{\{y_0^0 \in \Gamma_{02}\}} \left[Q(x; y_0^0, x_0^1, \ldots, x_{N_0})\right]$$

$$\times M_{\{x_0' \in \Gamma_{12}\}} \left[Q(y_0^0; x_0^1, \ldots, x_{N_1}^0) \ldots M_{\{x_0^k \in \Gamma_{k2}\}}\right.$$

$$\times \left. \left[Q(y_0^{k-1}; x_0^k, \ldots, x_{N_k}^k)\delta_0(x_0^k)|y_0^{k-1})\ldots\right]\right] . \tag{3.57}$$

Consequently,

$$\left| M_{\{x_0^k \in \Gamma_{02}\}} \left[y_0^{k-1}; x_0^k, \ldots, x_{N_k}^k)\delta_0(y_0^k)|y_0^{k-1}\right] \right|$$

$$= |\delta_1(y_0^{k-1}) + \delta_0(y_0^{k-1})| \leq \sup_{x \in G} |u(x)|\varepsilon(q+\varepsilon) ,$$

since

$$|\delta_1(y_0^{k-1})| \leq q\varepsilon \sup_{x \in G} |u(x)| .$$

The k-th term in (3.57) is not larger than $\varepsilon(\varepsilon + q)^{k+1}$ (by induction). Hence

$$|M\{\zeta_x; \kappa \leq a\} - u_a(x)| \leq \varepsilon \sup_{x \in G} |u(x)| \sum_{k=0}^{a} (\varepsilon + q)^k .$$

Choosing ε so that $\varepsilon + q < 1$ we obtain (3.56). Note that if $|Q| \leq 1$ (for example, this is the case when the walk on spheres algorithm is applied to solve the equation $\Delta u - cu = 0$), then the bias of the estimate ζ_x is of order $\omega(\varepsilon)$, and the variance of this estimate is finite.

To investigate the variance in the general case, we introduce an operator F in G_i according to the law

$$[FW](x) = M_{\{y_0 \in \Gamma_{i2}\}} \left[Q^2(x; y_0, x_0, \ldots, x_n)W(y_0)\right] , \quad x \in G_i .$$

Let

$$g_1(x) = \left\{ M_{\{y_0 \in \Gamma_{i1}\}}[Q^2(x; y_0, x_0, \ldots, x_N)g^2(y_0)]\right\}^{1/2} .$$

Then

$$M\{\xi_x^2; \kappa = j\} = F^j q_1^2(x) .$$

Now we show that

$$\|F\| = \sup_{x \in G_i} |FW| < 1 .$$

It is clear that $\sup_{x \in G_i} |FW|$ is reached at $W = 1$ on Γ_{i2}.

We shall assume that $M\xi_x \equiv 1$ if $g(y) = 1$. Hence $M\xi_x^2 = 1$, since in this case $D\xi_x = 0$.

For the problem (3.52) in G_i with boundary conditions

$$g(y) = \begin{cases} 1 & \text{if } y \in \Gamma_{i2} \\ 0 & \text{if } y \in \Gamma_{i1} \end{cases}$$

we have

$$\begin{aligned} [Fg^2](x) &= M(g(y_0)Q(x; y_0, x_0, \ldots, x_N))^2 \\ &< M[Q(x; y_0, x_0, \ldots, x_N)]^2 = 1 . \end{aligned}$$ □

Note that construction of estimates for complicated domains on the basis of iteration on standard subdomains can be carried out by using all known random walk algorithms, e.g., the walk on spheres, on ellipsoids, on the boundary, on a grid, etc. In the case of the walk on boundary algorithm it is natural to use convex subdomains G_i. Note that it is also possible to generalize the iterative process described in such a way that the subdomains do not intersect.

3.2.2 Non-homogeneous Equations and Global Walk on Spheres Algorithm for Calculating the Solution and Derivative Fields

Consider a non-homogeneous equation of the second order in a bounded simple-connected domain $G \subset \mathbb{R}^n$ with a boundary ∂G:

$$Lu(x) = f(x) , \quad x \in G \tag{3.58}$$

with the boundary conditions

$$u(y) = g(y) , \quad y \in \partial G . \tag{3.59}$$

Suppose that there exists a random estimate of the type (3.53) on a random trajectory $\{x_k\}_{k=0}^N$ inside the domain. This estimate permits one to calculate an approximate solution at a fixed point $x_0 \in G$. Note that there exists a modification of the walk on spheres algorithm which permits one to calculate the estimates for a set of points lying inside the first sphere $S(x_0, d(x_0))$ by using one trajectory of the ε-spherical process [3.17]. However, calculations schow that this technique works well only if the points $\{x_k\}_{k=1}^m$ lie close enough to the point x_0, because the variance increases rapidly as the distance between x_k and x_0 increases.

We describe here a different approach which permits one to calculate the solution and its derivatives at arbitrary points of the domain using one trajectory of the ε-spherical process.

Thus, suppose that it is desired to calculate the solution (3.58,59) at arbitrary points x_0, x_1, ..., x_m. Let $V(x_0, x)$ be a fundamental solution of (3.58), i.e., V is a known funcion of x_0 and x, satisfying $LV = \delta(x - x_0)$. Denote by $\pi(x)$ some appropriate distribution density of a point x in the domain G. First, consider the case when $g(y) \equiv 0$. Let $W(x_0, x)$ be the solution to the problem

$$LW(x_0, x) = 0 , \quad W|_{\partial G} = -V(y, x) , \quad y \in \partial G , \tag{3.60}$$

and let $\xi(x_0, x)$ be an ε-biased estimate of $W(x_0, x)$.

Theorem 3.4. Denote by E_x the mathematical expectation in the space of trajectories of the ε-spherical process starting at the point x, and $J = M_\pi F$ means that

$$J = \int_G F(x)\pi(x)dx .$$

Then

$$u(x_i) \cong M_\pi E_x[\eta(x_i, x)|x] . \tag{3.61}$$

Here

$$\eta(x_i, x) = \frac{f(x)}{\pi(x)}\{V(x_i, x) + \xi(x_i, x)\} , \quad i = 0, 1, \ldots, m , \tag{3.62}$$

$$Bu(x_i) \cong M_\pi E_x[\xi(x_i, x)|x] , \quad i = 0, 1, \ldots m , \tag{3.63}$$

if $|(f, B_p V(P, P_0))| < \infty$, where

$$\zeta(x_i, x) = \frac{f(x)}{\pi(x)}\{BV(x_i, x) + \xi_1(x_i, x)\} , \quad i = 0, 1, \ldots, m , \tag{3.64}$$

In (3.63, 64) B is a differential operator, $\xi_1(x_i, x)$ is an ε-biased estimate for the solution to the boundary value problem (3.60) where $V(y, x)$ is replaced by $BV(y, x)$.

Proof. First, we obtain (3.61). We start with construction of the direct estimate for the solution at a fixed point $x_0 \in G$ [the estimates (3.62, 64) are called adjoint estimates]. Solution to (3.58, 59) is represented as follows ($g \equiv 0$):

$$u(x_0) = (u_\delta, f) = \int_G u_\delta(x, x_0)f(x)dx \tag{3.65}$$

where u_δ is the Green's function, i.e.,

$$Lu_\delta(x, x_0) = \delta(x - x_0) , u_\delta|_{\partial G} = 0 . \tag{3.66}$$

We seek the solution to (3.66) in the form

$$u_\delta(x, x_0) = V(x, x_0) + W(x, x_0) .$$

Then

$$LW = 0 , \quad W|_{\partial G} = -V|_{\partial G} .$$

We denote by $\xi_0(x_0, x)$ the ε-biased estimate for the solution W. Substitution of the randomized representation of u_δ into (3.65) yields

$$u(x, x_0) \cong M_\pi E_{x_0}[\eta_0(x_0, x)|x] , \tag{3.67}$$

where

$$\eta_0(x_0, x) = \frac{f(x)}{\pi(x)} \{V(x, x_0) + \xi_0(x, x_0)\} . \qquad (3.68)$$

Thus by using the double randomization principle (Chap. 1), we can describe a direct algorithm for calculating the solution at the point x_0 as follows. A point x is chosen at random in G according to the distribution density $\pi(x)$, and a δ-source is placed at x. Then the ε-spherical process starting from x_0 is constructed and estimate (3.68) is calculated for this process (for brevity, we say that a contribution from a δ-source is calculated). Next a new random point $x = x'$ is sampled according to the density $\pi(x)$ and the contribution from the δ-source located at $x = x'$ is calculated, etc. The result is obtained by averaging the contributions over a sufficiently large ensemble of trajectories.

We described here a double randomization technique "1–1", what means that, for each random point x, one random trajectory is constructed. In general, for n random points x, m trajectories can be used (a scheme "n–m"), subject to the computational cost of simulation of the trajectories and of sampling random points according to $\pi(x)$, as in the splitting technique.

Consider the adjoint estimate (3.62). The proof of (3.61) follows immediately from (3.67, 68) since the Green's function is symmetric : $U_\delta(x, x_0) = u_\delta(x_0, x)$.

Let us describe the adjoint algorithm for calculating $u(x_i)$, $i = 0, \ldots, m$. The δ-sources are located at points x_i, $i = 0, 1, \ldots, m$. A random point x is sampled from the density $\pi(x)$. The ε-spherical process starting from the point x is simulated, and contributions from the δ-sources Q_i, $i = 0, 1, \ldots, m$ are calculated, i.e., estimates $\eta(x_i, x)$ are calculated simultaneously for $i = 0, 1, \ldots, m$ on one ε-spherical process according to (3.62). Next a new random point $x = x'$ is sampled according to the density $\pi(x)$, the ε-spherical process starting from this point is simulated, the contributions $\eta(x_i, x)$ $i = 0, 1, , \ldots, m$) are calculated. The result is obtained by averaging the contributions over a sufficiently large ensemble of trajectories.

Finally, consider estimate (3.64) for the differential expression $Bu(x_i)$. We introduce a generalized function

$$A_{P_0}^h = \sum_{i=1}^{1} c_i \delta(x - P_i)$$

such that

$$Bu(P_0) = \lim_{h \to 0} (A_{P_0}^h, u) ,$$

i.e., $A_{P_0}^h$ is an approximation of the differential operator B. We have $(A_{P_0}^h, u) = (f, W_h)$ where W_h is the solution to the problem

$$LW_h = A_{P_0}^h , \qquad W_h|_\Gamma = 0$$

in view of the homogeneity of the boundary conditions. Now if we assume that the Lebesque-dominated convergence theorem is applicable here, then

$$Bu(P_0) = \lim_{h \to 0} (A_{P_0}^h, u) = (f, W) \qquad (3.69)$$

where $W = \lim_{h \to 0} W_h$.

Let $S(P, P_0) = B_P V(Q, P)$, where the subscript shows that the operator acts with respect to P, and Q is a parameter. Then

$$W(P, P_0) = S(P, P_0) + W_1(P, P_0) ,$$

where W_1 solves the problem

$$LW_1(P, P_0) = 0 \, [P \in G] , \quad W_1|_\Gamma = -S(P, P_0) \, [P \in \Gamma] .$$

Using this representation and (3.69) we obtain (3.63, 64). The theorem is proved.

□

The adjoint algorithm thus permits us to calculate the solution and its derivatives at arbitrary points $x_0, x_1, \ldots, m$ simultaneously. Note that the variances of the estimates $\eta(x_i, x)$ and $\zeta(x_i, x)$ are completly determined by the variances of the estimates $\xi(x_i, x)$, $\xi_1(x_i, x)$. This follows from the variance representation

$$D\eta = D_\pi E_x \{\eta(x_i, x)|x\} + M_\pi D_x \{\eta(x_i, x)|x\}$$

and the properties of V and BV provided that $f \in L_2(G)$.

In the global direct and adjoint methods described it was assumed that $g \equiv 0$. It is possible to pass from non-homogeneous boundary conditions to the case $g \equiv 0$, using R-function technique [3.6]. Indeed, we seek the solution to (3.58, 59) in the form $u(x) = U(x) + \varphi(x)$, where φ is smooth enough and such that $\varphi|_{\partial G} = g$. then U satisfies the conditions

$$Lu = f - L\varphi , \quad U|_{\partial G} = 0 .$$

In general, the problem of construction of the function $\varphi(x)$ is difficult. However, using the R-function technique, it is possible to construct $\varphi(x)$ for a wide class of domains and even for discontinuous $g(x)$

Let us consider an example which can be used to solve electrooptics problems. Let G be a cylinder $x_1^2 + x_2^2 = R^2$, $|x_3| \leq L$. Consider the boundary value problem

$$\Delta u(x) = 0 , \quad x \in G \qquad (3.70)$$

$$u|_{\partial G} = g(y) \equiv \begin{cases} g_0 , & x_1^2 + x_2^2 = R^2 , & |x_3| \leq l \\ 0 , & x_1^2 + x_2^2 = R^2 , & l < |x_3| \leq L \\ 0 , & x_1^2 + x_2^2 < R^2 , & |x_3| = L \end{cases}$$

In this case, the function φ has the form

$$\varphi(x) = \tfrac{1}{2}\{(l^2 - x_3^2)[(l^2 - x_3^2)^2 + (R^2 - x_1^2 - x_2^2)(L^2 - x_3^2)]^{-1/2} + 1\}g_0 .$$

3.2.3 The Walk on Small Spheres and on Other Standard Domains

Consider a Dirichlet problem

$$\Delta u - k^2 u = -f , \quad x \in G \subset \mathbf{R}^3 , \quad u(y) = g(y) , \quad y \in \partial G \tag{3.71}$$

where k^2 is a complex number. The walk on maximal spheres for solving this problem when $c = \operatorname{Re}\{k^2\} \geq 0$, $\operatorname{Im}\{k^2\} = 0$ in a bounded domain G is described in detail [3.11]. Note that if $k = i\kappa$, κ real, then serious difficulties arise, especially if the domain G is unbounded. Note also that the method [3.11] which takes into account the right-hand side of equations on the basis of the "single random node" technique can be improved by using an optimal radius in the ε-spherical process.

First consider the case when G is the half-space $\mathbf{R}_{(+)}^3 = \{(x, y, z) : z > 0\}$. We shall study the applicability of the walk on maximal spheres algorithm. We first obtain the corresponding local integral equation of type (1.24) for the problem (3.71) in the case of $f \equiv 0$. To this end, we prove an auxiliary proposition. Let $J_\nu(z)$ be the Bessel function. It is known that [3.18]

$$J_\nu(z) = \sum_{k=0}^{\infty} \frac{(-1)^k}{\Gamma(k + \nu + 1)\Gamma(k + 1)} \left(\frac{z}{2}\right)^{2k+\nu} .$$

We introduce the function

$$W_n(z) = \Gamma\left(\frac{n}{2}\right) \left(\frac{2}{z}\right)^{n/2-1} J_{n/2-1}(z) ,$$

and put $w_n(r, \lambda) = W_n(ir(\lambda)^{1/2})$. Using $I_\nu(z) = J_\nu(iz)i^{-\nu}$ we can write

$$w_n(r, \lambda) = W_n(ir(\lambda)^{1/2}) = \Gamma\left(\frac{n}{2}\right) \left(\frac{2}{r\sqrt{\lambda}}\right)^{n/2-1} J_{n/2-1}(r\sqrt{\lambda})$$

$$= \Gamma\left(\frac{n}{2}\right) \sum_{k=0}^{\infty} \frac{\lambda^k r^{2k}}{2^{2k} k! \Gamma(k + n/2)}$$

$$= \sum_{k=0}^{\infty} \frac{\lambda^2 r^{2k}}{2^{2k} k! n(n + 2) \dots (n + 2k - 2)} ,$$

since

$$\Gamma\left(k + \frac{n}{2}\right) = \Gamma\left(\frac{n}{2}\right) n(n + 2) \dots \frac{n + 2k - 2}{2^k} .$$

In Chap. 4 we shall also use the derivatives

$$\frac{\partial^p w_m(r, \lambda)}{\partial \lambda^p} = \sum_{k=p}^{\infty} \frac{k(k - 1) \dots (k - p + 1)\lambda^{k-p} r^{2k}}{2^k k! m(m + 2) \dots (m + 2k - 2)}$$

$$= \sum_{j=0}^{\infty} \frac{r^{2(j+p)}\lambda^j}{2^{j+p}} \left[\frac{(j+p)(j+p-1)\ldots(j+1)}{(j+p)!} \right]$$

$$\times \frac{1}{m(m+2)\ldots[m+2(j+p)-2]} = \frac{r^{2p}}{2^p} \sum_{j=0}^{\infty}$$

$$\times \frac{\lambda^j r^{2j}}{2^j j! m(m+2)\ldots(m+2p-2)(m+2p)\ldots[(m+2p)+2j-2]}$$

$$= \frac{r^{2p}}{2^p m(m+2)\ldots(m+2p-2)} w_{m+2p}(r,\lambda) \, .$$

Theorem 3.5. Let $N_x^r(u)$ be the spherical mean in the space $\mathbf{R}^n$

$$N_x^r(u) = \int_{S(x,r)} u(y)d\sigma(y) = \frac{1}{\omega_n} \int u(x+re)d\Omega(e)$$

and $u(x)$ is a real valued analytic function, then

$$N_x^r(u) = \{W(r\sqrt{-\Delta})\}u(x) = \{W_n(ir\sqrt{\Delta})\}u(x) \tag{3.72}$$

for x, r, u, for which the left-hand side is defined and the expression in the right-hand side of (3.72) exists. This expression is defined as the value at the point x of the operator (in braces) acting on the function u.

Proof. Let η be a measure in $\mathbf{R}^n$ with compact support. Consider the Fourier transform of η,

$$h(y) = \int_{\mathbf{R}^n} \exp\{-i(y,s\}d\eta(\sigma)$$

and put

$$D = \left(-i\frac{\partial}{\partial x_k}\right) 1 \leq k \leq n \, .$$

Then [3.19]:

$$\int u(x+ry)d\eta(y) = \{h(-rD)u\}(x) \tag{3.73}$$

if the left-hand side is defined for x, r, u and the right-hand side of (3.73) exists [the latter expression is defined as the right-hand side of (3.72)]. To obtain (3.72) from (3.73), it is now sufficient to take for η the uniform measure $d\Omega(e)/\omega_n$ since the Fourier transform of $d\Omega(e)/\omega_n$ is given by $W_n(|y|)$ and $|D| = (-\Delta)^{1/2}$. $\square$

Consider, for example, the three-dimensional case:

$$h(iy) = \frac{1}{4\pi} \int_0^{2\pi} d\varphi \int_0^{\pi} d\theta \sin(\theta) \exp\{|y|\cos(\theta)\} = \frac{\sin(|y|)}{|y|} \, .$$

Equation (3.72) now takes the form

$$N_x^r(u) = \left\{ \frac{\sin(r\sqrt{\Delta})}{r\sqrt{\Delta}} \right\} u(x) .$$ (3.74)

From (3.72) we can obtain the following mean value relation. Solution of the equation

$$\Delta u(x) - \lambda u(x) = 0 , \quad x \in G \subset \mathbf{R}^n$$

satisfies the relation

$$u(x) = w_n^{-1}(r, \lambda) N_x^r(u) \quad \forall x, r : S(x, r) \subseteq G ,$$

for λ such that $r\sqrt{\lambda} \neq \xi_l$, ξ_l is the l-th root of the equation $W_n(i\xi) = 0$. The proof for λ such that $\mathrm{Re}\,\{\lambda\} \geq 0$ is obvious. Generalization to other values of λ is carried out by analytic continuation with respect to λ.

Let θ be the Heaviside function, and $d(x)$ is the distance from the point x to the boundary ∂G.

Theorem 3.6. The solution of (3.71) for $f \equiv 0$ satisfies the integral equation

$$u(x) = u_0(x) + \int_G k(x, x') u(x') dx'$$ (3.75)

where

$$k(x, x') = \frac{k\delta(|x - x'| - d(x))\theta(d(x') - \varepsilon)}{4\pi d(x) \sinh kd(x)}$$

and

$$u_0(x) = \frac{k}{4\pi d(x) \sinh kd(x)} \int_G \delta(|x - x'| - d(x))\theta(\varepsilon - d(x'))g(x')dx' .$$

Proof. The right-hand side of (3.74) can be rewritten as a power series in $r\sqrt{\Delta}$. Then, using the relation $\Delta u = k^2 u$, we obtain

$$N_x^r(u) = \frac{\sinh kr}{kr} u(x)$$ (3.76)

which is a generalization of the well-knownmean value theorem for harmonic functions.

Let $re = x_1$ and rewrite the integral in (3.76) in the form of a three-dimensional integral $[dx_1 = x_1^2 dx_1 d\Omega(e)]$:

$$u(x) = \frac{kr}{4\pi \sinh kr} \int_G u(x + x_1)\delta(x_1 - r)\frac{dx_1}{r^2} .$$

If $r = d(x)$, then $(x_1 = x - x')$:

$$u(x) = \frac{k}{4\pi d(x) \sinh kd(x)} \int_G \delta(|x - x'| - d(x))u(x')dx' .$$

From this we have

$$u(x) = \frac{k}{4\pi d(x) \sinh kd(x)} \left\{ \int_G \delta(|x - x'| - d(x))\theta(d(x') - \varepsilon)u(x')dx' \right.$$

$$\left. + \int_G \delta(|x - x'| - d(x))\theta(\varepsilon - d(x'))u(x')dx' \right\} \qquad (3.77)$$

since $1 \equiv \theta(d(x') - \varepsilon)) + \theta(\varepsilon - d(x'))$. In the second integral [where $d(x') \leq \varepsilon$] the solution is assumed to be known: $u(x') = \tilde{g}(x')$ [in $\partial G_\varepsilon \tilde{g}(x') \approx g(x_*(x'))$ where $x_*(x')$ is the point of ∂G nearest to x', since ε is small], and (3.77) has the form of (3.75). □

Theorem 3.7. In the case when $G = R_3^+$, the Neumann series for (3.75) converges if

$$\varepsilon \text{Im} \{k\} > 2\kappa \left\{ \ln[\kappa + (\kappa^2 + 1)^{1/2}] - \frac{\kappa}{(\kappa^2 + 1)^{1/2}} \right\} \qquad (3.78)$$

where $\kappa = \text{Im} \{k\}/\text{Re} \{k\}$.

Proof. Consider the norm

$$J \equiv \|K^*\| = \sup_x \int_G |k(x, x')|dx' .$$

Now

$$J = \sup_x \frac{|k|}{4\pi d(x) \sinh kd(x)} \left\{ \int_G \delta(|x - x'| - d(x))\theta(d(x') - \varepsilon)dx' \right\} .$$

In terms of $x_1 = x - x'$, $dx_1 = x_1^2 dx_1 d\Omega(e)$, we have

$$J = \sup_x \left| \frac{k}{\sinh kd(x)} \right| \int \theta(d(x - ed(x)) - \varepsilon) \frac{d\Omega(e)}{4\pi} .$$

Let

$$\nu_\varepsilon(x) \equiv \int \theta(d(x - ed(x)) - \varepsilon) \frac{d\Omega(e)}{4\pi} ,$$

then $0 \leq \nu_\varepsilon < 1$. For the half-space R_3^+, $\nu_\varepsilon(x) \leq 1 - \varepsilon/2d(x)$. Let $kd(x) = \varphi(x) = \varphi_1 + i\varphi_2$. Then

$$|\sinh(\varphi_1 + i\varphi_2)|^2 = \sin^2 \varphi_2 + \sinh^2(\varphi_1) .$$

Hence,

$$J = \sup_x \left[\frac{\varphi_1^2 + \varphi_2^2}{\sin^2 \varphi_2 + \sinh \varphi_1^2} \right]^{1/2} \nu_\varepsilon(x) \leq \sup_x \frac{(\varphi_1^2 + \varphi_2^2)^{1/2}}{\sinh \varphi_1(x)} \nu_\varepsilon(x) .$$

Let $\varphi_i = k_i d$, $i = 1, 2$; $\nu = 1 - \varepsilon/2d$. Then $J = \sup\limits_{d} f(d)$, where

$$f(d) = [k_1^2 + k_2^2]^{1/2}\frac{d - \varepsilon/2}{\sinh k_1 d} \, .$$

We shall find the maximum of the function $\ln[f(d)]$ from the equation

$$\tanh k_1 d_0 = k_1 d_0 - k_1\frac{\varepsilon}{2} \, .$$

Then

$$f(d_0) = \left[1 + \left(\frac{k_2}{k_1}\right)^2\right]^{1/2}\frac{k_1 d_0 - k_1\varepsilon/2}{\sinh k_1 d_0} = \left[1 + \left(\frac{k_2}{k_1}\right)^2\right]^{1/2}\frac{1}{\cosh k_1 d_0} \, .$$

We require

$$\left[1 + \left(\frac{k_2}{k_1}\right)^2\right]^{1/2}\frac{1}{\cosh k_1 d_0} < 1, \quad \text{i.e.,} \quad \sinh k_1 d_0 > \frac{k_2}{k_1} \, .$$

Since

$$k_1\varepsilon = 2[k_1 d_0 - \tanh k_1 d_0]$$

increases monotonically, we obtain $J < 1$ if

$$k_1\varepsilon > 2\kappa\left\{\ln[\kappa + (\kappa^2 + 1)^{1/2}]\frac{\kappa}{(\kappa^2 + 1)^{1/2}}\right\}$$

[note that $\mathrm{arcsinh}\, x = \ln(x + [x^2 + 1]^{1/2})$]. $\qquad\square$

In Table 3.2 we show when the condition $J < 1$ is satisfied. Thus for fixed ε the following ε-biased Monte Carlo estimate can be constructed

$$\eta = \prod_{i=1}^{n_\varepsilon}\frac{kd(x_i)}{\sinh kd(x_i)}\, g(x_*(x_{n_\varepsilon})) \tag{3.79}$$

provided that the condition (3.78) is satisfied. Using (3.79), it is possible to construct ε-biased estimates for solving exterior boundary value problems of the type (3.71) for convex polyhedrons on the basis of iterations on half-spaces R_+^3 according to the scheme described above in Sect. 3.2.1 (in Fig. 3.3 we show schematically the subdomains G_i – the half-spaces and one possible trajectory).

Table 3.2. Conditions for $J < 1$

cond.	$\kappa = 1$	1.5	10	100	10^3	10^6	10^7	0.1	0.01
$\dfrac{\varepsilon k_1}{2\pi} >$	0.055	0.173	6,38	136.8	211	$4\cdot 10^6$	$5\cdot 10^7$	10^{-5}	$7.5\cdot 10^{-10}$

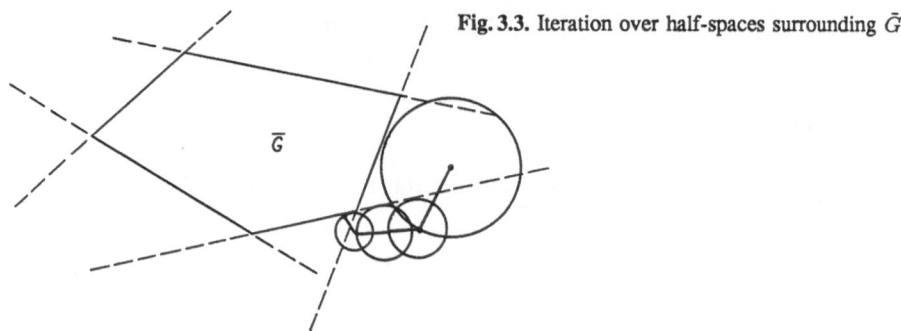

Fig. 3.3. Iteration over half-spaces surrounding $\bar{G}$

Note that when G_i are half-spaces or dihedral angles then the distribution of points $y_k \in \partial G_i$ can be constructed on the basis of the Green's functions for these domains (the explicit form of the Green's function is known [3.3]). The algorithm of iteration on dihedral angles in this case is based on simulation of the random walk $\{y_k\}$, where the process is stopped if $y_k \in \partial G_i^r$ (Fig. 3.2), while the transition $y_k \to y_{k+1}$ is simulated if $y_k \in \partial G_i^f$.

Up to now, we have considered the case $f = 0$. If in the walk on spheres algorithm for $f \neq 0$ the integral

$$J(r) = \int f G^r dx \quad (G^r \text{ is the Green function})$$

is calculated using a single random node in the ball $K(x_0, r)$, then it may appear that the optimal radius is not maximal although the average number of steps $n(r)$ is minimal (Sect. 1.3.1). Indeed, if $\underset{K}{\text{osc}} f$ increases rapidly as r increases then it is more effective to use the walk on small spheres, and the optimal value of r can be obtained from the condition that the quantity $\alpha(r) = D\xi(r)/g(r)$ is minimal; here $\alpha(r)$ is an estimate of $J(r)$, and $g(r)$ is the function from (1.45). In Fig. 3.4 we show such a simulation for a unit cube, $\varepsilon = 0.01$, the trajectories were started at the center of the cube; $n_1(r)$ is the average number of steps of the process of type I (calculations), $n_2(r)$ is the average number of steps of the process of type II (calculations); $n_2^T(r)$ is the average number of steps for a half-space $R_{(+)}^3$ (1.45); the points are the calculated values. Note that $n_2^T(r) = 0(r^{-2})$ as $r \to 0$. It can be seen that in the walk on small spheres of type II the boundary "is felt" beginning with $r_0 \approx 0.1$–0.2. When the variance $D\xi(r))$ increases rapidly, the near optimal value lies in the region $r_0 \geq r_c$. When the increase of $D\xi(r)$ is slow, then the optimal value of r_0 is close to 1. It should be noted that if the coefficients of the equation to be solved are not constants, then it is convenient to use the walk on small spheres algorithm; the radii of the random walk are then determined on the basis of the information about the variation of the coefficients. The case of variable coefficients is considered in Chap. 4.

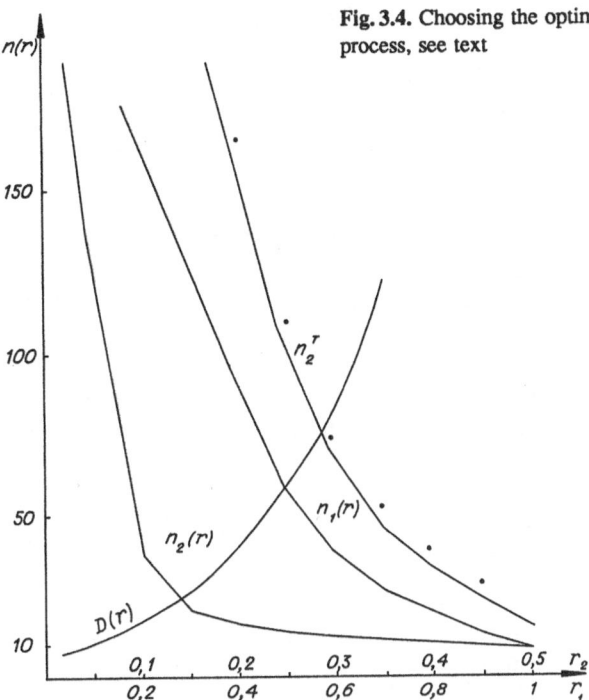

Fig. 3.4. Choosing the optimal radius for the walk on spheres process, see text

3.3 Numerical Solution of Some Test and Applied Problems of Potential Theory in Deterministic and Stochastic Formulation

In Sect. 3.3.2, problems concerning motions of particles [aerosol and charged particles] in stochastic and electromagnetic fields are solved. Various functionals of the particle trajectories are calculated, for example, the capture coefficient, the size distribution function for particles captured, the probability that some functional [e.g., a concentration] exceeds a critical value.

In Sect. 3.3.1 results of numerical experiments carried out to investigate practical efficiency of the algorithms developed are given. We present also results of comparison with different numerical methods.

3.3.1 Numerical Experiments: Solution of Test Problems of Potential Theory

1) We have used the walk on boundary algorithm in calculations for a cylindrical capacitor. This is equivalent to solving a two-dimensional Dirichlet problem in the domain

$$G = \{(x,y) : r < [x^2 + y^2]^{1/2} < R\}$$

with the boundary conditions $g(r) = 1$, $g(R) = 0$. The exact solution is known:

$$u = g(r) + [g(R) - g(r)] \frac{\ln\left\{\dfrac{x^2 + y^2}{r}\right\}}{\ln\left(\dfrac{R}{r}\right)}.$$

In Table 3.3 the numerical results for various points lying on the circle $[x^2 + y^2]^{1/2} = 1.5$ are presented. The calculations were carried out by (I) the multiplication method; (II) the method using the transformation $\lambda = \eta/(1 - \eta)$; (III) the Lindelef method, $\delta = 10^{-3}$, and (IV) the method using the continued fraction. The total number of trajectories was 10^5. The exact solution is $u(1.5) \cong 0.415$. In methods I and II 6 terms of the series were taken, in method III we used 30 terms, in method IV 8 iterations were made. The relative error varied from 0.1% to 1%.

Table 3.3. Comparison of estimates I–IV in calculations of a cylindrical capacitor

x	y	I	II	III	IV
−0.3	1.47	0.419	0.417	0.420	0.416
−1.2	0.90	0.418	0.417	0.419	0.415
−0.9	1.20	0.420	0.417	0.421	0.415
−1.1	1.01	0.419	0.418	0.420	0.416

2) Methods I–IV were applied to solve the problem $\Delta u(x) = 0$, $x \in G$, G is a hemisphere centered at the origin, with the following boundary conditions:

$$u(y) = [2(y_3 + 1)]^{-1/2} \quad \text{for} \quad |y| = 1,$$

and

$$u(y) = [(y_1^2 + y_2^2 + 1)]^{-1/2} \quad \text{for} \quad y_3 = 0.$$

The exact solution is known:

$$u(x) = [x_1^2 + x_2^2 + (x_3 + 1)^2]^{-1/2}.$$

In Table 3.4 we show the results obtained in 12 minutes of computing time.

Table 3.4. Comparison of estimates I–IV for the solution and its derivatives

Functional	Exact solution	I	II	III	IV
$u(0,0,0.5)$	0.666	0.665	0.665	0.665	0.666
$u_{x_3}(0,0,0.5)$	−0.444	−0.438	−0.435	−0.434	−0.440
$u(0,0,0.9)$	0.526	0.519	0.521	0.520	0.525
$u_{x_3}(0,0,0.9)$	−0.277	−0.348	−0.350	−0.321	−0.279

3) Methods I–IV were also used to solve the exterior Neumann problem

$$\Delta u(x) = 0 , \quad x \in \bar{G} = \{x : |x| > 1\} ,$$

$$\frac{\partial u}{\partial n}\bigg|_{\partial G} = 0 , \quad \lim_{|x| \to \infty} \frac{\partial u}{\partial l}(x) = (v_\infty, l) .$$

This problem describes the laminar flow of an ideal incompressible fluid around a sphere, where v_∞ is the velocity of the flux.

We seek the solution in the form

$$u(x) = (v_\infty, x) + W(x) .$$

Then $W(x)$ is harmonic in G such that

$$\frac{\partial W}{\partial n}(y) = -(v_\infty, n(y)) , \quad y \in \partial G .$$

The exact solution to this problem is known:

$$u(x) = (v_\infty, x)\left(1 + \frac{|x|^{-3/2}}{2}\right) .$$

In methods I–IV the initial points of the Markov chain used to calculate the iterations of the boundary integral operator was sampled according to the density $p_0(y_0) = 1/4\pi$, i.e., uniformly in the surface. For example, the estimate of method I was taken as

$$\xi = 2(v_\infty, y_0)\left\{\sum_{i=0}^{5}(-1)^i|x - y_i| + \tfrac{1}{2}|x - y_0|^{-1}\right\} .$$

In Table 3.5 the solution and its derivatives calculated by methods I–IV at the point (1,2,2) for $v_\infty = (0,0,1)$ are given.

4) The results of numerical solution of the problem

$$\Delta u(x) = 0 , \quad x \in G_1 = \left\{\frac{x_1^2}{a^2} + \frac{x_2^2}{b^2} + \frac{x_3^2}{c^2} > 1\right\} , u(y) = 1 , \quad y = \partial G_1$$

are given in Table 3.6.

Table 3.5. Comparison of estimates I–IV for the exterior Neumann problem at the point $(1,2,2)$ for $v_\infty = (0,0,1)$

Functional	Exact solution	I	II	III	IV
u	2.0370	2.0367	2.0368	2.0368	2.0370
u_{x_1}	−0.0123	−0.0121	−0.0122	−0.0124	−0.0123
u_{x_2}	−0.0247	−0.0247	−0.0247	−0.0248	−0.0247
u_{x_3}	0.9938	0.9938	0.9937	0.9939	0.9938

Table 3.6. Solution to the Neumann problem

Point	Exact solution	Method I	II	III	IV
(0.1, 2, 0.9)	0.544	0.545	0.546	0.545	0.544

The solution was obtained by using the representation

$$u(x) = \int_{\Gamma} \mu(y) \frac{1}{2\pi} \frac{(n(y), x - y)}{|x - y|^3} d\sigma(y) + \frac{c}{|x|} .$$

The constant c was calculated as described in Sect. 3.1.3, $c \approx 0.8277$ (the ellipsoid with $a = b = 1$, $c = 0.5$ was taken). The exact solution is known:

$$u(x) = \arctan \left\{ \frac{a^2 - c^2}{\alpha + c^2} \right\}^{1/2} \left[\arctan \left\{ \frac{a^2 - c^2}{c^2} \right\} \right]^{-1} ,$$

where α is obtained from the equation

$$\frac{x_1^2 + x_2^2}{a^2 + \alpha} + \frac{x_3^2}{c^2 + \alpha} = 1 .$$

Again, four methods were used to obtain solutions: (I) by the method of Sect. 3.2.3, with 4 iterations; (II) by acceleration of the Neumann series on the basis of the Euler transformation, with 4 iterations; (III) by acceleration of the Neumann series using the transformation of Sect. 3.2.1, with 6 iterations; (IV) by continued fraction, with 8 iterations. The computations required 32 sec ($\sigma = 0.004$) for method I, 30 sec, 23 sec, 40 sec for II–IV. Note that the result $u = 0.542$ was obtained in 2 min by the method of integral equations [3.20]. These calculations were carried out by *Simonov*: in example 2 by method II and in example 3 and 4 by method I.

3.3.2 Calculation of the Capture Coefficient of Highly Dispersed Aerosols (3D)

Three problems connected with aerosol technology have been solved in cooperation with the Institue of Chemical Kinetics and Combustion of Siberian Branch of Academy of Science of USSR [IChKC]. The first problem is the calculation of aerosol transport in turbulent velocity fields in the presence of vegetation. In practice, it is interesting to calculate characteristics such as the distribution of concentration of aerosols in the vegetation, the probability of exceeding a critical dose, the intensity of concentration fluctuations as a function of the distance to the source. This problem is considered in Sect. 5.4, since it is solved in the nonstationary formulation. The second problem is the calculation of the aerosol capture coefficient for insects on a leaf, i.e., the evaluation of deposition of highly dispersed particles on 3D-bodies. The third problem is the calculation of characteristics of electron microscopes.

Note that in all these processes there occurs coagulation of particles. Of course, accurate solution of the problem which takes this phenomenon into account is not yet possible since in this case it is necessary to solve a system of nonlinear kinetic equations together with the hydrodynamics equations and the equations of motion. An additional difficulty lies in the large difference in the characteristic scales of the processes under study. Thus, it is necessary to investigate approximate techniques of taking into account the processes of coagulation. The construction of approxiamte solutions of coagulation equations and the investigation of the influence of the coagulation on various functionals under study is required.

Consider now the problem of calculating the capture coefficient, first neglecting coagulation and Brownian motion. A reference list on this subject can be found in [3.21, 22]. However, as noted in [3.21] many questions of the mechanics of aerosols, especially quantitative estimations, are very difficult to answer. Thus, let us consider an incompressible gas including aerosol particles flowing around a finite body. Assume that the flow is potential whose velocity at infinity is defined by a constant vector v_∞. Generally we suppose that the aerosol particles are well mixed in the flow. The size distribution of the aerosol particles is given.

We assume that the spherical aerosol particles deviate from the streamline due to their inertia. The inertial motion is characterized by the Stokes number $k = 2\varrho_a R^2 v_\infty / 9\eta L$ where ϱ_a is the mass density of the aerosol particle, R is its radius, L is the characteristic spatial scale of inhomogeneities of the flow. This may lead to capture of aerosol particles by the body (pure inertial capture). The capture coefficient ε is defined as follows [3.21]. We denote by Q_∞ the number of particles which would intersect the body (or a part of the body) during a unit time interval if they had moved parallel to the vector v_∞. In this definition, the aerosol particles are taken from a certain region of the size distribution. Let Q_k be the number of aerosol particles which in fact intersect the body (or a fixed part of the body) during a unit time interval. Then $\varepsilon = Q_k / Q_\infty$.

In general, the capture coefficient depends on many factors, in particular, on the parameters determining the flow regime, on the properties of aerosol particles, on the forces determining interactions between the particles and the body, on the Stokes number, etc. Of course, this dependence is very complicated. Often the capture depends on the electrostatic forces between a particle and the body. The deposition of highly-dispersed aerosol particles ($r \approx 1\ \mu$m) strongly depends on the Brownian motion since these particles "feel" microscopic imhomogeneities of the flow.

We make one more assumption; that the Reynolds number of the flow is small: Re $\ll 1$. This assumption is satisfied in the problem of deposition of particles of aerosol insecticide on a leaf surface and on the insects. If Re is large, the problem becomes very difficult because the tail and the hydrodynamic boundary layer must be taken into account. However there exist models which include the exterior potential flow and viscuous boundary layer (Re $> 10^5$) [3.23]. Consider the equation of motion of an aerosol particle moving around a body G:

$$k\frac{dv}{dt} + v = u + F , \quad v(t_0) = v_0 , \tag{3.80}$$

where $t \approx L/|v_\infty|$ is a dimensionless time, L is the characteristic scale of flow inhomogeneities, k is the Stokes number, $F = F^*/6\pi\eta R$, F^* is the external force, u is the velocity of the flow, v is the particle velocity, η is the viscosity, R is the radius of the aerosol particle. We suppose that u is a potential flow and seek it in the form $u(y) = (v_\infty, y) + V(y)$. Then V is found from the exterior Neumann problem

$$\Delta V(y) = 0 , \quad y \in \mathbf{R}^3/\bar{G} ; \quad \frac{\partial V}{\partial n}(z) = -(v_\infty, n(z)) , \quad z \in \Gamma . \tag{3.81}$$

Here $n(z)$ is a interior normal vector at $z \in \Gamma$.

There exist two approaches to solve this problem. The first approach is based on (1) numerical construction of the velocity field by solving (3.81), (2) solution of the motion equation (3.80) for an ensemble of particles; then $\varepsilon = Q_k/Q_\infty$.

The second approch is based on the Euler description of the motion of particles

$$k\left[\frac{\partial}{\partial t} + (v\nabla)\right] v + v = u + F ,$$
$$\frac{\partial c}{\partial t} + \mathrm{div}(cv) = 0 \tag{3.82}$$

where $c = c_*/c_\infty$; c_* is the number of particles in unit volume, c_∞ is a characteristic value of the particle concentration. For relatively large Stokes numbers k (3.82) can be linearized by changing $v\nabla$ with $e\nabla$ where $e = (1,0,0)$ is a basis vector. Under the above assumptions the following approximate formula can be obtained [3.21]:

$$\varepsilon = -\frac{1}{S}\int_T \exp\{x - x_0\}u_\tau dT . \tag{3.83}$$

Here T is the side surface of a cylinder oriented parallel to v_∞ whose base surface is defined as the part of the body cut by this cylinder. If this part is small (compared with the body size) then (3.83) defines a local capture coefficient ε of this part. If the cylinder cuts exactly the entire body then (3.83) defines the integral capture coefficient of the body. In (3.83) S is the area of the projection of T on a plane perpendicular to φ_∞, x_0 is the x_1-coordinate of the contour of the cut part of the body, u_τ is the velocity of the flow in the direction τ, where τ is the normal vector to T, $v_\infty = (1,0,0)$.

Thus the coefficient ε is represented as a linear integral functional (3.83) of the solution of exterior Neumann problem (3.81). Consequently, to calculate this functional, the walk on boundary algorithm described in Sect. 3.1.2 can be applied.

We define a critical Stokes number as $k_{cr} = \sup\{k : k \in J\}$ where the set J consists of Stokes numbers such that the particles having inertia parameters $k \in J$ can not be captured by the body [3.21]. The existence of k_{cr} in the problem of inertial deposition of particles was first discovered for simple flows by *Albert* and *Taylor* (bibliography in [3.21]).

Thus $\varepsilon = 0$ for $k \leq k_{cr}$, and the problem is to determine k_{cr} and the coefficient ε. We now describe the Monte Carlo estimate for calculating ε from (3.83) by the walk on boundary algorithm. Let $\{y_i\}_{i=1}^n$ be the isotropic walk on bondary process starting at $y_{-1} \in \Gamma$, and let $q(y_i)$ be the number of intersections of Γ by the line with direction $\overrightarrow{y_{i-1}y_i}$. Then if we seek the solution V in the form of a simple layer potential, the adjoint estimate of Sect. 3.1.2 for $u_\tau(y)$ will take the form

$$\xi^* = \sum_{i=0}^{n} a_i Q_i^* f(y_i) ,$$

$$Q_0^* = \frac{(\tau, y_0 - y)}{|y_0 - y|^3} \frac{|y_0 - y_{-1}|^3}{(n(y_0), y_{-1} - y_0)} ,$$

$$Q_i^* = Q_{i-1}^* \left[-\frac{(n(y_{i-1}), y_i - y_{i-1})}{(n(y_i), y_{i-1} - y_i)} \right] q(y_i) ,$$

$$f(y_i) = -(v_\infty, n(y_i)) .$$

(3.84)

Here y_{-1} is the initial point of the walk which is taken as a point of the boundary nearest to y; a_i are defined by the transformation method used.

Replacing in (3.83) (according to the double randomization scheme) $u_\tau = (\mathrm{grad}\, V, \tau)$ by the random estimate (3.84) we obtain

$$\varepsilon = M \left\{ \frac{L}{S p_1(x)} \xi^* \exp[\frac{x - x_0}{k}] \right\} + \delta .$$

(3.85)

Here δ is the deterministic error of the algorithm, $p(x)$ is the probability density of the x-distribution. Let $L_T(x)$ be a curve obtained by the intersection of T with the plane $x = \mathrm{const}$, and let L be the length of L_T. Then the point on the surface T [where (3.85) is calculated] is chosen as follows: its x-coordinate is sampled according to $p_1(x)$, next the desired point is sampled on $L_T(x)$ according to the density estimate IV in Table 3.7.

Table 3.7. The capture coefficient as a function of Stokes number for a sphere

k	Exact solution	Method I	Method IV	k	Exact solution	Method I	Method IV
0.01	0	0	0	1	0.46	0.44	0.45
0.083	0	10^{-3}	10^{-3}	2	0.65	0.64	0.64
0.5	0.24	0.23	0.24				

We now describe the results for two bodies: a sphere $G_1 = \{x \in \mathbf{R}^3 : |x| \leq 1\}$, and two intersected spheres: $G_2 = \{x \in \mathbf{R}^3 : |x| \leq 1\} \bigcup \{|x - x_T| < 0.1\}$, where $x_T = (0, 0, 1)$. The numerical results are shown in Table 3.7 for G_1.

The results were obtained in 7:46 min by the multiplication method (I) of Sect. 3.3.1, (Table 3.3) using 9×10^4 samples, the mean squared error $\sigma = 0.04$; and by method (IV) of continued fraction with $\sigma = 0.02$, in a computing time of 6:20 min. The body G_2 is a rough model of a leaf–insect system. Using experimental data obtained by the physics department of IChKC on the size of the δ-layer, $\delta \approx k_{\mathrm{cr}} R$, we obtained the results for G_2 shown in Fig. 3.5. Here ε_0 is the integral capture coefficient of the entire body, $\varepsilon(y_0, y_1)$ is the local coefficient calculated for the part of G_2 where $y_0 \leq y \leq y_1$, $0 \leq y_0, y_1 < 1.1$. The integral coefficient was calculated on the basis of (3.85). The local coefficient

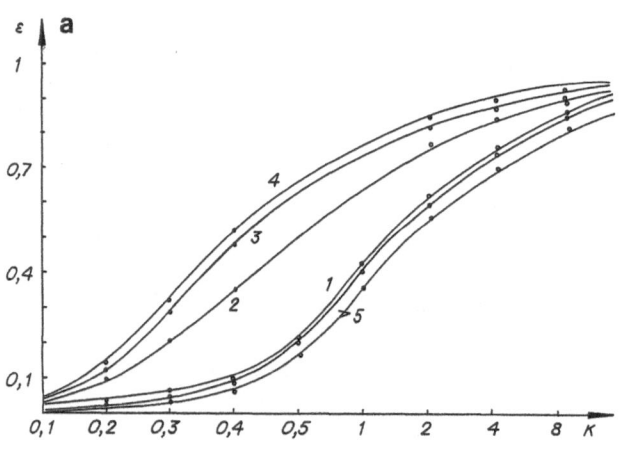

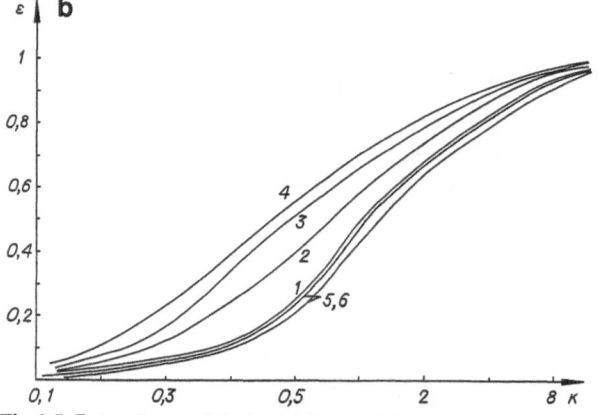

Fig. 3.5. Dependence of the integral ε_0 and the local ε capture coefficient for the body G_2 on k: (a) fixed position of the small sphere, (b) averaged results for a random position of the small sphere. Curves $5, 6$ – the local capture coefficient for the large sphere: $5 - \varepsilon(0, 0.1)$, $6 - \varepsilon(0.1, 0.2)$. Curve $1 - \varepsilon_0$. Curve $2 - \varepsilon(0.9, 1)$. Curve $3 - \varepsilon(1.05, 1.1)$. Curve $4 - \varepsilon(1, 1.05)$

was calculated by solving the equation of motion (3.80) where the velocity was calculated only along the trajectories. It is an important advantage of the walk on boundary algorithm that in calculations of the velocities along the trajectories a fixed set of samples of boundary values stored beforehand is used.

As the results of Fig. 3.5a show, k_{cr} for G_1 and G_2 is approximately the same. However, there exists a region of k where the local coefficient for the small sphere is larger than that of the large sphere. This can apparently explain following experimental results. The upper picture in Fig. 3.6 shows the deposited aerosol particles of $d = 25\,\mu$m diameter on an insect placed on a leaf. It is seen that the deposited particles are approximately uniformly distributed over the insect and the leaf surface. The data in the lower picture are different only in the size of the particles: $d = 11\,\mu$m. Here the particles are deposited mainly on the insect.

It is interesting to investigate the dependence of the behavior of the curves in Fig. 3.5 on various parameters. Consider the following modifications of the basic model:

I) The position of the small sphere on the large sphere is sampled uniformly (keeping axial symmetry).
II) The particles are sampled from a lognormal size distribution density

$$g(r) = \frac{1}{[2\pi]^{1/2} r \ln(\sigma)} \exp\left\{\left[\frac{\ln(r) - \ln(r_m)}{2^{1/2} \ln(\sigma)}\right]^2\right\}$$

where $\ln(r_m)$ and $\ln(\sigma)$ are the average and the mean squared deviation of the logarithm of the relative radii of the particles, respectively.
III) In (3.82) $F \not\equiv 0$ is the electrostatic force between the particles and the body.

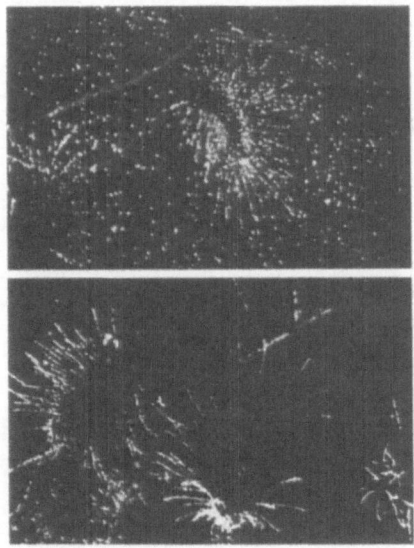

Fig. 3.6. Aerosol deposition on an insect. *Upper* photograph, particle diameter 25 μm; *lower* photograph 11 μm

135

IV) For small values of k a diffusion mechanism is taken into account with diffusion coefficient $D(r) = ar^{-1} + br^{-2}$, where a, b are known constants.

V) For arbitrary values of k, diffusion in a turbulent velocity field $u = \bar{u} + u'$ is present; here $\bar{u}$ is the mean flow, u' are the fluctuations.

In real problems all the situations I–IV may appear. The problem is to study the influence of all the factors on the capture coefficient.

In the first model, by using the double randomization scheme, we first simulate the position of the small sphere, next we calculate the capture coefficient as in the basic model by using a low number of trials. Then a new position of the small sphere is sampled and the entire procedure is repeated. The result is obtained by averaging over all trials. In Fig. 3.5b the results are shown for 4×10^4 trials, $\sigma = 0.002$. Comparison of Fig. 3.5a and b shows that the integral capture coefficient for a fixed G_2 and a random deposition of the small sphere do not really differ. The local coefficients for the middle part of the large sphere are also slightly changed. The change of local coefficients of the small sphere at $k \approx 0.3$–0.4 was about 10–15% (curves 2–4).

Model II was implemented for a fixed small sphere (as in the basic model) where the mean squared deviation was taken to be 1.5; the results are close to that of Fig. 3.5b with the difference that curve 2 at $k \approx 1$ is 8–10% higher.

Note that often coagulation processes change the lognormal distribution. Below we describe the coagulation processes in more detail.

In model III, similar behavior of the curves was obtained only for electrostatic forces which were too large to be realistic. In model IV (Fig. 3.7) the numerical results show that k_{cr} disappears, i.e., ε changes more smoothly. Analogously, in model V, $\varepsilon(k)$ is also smoothed out (curve 7, Fig. 3.7).

We now describe the results of simulation of coagulation processes. Examples of dispersed systems where coagulation processes play an important role are highly dispersed atmospheric aerosol ($d < 1\,\mu$m), clouds ($d \leq 100\,\mu$m), aerosol products of explosive combustion, etc.

Aerosol formation consists usually of three main stages: nucleation of a condensed phase (particles), condensation and coagulation growth of particles and unimolecular evaporation. The two last stages can be described by the Smolukhovsky equation

$$\frac{dn_l}{dt} = \frac{1}{2} \sum_{l=i+k} a_{ik} k_{ik} n_i n_k - n_l \sum_{l=1}^{\infty} a_{il} k_{il} n_i + b_{l+1} n_{l+1} - b_l n_l , \qquad (3.86)$$

where k_{ij} is the rate constant for two particles (consisting of i and j atoms) to coagulate, b_l is the constant of unimolecular evaporation, a_{ij} is the effective coefficient of coagulation, n_i is the concentration of particles consisting of i atoms.

The explicit solution of this equation is known only for particular cases, for example, when $b_i = 0$, $a_{ij} = 1$ and $k_{ij} = 1$, $k_{ij} = i + j$, $k_{ij} = ij$. In [3.24] a numerical method was proposed based on subdivision of (3.86) into classes, one of which describes coagulation of particles of "mean" size. As a consequence, it is

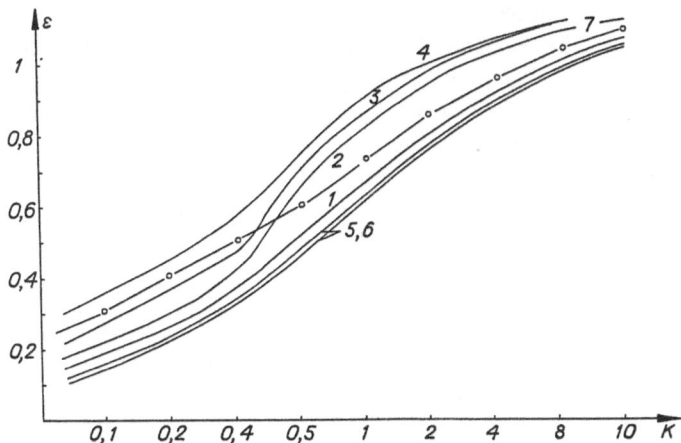

Fig. 3.7. Curves *1–6* are the same as on Fig. 3.5, but using model 4. Curve 7 is the integral capture coefficient for model 5

not possible to calculate narrow size distributions by this averaging procedure. Let us consider a particle in the gas phase. We assume that the collision frequencies between two particles are described by coagulation constants k_{ij}. The number of collisions per time interval is approximately given by $m \approx kn_1^2\Delta t$ where k is an average collision constant, n_1 is the average particle concentration. The average time interval between two collisions is $t \approx 1/kn_1$. The corresponding mean distance between particles is then on the order of $l \approx [2Dt]^{1/2}$ where D is the diffusion coefficient of a particle moving in the gas phase. The average number of particles in a ball of radius l is then on the order of $N \approx l^3 n_1$ or $N \approx n_1[D/kn_1]^{3/2}$. Assume that $D \approx V_1/3n_2\sigma_{12}$ and $k \approx V_1\sigma_{11}$, where V_1 is the mean relative velocity of the colliding particles, σ_{11} is the collision cross section of the aerosol particles, and σ_{12} is the cross section determining the collision between the aerosol particles and the gas molecules whose concentration is denoted by n_2. Thus

$$N \approx [\sigma_{12}\sigma_{11}n_2]^{-3/2}n^{-1/2} .$$

Thus for particles whose size is on the order of $d \approx 10^{-9}$ m, and $n_1 > 10^{10}$ cm^{-3}, $n_2 \approx 3 \times 10^{19}$ cm^{-3}, the mean number of gas particles which collide with a fixed aerosol particle between two collisions is on the order of several tens of thousands. For example, if $d \approx 10^{-7}$ m, $n_1 \approx 10^2$ cm^{-3}, $N < 10^4$.

Let us describe the algorithm for solving (3.86) (Fig. 3.8). Consider m arrays $x_1[1 : N], \ldots, x_m[1 : N]$, where n is the number of particles in a volume V. Two particles of i and j atoms chosen at random collide with a given (normalized) probability and form a new particle consisting of $i + j$ atoms. One cell of the two arrays becomes empty. After k collisions, the particles of the formed arrays are mixed filling the empty cells. The last procedure prevents emptying the entire array and is repeated k_0 times (k_0 combinations). The time is measured in

Monte Carlo steps, i.e., in the number of collisions of two particles having the minimum collision frequency. To evaluate the process in real time an invariant factor is introduced: $\text{Inv} = N_{\text{rand}}/k_{ij}(\text{max})$. This quantity is constant to within 0.5–1% depending on the normalization factor. Comparison of this method with the method of *Rosinsky–Snow* (Fig. 3.9) shows good agreement in the case of free molecular collisions. Another test is shown in Fig. 3.10 when an exact solution is known for $k_{ij} = \text{const}$. The error of the Monte Carlo calculations was about

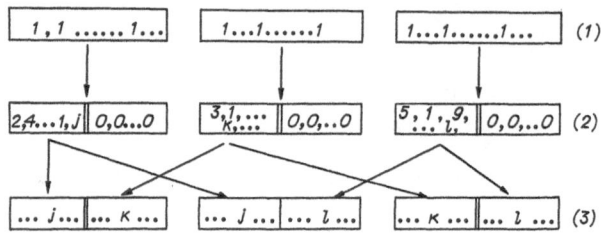

Fig. 3.8. Flow diagram of the algorithm for simulating coagulation

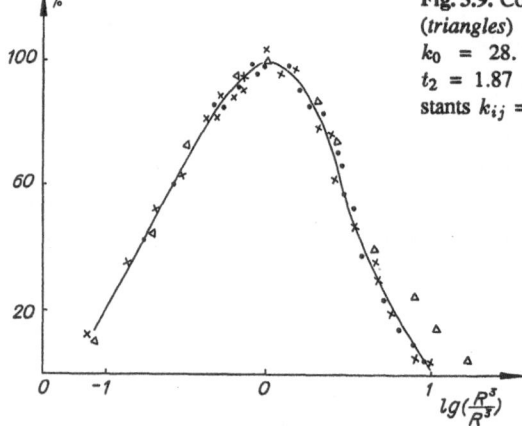

Fig. 3.9. Comparison of the Rosinsky–Snow algorithm (*triangles*) with our results, *crosses* $k_0 = 19$, *points* $k_0 = 28$. $\theta = 1/k_0 n_0$, $t_1 = 1.44 \times 10^{-4}$ and $t_2 = 1.87 \times 10^6 \theta_0$ for free molecular collision constants $k_{ij} = k_{fm} \equiv (i^{1/3} + j^{1/3})(i^{-1} + j^{-1})^{1/2}$

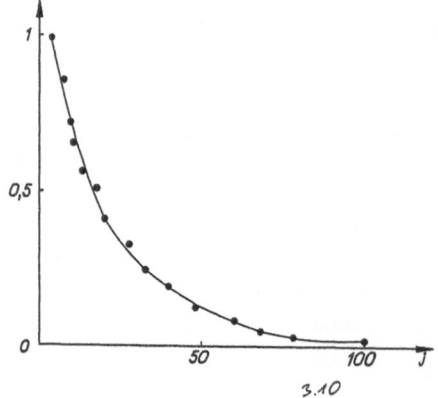

Fig. 3.10. Comparison of results (*points*) with exact solution to the Smolukhovsky equation (*solid line*) $k_{ij} = \text{const}$ $t = 32\,\theta_0$

3.10

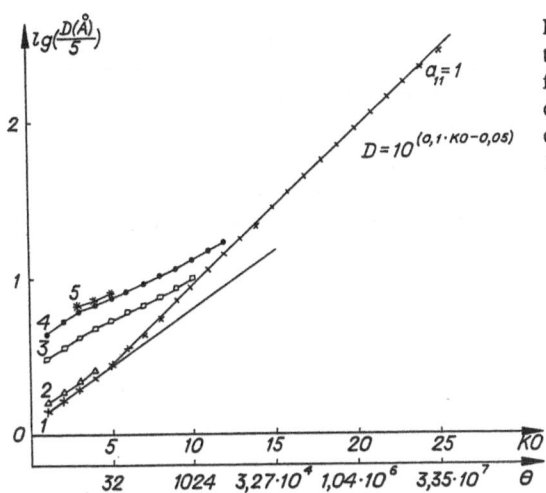

Fig. 3.11. Dependence of the mean particle size on dimensionless time for different values of a_{11}. Curve *1*, $a_{11} = 1$; curve *2*, $a_{11} = 0.1$; curve *3*, $a_{11} = 10^{-3}$; curve *4*, $a_{11} = 10^{-4}$; curve *5*, $a_{11} = 10^{-5}$; $\log(D) = 0.1 k_0 - 0.05$

1%. The dependence of the mean particle size on dimensionless time for various values of a_{11} is shown in Fig. 3.11. The results show that a decrease of a_{11} leads to delay of stabilization. A qualitative estimate of the stabilization time is $t_{st} \approx 1/a_{11}$.

Size distributions of particles experimentally obtained in the process of explosive combustion Fig. 3.12 are shown in Fig. 3.17. An example of size distribution for free molecular coagulation is shown in Fig. 3.13 for $a_{11} = 10^{-3}$; calculations were carried out for $k_0 = 5$ combinations of half-emptied arrays. It is clear from this that for $k_0 = 5$ size distribution does not depend on the presence of monomers.

In experiments there often arises the question of what parameters must be measured to determine the collision constants, especially the constant a_{11}. As the results of Fig. 3.14 show, the mean size does not strongly depend on a_{11}. Comparison with the results of Fig. 3.16 and 3.10 shows that the second moment of the size distribution α contains far more information about a_{11}. Therefore, second moments of the size distribution in experiments (on AgI particles) were

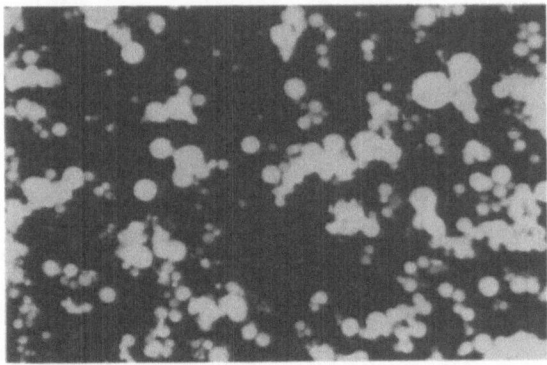

Fig. 3.12. Spherical particles of Al_2O_3 obtained in experiments

obtained. The experimental results agree well with the calculations carried out for $a_{11} \approx 10^{-5}$ but deviate strongly from $a_{11} \approx 1$ (Fig. 3.15). Size distributions for $a_{11} = 10^{-3}$ and $a_{11} = 10^{-5}$ are shown in Fig. 3.16. The difference in the second moment for the two curves is about $\Delta\alpha \approx 0.1$.

An old question discussed in the literature is what is the influence of the constants k_{ij} on the form of the steady-state size distribution. In experiments, one usually measures the mean size R and the variation α. Therefore we also tried

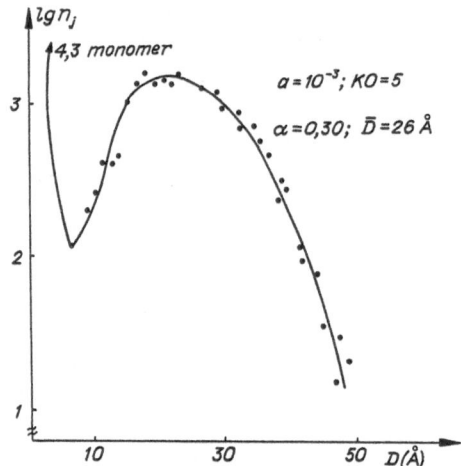

Fig. 3.13. Size distribution calculated for free molecular coagulation for $a_{11} = 10^{-3}$; $\alpha = 0.30$, $\bar{D} = 26\,\text{Å}$; $k_0 = 5$; $k_{ij} = k_{fm}$

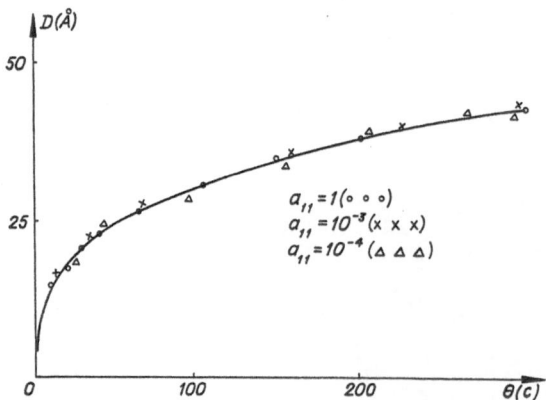

Fig. 3.14. Dependence of the mean particle size on the time for different values of a_{11} : $n_1 = 10^{10}\,\text{cm}^3$, $k_{ij} = k_{fm}$. *Points* $a_{11} = 1$; *crosses* $a_{11} = 10^{-3}$; *triangles* $a_{11} = 10^{-4}$

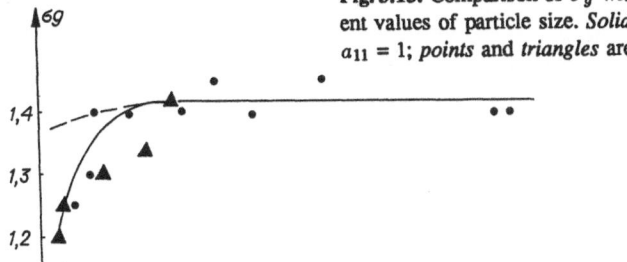

Fig. 3.15. Comparison of σ_g with experimental data for different values of particle size. *Solid line* $a_{11} = 10^{-5}$; *dashed line* $a_{11} = 1$; *points* and *triangles* are experimental data [5.26]

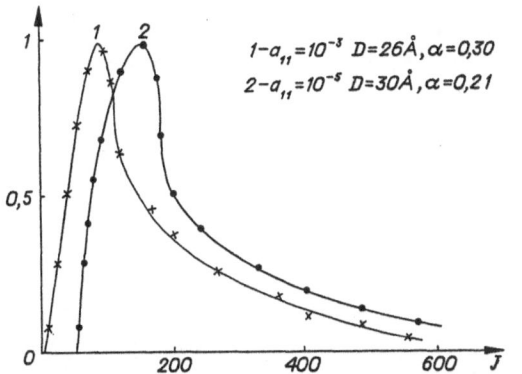

$$1-a_{11}=10^{-3}\ D=26\text{Å},\alpha=0,30$$
$$2-a_{11}=10^{-5}\ D=30\text{Å},\alpha=0,21$$

Fig. 3.16. Size distributions for similar values of $\bar{D}$ for different a_{11}; ($k_{ij} = k_{fm}$). *Crosses*, $a_{11} = 10^{-3}$; $\bar{D} = 26$ Å, $\alpha = 0.3$. *Points*, $a_{11} = 10^{-5}$; $\bar{D} = 30$ Å, $\alpha = 0.21$

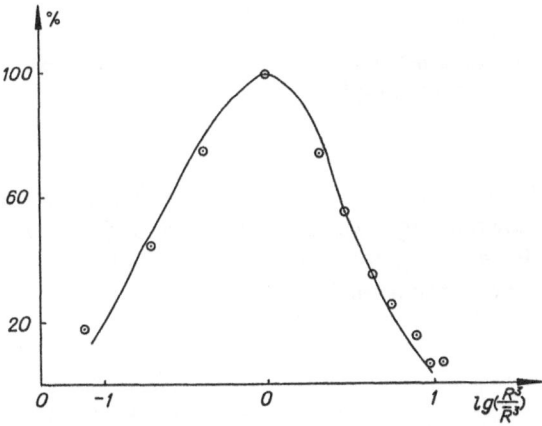

Fig. 3.17. Comparison of computational results of particles size distribution (*solid line*) with experimental data (*points*) of Fig. 3.12

141

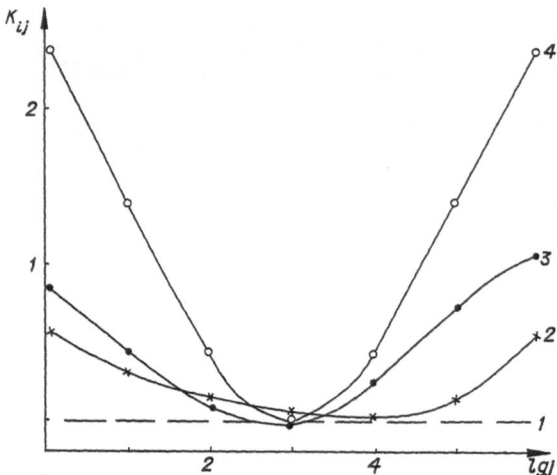

Fig. 3.18. Dependence of the collision constants on the particle size; curve 1, k_{ij} = const; curve 2, diffusion constant $k_{dif} = k_{ij} = k_{01}(i^{-1/3} + j^{-1/3})$; curve 3, $k_{ij} = k_{fm}$; *curve 4*, $k_m = k_{ij} = (i^{-1} + j^{-1})(i + j)$

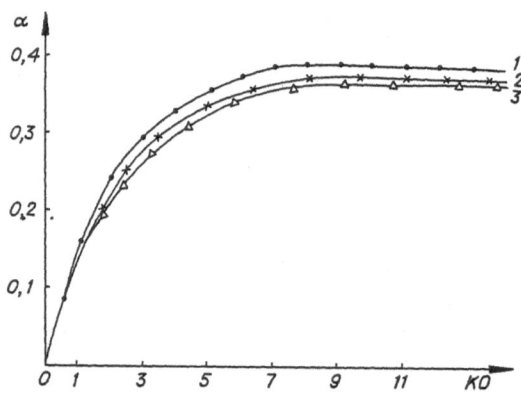

Fig. 3.19. Dependence of the variation coefficient on the total number of combinations: curve 1, $k_{ij} = k_m$; curve 2, $k_{ij} = k_{fm}$; curve 3, k_{ij} = const; $a_{11} = 1$

to study the influence of k_{ij} on R and α in the steady-state regime. The behavior of the different constants k_{ij} is shown in Fig. 3.18. Calculations show (Fig. 3.19) a slow dependence of α on the form of k_{ij}. Analysis of many calculations has shown that the dimensionless time θ depends on k_0 as $\theta = 2^{0.78k_0 - 1}$.

142

4. Monte Carlo Algorithms for Solving High-Order Equations and the Elasticity Problems

In this chapter we construct walk on spheres and walk on boundary algorithms for solving high-order equations and systems primarily related to elasticity problems: the biharmonic problem, general metaharmonic equations and the Lamé equation.

4.1 Biharmonic Problem

Consider the equation describing the bending of a thin elastic plate [4.1]

$$\Delta\Delta u(x,y) = \frac{f(xy)}{D}, \quad (x,y) \in G \subset \mathbf{R}^2 \tag{4.1}$$

where $u(x, y)$ is the normal displacement at a point (x, y), $f(x, y)$ is the magnitude of the normal load, and $D = Eh^3/12(1 - \sigma^2)$, where E is Young's modulus, σ is the Poisson's ratio of the plate, and $2h$ is its thickness.

Depending on how the plate boundary is fixed, one typically uses one of the following boundary conditions:

1) simply supported boundary of the plate:

$$u|_\Gamma = 0, \quad \left\{\Delta u - \frac{1-\sigma}{\varrho}\frac{\partial u}{\partial n}\right\}\Big|_\Gamma = 0. \tag{4.2}$$

2) rigid fixing of the boundary:

$$u|_\Gamma = 0, \quad \frac{\partial u}{\partial n}\Big|_\Gamma = 0. \tag{4.3}$$

Here, n is an exterior normal to the boundary $\partial G = \Gamma$ and ϱ is the curvature of the boundary. It is also possible to consider the case where different parts of the boundary ∂G are fixed differently. Then the appropriate combination of boundary conditions (4.2, 3) are used.

Let us construct a vector walk on spheres algorithm for (4.1) with boundary conditions (4.2, 3). If the plate lies on an elastic base surface, then $u(x, y)$ satisfies the equation

$$\Delta\Delta u(x,y) + \frac{k}{D} u = \frac{f(x,y)}{D} \tag{4.4}$$

143

where k is a coefficient characterizing the underlying surface. If one considers, in addition, a crosswise bending of a circular plate on an elastic base surface, then $u(x, y)$ satisfies the equation

$$\Delta\Delta u(x, y) + \frac{k}{D} u + R\Delta u = \frac{f(x, y)}{D} , \tag{4.5}$$

where R is the intensity of a longitudinal load uniformly distributed over the boundary. Note that (4.4, 5) are special cases of the metaharmonic equation which will be investigated in Sect. 4.2.

Hereafter we assume that there exist unique classical solutions of the formulated problems.

It should be noted that one should consider various forms of descriptions of boundary conditions. For example, the simply supported boundary of the plate is often described by the conditions

$$u|_\Gamma = 0 \quad \left\{ \frac{\partial^2 u}{\partial n^2} + \sigma \left(\frac{\partial^2 u}{\partial s^2} + \frac{1}{\varrho} \frac{\partial u}{\partial n} \right) \right\} \bigg|_\Gamma = 0 \tag{4.2a}$$

or

$$u|_\Gamma = 0 , \quad \left\{ \frac{\partial^2 u}{\partial n^2} + \sigma \frac{\partial^2 u}{\partial t^2} \right\} \bigg|_\Gamma = 0 , \tag{4.2b}$$

or

$$u|_\Gamma = 0 , \quad \left\{ \sigma\Delta u + (1 - \sigma)\frac{\partial^2 u}{\partial n^2} \right\} \bigg|_\Gamma = 0 . \tag{4.2c}$$

Here t is a tangent to the boundary, S is arc of Γ. All these conditions are equivalent to (4.2) because they describe the vanishing of the bending moment

$$M_n = M_x \cos^2(\alpha) + M_y \sin^2(\alpha) + H \sin(2\alpha)$$

where α is the angle between the axes x and t, and

$$M_x = -D \left(\frac{\partial^2 u}{\partial x^2} + \sigma \frac{\partial^2 u}{\partial y^2} \right), \quad M_y = -D \left(\frac{\partial^2 u}{\partial y^2} + \sigma \frac{\partial^2 u}{\partial x^2} \right),$$

$$H = -D(1 - \sigma)\frac{\partial^2 u}{\partial x \partial y} .$$

Indeed, equivalence of (4.2b) and (4.2c) follows from the equality

$$\Delta u = \frac{\partial^2 u}{\partial n^2} + \frac{\partial^2 u}{\partial t^2} = \frac{\partial^2 u}{\partial x^2} + \frac{\partial^2 u}{\partial y^2} .$$

To prove equivalence of (4.2a) and (4.2b) we use the representation

$$\frac{\partial^2 u}{\partial s^2} = -\frac{1}{\varrho} \frac{\partial u}{\partial n} + \frac{\partial^2 u}{\partial t^2} ;$$

then, adding and substracting $\frac{\partial^2 u}{\partial t^2}$ in (4.2b) we obtain

$$\left\{ \Delta u - (1-\sigma)\frac{\partial^2 u}{\partial t^2} \right\}\bigg|_\Gamma = 0 \, ,$$

consequently

$$\left\{ \Delta u - (1-\sigma)\left(\frac{\partial^2 u}{\partial s^2} + \frac{1}{\varrho}\frac{\partial u}{\partial n} \right) \right\}\bigg|_\Gamma = 0 \, .$$

Recalling that $u|_\Gamma = 0$ we obtain (4.2).

Sometimes one also considers the following condition:

$$u|_\Gamma = 0 \, , \quad \left\{ \Delta u + \left(\frac{1-\sigma}{\varrho} + k_0 \right)\frac{\partial u}{\partial n} \right\}\bigg|_\Gamma = 0$$

which describes an elastic support, where k_0 is a coefficient of rigid sealing of the boundary.

Although all the boundary conditions are homogeneous (with zero right-hand side), we shall consider the general case of nonhomogeneous boundary conditions. The reason is two-fold: (1) problems of thin plates with nonhomogeneous conditions exist; (2) even in the case of homogeneous boundary conditions the walk on circles method requires consideration of the nonhomogeneous case.

4.1.1 Vector Walk on Circles Algorithm for Solving the Plate Bending Problem for Simply Supported Plates

Consider the following simple case of (4.1, 2) when $f = 0$ and the boundary conditions are given on a piecewise linear boundary in the form

$$u|_\Gamma = \varphi_1 \, , \quad \Delta u|_\Gamma = \varphi_2$$

(because in this case $\varrho = \infty$). Suppose that we need to calculate the solution of this problem at a fixed point $x = (x_1, x_2)$. In accordance with the general scheme described in Chap. 1, it is necessary to derive a mean value theorem which relates the vector $v = (u, \Delta u)^\mathrm{T}$ at a point x to the integral of v over $S(x, d(x))$. Since $\Delta^2 u = 0$, we obtain from (3.72) for $n = 2$, $r = d(x)$:

$$u(x) = N_x^r(u) - \frac{r^2}{4} N_x^r(\Delta u) \tag{4.6}$$

$$\Delta u(x) = N_x^r(\Delta u)$$

where N_x^r is an averaging operator over $S(x, r)$; that is,

$$N_x^r(u) = \frac{1}{2\pi r}\int_{S(x,r)} u \, ds = \frac{1}{2\pi}\int_0^{2\pi} u(x_1 + r\cos\varphi, x_2 + r\sin\varphi)d\varphi \, .$$

Let $\{\xi_k\}_k^\infty = 0$ be an ε-spherical process starting at a point $\xi_0 = x$,

$$\xi_{k+1} = \xi_k + d(\xi_k)\omega_k , \quad k = 0, 1, \ldots$$

and put

$$C(i) = \begin{pmatrix} 1 & -\frac{1}{4}d^2(\xi_i) \\ 0 & 1 \end{pmatrix}, \quad C^{(n)} = \prod_{i=0}^{n} C(i) .$$

Then the vector ε-biased Monte Carlo estimate for the formulated problem is constructed as described in Sect. 1.1.2:

$$\eta_x = C^{(n)}\varphi(Q_*) \tag{4.7}$$

where $\varphi = (\varphi_1, \varphi_2)^{\mathrm{T}}$; Q_* is the point of Γ nearest to the point $Q = \xi_{N_\varepsilon} \in \Gamma_\varepsilon$, i.e., to the last state of the ε-spherical process.

In Sect. 4.2.2 we will give the proof of ε-biasedness and of the uniform (with respect to ε) boundedness of the variance of the estimate η_x.

Consider now equation (4.1) with non-zero right-hand side $f/D = g$ and derive the corresponding mean value theorem in a circle $K(x, r)$ with boundary $S(x, d(x))$. To do this we use the equality [4.2]

$$N_x^r(u) = \sum_{\nu=0}^{m} \left(\frac{r}{2}\right)^{2\nu} \frac{\Delta^\nu u(x)}{(\nu!)^2} + B_x^r\left(v_m \Delta^{m+1} u\right) \tag{4.8}$$

where $B_x^r(u)$ is the averaging operator of $u(x)$ over the circle $K(x, r)$, i.e.,

$$B_x^r(u) = \int_{K(x,r)} u d\sigma_x^r .$$

The functions v_m are determined from the recurrence formulas

$$v_0(\varrho) = \frac{1}{2\pi} \ln \frac{r}{\varrho} , \quad v_{\nu+1} = \int_{r_1}^{r} \varrho v_\nu(\varrho) \ln \frac{\varrho}{r} d\varrho . \tag{4.9}$$

Hence, using $\Delta^2 u = g$ we obtained from (4.8) for $m = 0$:

$$\Delta u(x) = N_x^r(\Delta u) + B_x^r\left(-\frac{g}{2\pi} \ln \frac{r}{r_1}\right) . \tag{4.10}$$

Integrating (4.9) by parts, we find

$$v_1 = \int_{r_1}^{r} \varrho v_0(\varrho) \ln \frac{\varrho}{r_1} d\varrho$$

$$= \frac{1}{2\pi} \left[\frac{r^2 + r_1^2}{4} \ln \frac{r}{r_1} - \frac{r^2 - r_1^2}{4} \right] ;$$

from this we get, using (4.8) for $m = 1$, the identity

146

$$u(x) = N_x^r(u) - \frac{r^2}{4} N_x^r(\Delta u)$$

$$+ B_x^r \left(\frac{1}{2\pi} \left[\frac{r^2 - r_1^2}{4} - \frac{r_1^2}{4} \ln \frac{r}{r_1} \right] g \right). \tag{4.11}$$

Hence, (4.10, 11) present a generalized mean value relation in the case of $g \neq 0$. We rewrite them in the form

$$u(x) = N_x^r(u) - \frac{r^2}{4} N_x^r(\Delta u) + \int_0^{2\pi} \frac{d\varphi}{2\pi} \int_0^r \frac{r^2 - r_1^2 \left(\ln \frac{r}{r_1} + 1 \right) r_1}{4}$$

$$\times \, g(x_1 + r_1 \cos \varphi, x_2 + r_1 \sin \varphi) dr_1$$

$$\Delta u(x) = N_x^r(\Delta u) + \int_0^{2\pi} \frac{d\varphi}{2\pi} \int_0^r \tag{4.12}$$

$$\times \left[-\ln \left(\frac{r}{r_1} \right) r_1 g(x_1 + r_1 \cos \varphi, x_2 + r_1 \sin \varphi) \right] dr \, .$$

The integrals in the right-hand side of (4.12) over the circle $K(x, r)$ [we denote the integrals of u and Δu by $J_1(x)$ and $J_2(x)$, respectively] can be estimated using the single node technique. The polar coordinates of this node are sampled as follows. First, an isotropic direction φ is sampled, then the random distance ϱ is chosen with respect to the density

$$p(r_1) = \frac{1}{r^2} \left[4r \ln \frac{r}{r_1} \right], \quad r_1 \in [0, 1] \, .$$

Then the random estimates for the integrals take the form

$$\eta_1 = \frac{r^2 - \varrho^2 \left[\ln \left(\frac{r}{\varrho} \right) + 1 \right]}{4} \frac{r^2}{4 \ln \left(\frac{r}{\varrho} \right)} g(x + \varrho \omega)$$

$$\eta_2 = -\frac{r^2}{4} g(x + \varrho \omega) \tag{4.13}$$

where ω is a unit isotropic vector. The simulation of the distance in an arbitrary circle $K(x, r)$ is not difficult. For example, it is possible to use the Neumann method (rejection method) choosing $4/er$ as a majorant. The rejection probability p is then the same in all circles, namely $p = 1 - e/4$. Note that for many realistic loads the integrals $J_1(x)$ and $J_2(x)$ can be evaluated explicitly. Using the double randomization principle, (4.12), and the estimates given in (4.13), it is easy to derive a vector estimate ξ for $(u, \Delta u)^T$:

$$\xi = \sum_{n=0}^{N_e - 1} Q_n^* \eta(\xi_n) + Q_{N_e}^* \varphi(Q_*) \tag{4.14}$$

where $Q_0^* = I$, $Q_i^* = Q_{i-1}^* C(i)$ are matrix weights. To ensure the finiteness of the variance of estimate (4.14), we assume that the function $g(x)$ satisfies the following inequality in an arbitrary circle $K(x, r)$:

$$\int_{K(x,r)} \frac{F_i^2(g)(r_1)}{p(r_1)} \, dr_1 < \infty \, ,$$

where $F_i(g)$, $i = 1, 2$, are the corresponding integrands in (4.12).

4.1.2 Plates with Arbitrary Boundaries

Consider now the general case of boundary value problems (4.1–3). An arbitrary solution of (4.1) satisfies the local system of integral equations generated by (4.12). The right-hand side of this system consists of two terms, namely, of integrals $(J_1(x), J_2(x))^{\mathrm{T}}$ and the vector $(u(x), \Delta u(x))^{\mathrm{T}}$ in Γ_ε. The latter term is unknown and is therefore approximated by the vector $(\varphi_1(Q_*), \varphi_2(Q^*))^{\mathrm{T}}$. This approximation leads to a bias of the constructed vector Monte Carlo estimate of the order $0(\omega(\varepsilon))$, where $\omega(\varepsilon)$ is the minimal (in the order of ε) of the functions $\omega_i(\varepsilon)$, $i = 1, 2$, i.e., of the continuity moduli of the functions u and Δu in Γ_ε, respectively.

We use this approach in the general case when the following boundary conditions are considered:

$$u|_{\partial G} = \varphi_1 \, , \quad \left\{ \Delta u - \frac{1-\sigma}{\varrho} \frac{\partial u}{\partial n} \right\} \bigg|_{\partial G} = \varphi_2 \, ,$$

or

$$u|_{\partial G} = \varphi_1 \, , \quad \frac{\partial u}{\partial n} \bigg|_{\partial G} = \varphi_2 \, .$$

We obtain a mean value relation which connects $(u(x), \Delta u(x))^{\mathrm{T}}$ with $N_x^r(u)$, $N_x^r \left(\Delta u - \frac{1-\sigma}{\varrho} \frac{\partial u}{\partial r} \right)$ and $B_x^r(qg)$ in the last circle $K(x, r)$ of the ε-spherical process; $x = \xi_{N_\varepsilon} - 1$, $r = d(\xi_{N_\varepsilon} - 1)$, where q is known weight function. This relation permits us to obtain an approximation for $(u(x), \Delta u(x))^{\mathrm{T}}$ in Γ_ε using $\varphi_1(Q_*)$, $\varphi_2(Q_*)$, where Q_* is a point of Γ nearest to the last point of the ε-spherical process, $\xi_{N_\varepsilon} \in \Gamma_\varepsilon$. Moreover, the form of the Monte Carlo estimate is then the same as (4.14) with the difference that in the last term the matrix Q_{N_ε} is acting not on $\varphi(Q_*)$, but on another vector (we denote it by ψ) which is constructed as described above. First, take the following boundary conditions:

$$u|_\Gamma = \varphi_1 \, , \quad \frac{\partial u}{\partial n} \bigg|_\Gamma = \varphi_2 \, .$$

Theorem 4.1. An arbitrary solution of the equation $\Delta^2 u = g$ satisfies the mean value relations

$$u(x) = N_x^r(u) - \frac{r}{2} N_x^r \left(\frac{\partial u}{\partial r} \right) + B_x^r(g_1 g) \qquad (4.15)$$

$$\Delta u(x) = \frac{2}{r} N_x^r \left(\frac{\partial u}{\partial r} \right) + B_x^r(g_2 g) , \qquad (4.16)$$

where

$$q_1(r_1) = \frac{1}{2\pi} \left(\frac{r^2 + r_1^2}{8} - \frac{r_1^2}{4} \ln \left(\frac{r}{r_1} \right) \right)$$

$$q_2(r_1) = -\frac{1}{2\pi} \left(\ln \left(\frac{r}{r_1} \right) - \frac{1}{2} + \frac{r_1^2}{2r^2} \right)$$

and the integration B_x^r of the functions $q_1(r_1)g(r_1)$, $q_2(r_1)g(r_1)$ is performed over the circle $K(x, r)$.

Proof. Note that by the second Green's formula [4.2]

$$B_x^r(\Delta u) = N_x^r \left(\frac{\partial u}{\partial r} \right) 2\pi r .$$

Substituting this expression into Poisson's formula for $\Delta u(x)$,

$$\Delta u(x) = \frac{1}{\pi r^2} B_x^r(\Delta u) - \frac{1}{2\pi} B_x^r$$

$$\times \left[\left(r^2 \ln \frac{r}{r_1} - \frac{r^2 - r_1^2}{2} \right) g \right] , \qquad (4.17)$$

we obtain (4.16).

Substituting now (4.17) into (4.8) for $m = 1$, we obtain by using (4.16)

$$u(x) = N_x^r(u) - \frac{r^2}{4} \Delta u(x) - B_x^r(v_1 g) = N_x^r(u) - \frac{r}{2} N_x^r \left(\frac{\partial u}{\partial r} \right)$$

$$+ \frac{1}{4} \frac{1}{2\pi} B_x^r \left(\left[r^2 \ln \frac{r}{r_1} - \frac{r^2 - r_1^2}{2} \right] g \right)$$

$$- B_x^r(v_1 g) = N_x^r(u) - \frac{r}{2} N_x^r \left(\frac{\partial u}{\partial r} \right) + B_x^r(q_1 g) .$$

This proves (4.15). □

The unbiased estimate for the solution of

$$\Delta^2 u = f , \quad u|_{\partial G} = \varphi_1 , \quad \frac{\partial u}{\partial n} \bigg|_{\partial G} = \varphi_2$$

we obtained from (4.14), where $\varphi(Q_*)$ must be replaced with explicit representations (4.15, 16). The biased Monte Carlo estimate is obtained according to the general scheme from the unbiased estimate using a "drift" (i.e., a first-order approximation) of the boundary values in ∂G_ε:

$$\xi_\varepsilon(x) = \sum_{n=0}^{N_\varepsilon-1} Q_n^* \eta(\xi_n) + Q_{N_\varepsilon}^* \psi \tag{4.18}$$

where the vector ψ is an approximation of $(u, \Delta u)^{\mathrm{T}}$ in ∂G_ε, constructed from (4.15, 16). For example, it is possible to take

$$\psi_1 = \varphi_1(Q^*) - \frac{r}{2} \varphi_2(Q^*) + Q_1[q_1 g](x + \varrho\omega) \, ,$$

$$\psi_2 = \frac{2}{r} \varphi_2(Q^*) + Q_2[q_2 g](x + \varrho\omega) \, , \tag{4.19}$$

where the random weights Q_1, Q_2 and the random point ϱ and direction ω are constructed as in (4.13).

Assume that the continuity moduli in Γ_ε of the functions u and $\partial u/\partial r$ have the form $\omega_1(\varepsilon)$ and $\varepsilon\omega_2(\varepsilon)$, respectively, where $\omega_1, \omega_2 \to 0$ as $\varepsilon \to 0$. Then, obviously

$$\left| M\xi_\varepsilon^{(1)}(x) - u(x) \right| = 0(\omega(\varepsilon)) \quad \text{as} \quad \varepsilon \to 0 \, ,$$

where $\omega(\varepsilon)$ is the minimum (in the order of ε) of the functions $\omega_i(\varepsilon)$, $i = 1, 2$, i.e., of the continuity moduli of the functions u and Δu in Γ_ε, respectively.

It should be noted that this scheme is numerically stable for the problem (4.1, 2). Indeed, as we shall see below from Theorem 4.2, in this case the large weight r^{-1} [cf. (4.16)] can be eliminated.

Now we turn to (4.1, 2) with nonhomogeneous boundary conditions

$$u|_\Gamma = \varphi_1 \, , \quad \left\{ \Delta u - \frac{1-\sigma}{\varrho} \frac{\partial u}{\partial n} \right\}\bigg|_\Gamma = \varphi_2 \, .$$

Theorem 4.2. An arbitrary solution of the equation $\Delta^2 u = g$ satisfies the mean value relations

$$u(x) = N_x^r(u) - \frac{r^2}{4\left[1 - \frac{\alpha r}{2}\right]} N_x^r\left(\Delta u - \alpha \frac{\partial u}{\partial r}\right)$$

$$+ \frac{1}{1 - \frac{\alpha r}{2}} B_x^r(q_1' g) \, , \tag{4.20}$$

$$\Delta u(x) = \frac{1}{1 - \frac{\alpha r}{2}} N_x^r\left(\Delta u - \alpha \frac{\partial u}{\partial r}\right) + \frac{1}{1 - \frac{\alpha r}{2}} B_x^r(g_2' g) \, , \tag{4.21}$$

where α is an arbitrary constant, and

$$q_1' = \frac{1}{2\pi}\left(\frac{r^2 - r_1^2}{4} - \frac{r_1^2}{4}\ln\frac{r}{r_1}\right) - \frac{\alpha r}{2}q_1 \ ;$$

$$q_2' = -\frac{1}{2\pi}\ln\frac{r}{r_1} - \frac{\alpha r}{2}q_2 \ .$$

Proof. Multiplying (4.15) by $\alpha r/2$ and substracting this equality from (4.11), we get (4.20). Analogously, multiplication of (4.16) by $\alpha r/2$ and substraction of it from (4.10) yields (4.21). $\qquad\Box$

The biased estimate (4.18) remains in this case, where in (4.19) the functions ψ_1, ψ_2 are replaced by

$$\psi_1 = \varphi_1(Q_*) - \frac{r^2}{4\left[1 - \dfrac{\alpha r}{2}\right]}\varphi_2(Q_*)$$

$$+ \frac{1}{1 - \dfrac{\alpha r}{2}}Q_1'[q_1'g](x + \varrho\omega) \ , \qquad (4.22)$$

$$\psi_2 = \frac{1}{1 - \dfrac{\alpha r}{2}}\varphi_2(Q_*) + \frac{1}{1 - \dfrac{\alpha r}{2}}Q_2'[q_2'g](x + \varrho\omega) \ ,$$

where $\alpha = (1 - \sigma)/\varrho$, the weights Q_1', Q_2' and the random point $x + \varrho\omega$ in $K(\xi_{k-1}, d(\xi_{k-1}))$ are constructed as in (4.13, 19).

Note that if on one part of the boundary, say Γ_1, the conditions of type (4.2) are formulated, and on the other, say Γ_2, the conditions of the type (4.3) are used, then ψ_1, ψ_2 are calculated according to (4.22) if $\xi_{N_\varepsilon} \in \Gamma_{1\varepsilon}$, while (4.19) is used if $\xi_{N_\varepsilon} \in \Gamma_{2\varepsilon}$. If a part of Γ, say Γ_3, is a segment, then $\psi_1 = \psi_2(Q_*)$, if $\xi_{N_\varepsilon} \in \Gamma_3$.

We have used the mean value relations for the center of the circle. As in Chap. 3, we can construct the walk on circles algorithms for calculating the solutions simultaneously at a series of points lying inside the first circle, if we start from the mean value relation for an arbitrary point inside the sphere. We now derive such a general mean value relation for

$$\Delta^2 u = 0 \ , \quad u|_\Gamma = g_0 \ , \quad \Delta u|_\Gamma = g_1 \ .$$

Theorem 4.3. Any biharmonic function $u(x, y)$ satisfies the mean value relations

$$u(x) = \frac{r^2 - t^2}{2\pi r}\int_{S(x_0, r)}\frac{u(y)dS_y}{|x - y|^2}$$

$$+ \frac{r^2 - t^2}{2\pi r}\int_{S(x_0, r)}\left[\frac{r\cos^2\alpha}{|x - x_0|\sin\alpha}\arctan\left\{\frac{|x - x_0|\sin\alpha}{r - |x - x_0|\cos\alpha}\right\}\right.$$

$$\left. - \frac{1}{2} - \frac{r\cos\alpha}{2t}\ln\frac{|x - y|^2}{r^2}\right]\Delta u(y)dS_y \qquad (4.23)$$

for any point x lying inside the sphere $S(x_0, y)$, $t = |x - x_0|$, α is the angle between the vectors $\overrightarrow{x_0 x}$ and $\overrightarrow{x_0 y}$.

The function $\Delta u(x)$, as a harmonic function, satisfies the relation

$$\Delta u(x) = \frac{r^2 - t^2}{2\pi r} \int_{S(x_0, r)} \frac{u(y) dS_y}{|x - y|^2} \tag{4.24}$$

Proof. We seek the biharmonic function $u(x)$ in the form

$$u(x) = \left(|x - x_0|^2 - r^2 \right) v(x) + w(x) \tag{4.25}$$

where $v(x)$ and $w(x)$ are two harmonic functions. Since for $x \in \Gamma = S(x_0, r)$, $w(x) = u$, we get by the Poisson formula

$$w(x) = \frac{r^2 - t^2}{2\pi r} \int_{S(x_0, r)} \frac{u(y) dS_y}{\varrho^2}$$

where $t = |x - x_0|$, $\varrho = |x - y|$. It remains to define uniquely the function $v(x)$. Note that

$$\Delta u(x) = \Delta \left\{ \left(|x - x_0|^2 - r^2 \right) v(x) \right\} + \Delta w(x)$$
$$= v(x) \Delta \left(|x - x_0|^2 - r^2 \right) + 2 \frac{\partial v(x)}{\partial x_1} \frac{\partial}{\partial x_1} \left(|x - x_0|^2 - r^2 \right)$$
$$+ 2 \frac{\partial v(x)}{\partial x_2} \frac{\partial}{\partial x_2} \left(|x - x_0|^2 - r^2 \right)$$
$$+ \left(|x - x_0|^2 - r^2 \right) \Delta v(x) + \Delta w(x) .$$

Let $x = (x_1, x_2)$, $x_0 = (x_1^0, x_2^0)$. Using the obvious equalities

$$\frac{\partial}{\partial x_1} \left(|x - x_0|^2 - r^2 \right) = 2(x_1 - x_1^0) ,$$

$$\frac{\partial}{\partial x_2} \left(|x - x_0|^2 - r^2 \right) = 2(x_2 - x_2^0)$$

$$\Delta \left(|x - x_0|^2 - r^2 \right) = 4 ,$$

and the fact that the functions $v(x)$, $w(x)$ are harmonic, we obtain

$$\Delta u(x) = 4v(x) + 4(x_1 - x_1^0) \frac{\partial v(x)}{\partial x_1} + 4(x_1 - x_1^0) \frac{\partial v(x)}{\partial x_2}$$
$$= 4v(x) + 4r \left[\frac{x_1 - x_1^0}{r} + \frac{x_2 - x_2^0}{r} \frac{\partial v(x)}{\partial x_2} \right] .$$

From this we get for $x \in \Gamma$:

$$\Delta u(x) = 4v(x) + 4r \frac{\partial v}{\partial n} .$$

Thus the function $v(x)$ satisfies the third boundary value problem

$$\Delta v(x) = 0 , \quad x \in K(x_0, r) , \quad \frac{\partial v}{\partial n}(x) + \frac{1}{r} v(x) = \frac{\Delta u}{4r}\bigg|_\Gamma . \qquad (4.26)$$

Solution to (4.26) can be represented in polar coordinates ($x = \varrho e^{i\varphi}$):

$$v(x) = v\left(\varrho e^{i\varphi}\right) = \sum_{n=1}^{\infty} \frac{\varrho^n \left[A_n \cos(n\psi) + B_n \sin(n\psi) \right]}{r^{n-1}(n+1)} + A_0 \frac{r}{2} ,$$

where

$$A_n = -\frac{1}{4\pi} \int_0^{2\pi} \frac{\Delta u}{4r} \left(re^{i\theta}\right) \cos(n\theta)d\theta ,$$
$$n = 1, 2, \ldots$$
$$B_n = -\frac{1}{4\pi} \int_0^{2\pi} \frac{\Delta u}{4r} \left(re^{i\theta}\right) \sin(n\theta)d\theta .$$

Note that

$$A \cos(n\psi) + B \sin(n\varphi) = \frac{1}{4\pi} \int_0^{2\pi} \frac{\Delta u}{4r} \left(re^{i\theta}\right) \cos(\theta - \varphi)d\theta ,$$

$$A_0 = \frac{1}{2\pi} \int_0^{2\pi} \frac{\Delta u}{4} \left(re^{i\theta}\right) d\theta .$$

Then

$$v\left(\varrho e^{i\varphi}\right) = -\frac{1}{\pi} \int_0^{2\pi} \frac{\Delta u}{4} \left(\varrho e^{i\theta}\right) \left(\sum_{n=1}^{\infty} \frac{\varrho^n \cos[n(\theta - \varphi)]}{r^n(n+1)} + \frac{1}{2} \right) d\theta ,$$

since the series is uniformly convergent. Now we evaluate the series. Let $a = \varrho/r$, and let

$$R(a, \theta) = \sum_{n=1}^{\infty} \frac{a^n \cos(n\theta)}{n+1} ,$$

$$R_1(t, \theta) = \frac{1}{2} + \sum_{n=1}^{\infty} t^n \cos(n\theta) .$$

The last series converges uniformly for $t^2 < 1$, and

$$R_1(t, \theta) = \frac{1}{2} \frac{1 - t^2}{1 - 2t \cos(\theta) + t^2} .$$

Note that

$$R(a, \theta) = \frac{1}{a} \int_0^a R_1(t, \theta)dt - \frac{1}{2} .$$

Now evaluate the integral

153

$$I = \frac{1}{a} \int_0^a R_1(t, \theta) dt = \frac{1}{2a} \int_0^a \frac{1 - t^2}{1 - 2t\cos(\theta) + t^2} \, dt \ .$$

From standard handbooks the following integrals are known

$$\frac{1}{2a} \int_0^a \frac{dt}{1 - 2t\cos\theta + t^2} = \frac{1}{2a\sin(\theta)}$$
$$\times \left[\arctan\left\{ \frac{a - \cos\theta}{\sin\theta} \right\} - \arctan\{-\cot\theta\} \right] \ ,$$

$$\frac{1}{2a} \int_0^a \frac{t^2 \, dt}{1 - 2t\cos(\theta) + t^2} = \frac{1}{2} + \frac{\cos\theta}{2a} \ln|1 - 2a\cos(\theta) + a^2|$$
$$- \frac{\cos^2(\theta) - 1/2}{a\sin\theta} \left[\arctan\frac{a - \cos\theta}{\sin\theta} - \arctan(-\cot\theta) \right] \ .$$

From this we get

$$I = \frac{\cos^2(\theta)}{a\sin(\theta)} \left[\arctan\frac{a - \cos(\theta)}{\sin(\theta)} - \arctan(-\cot\theta) \right]$$
$$- \frac{1}{2} - \frac{\cos\theta}{2a} \ln|1 - 2a\cos(\theta) + a^2| \ .$$

Simple arguments show that

$$\alpha - \beta = \arctan[a\sin\theta/(1 - a\cos\theta)] \ ,$$

where

$$\alpha = \arctan\frac{a - \cos\theta}{\sin\theta} \ , \qquad \beta = \arctan(-\cot\theta) \ .$$

Thus

$$I = \frac{\cos^2(\theta)}{a\sin\theta} \arctan\frac{a\sin\theta}{1 - a\cos(\theta)} - \frac{1}{2} - \frac{\cos\theta}{2a} \ln|1 - 2a\cos\theta + a^2| \ .$$

Replacing θ by $\theta - \varphi$, we get

$$v\left(\varrho e^{i\theta}\right) = -\frac{1}{4\pi} \int_0^{2\pi} g_1\left(\varrho e^{i\theta}\right) \left[\frac{\cos^2(\theta - \varphi)}{a\sin(\theta - \varphi)} \arctan\frac{a\sin(\theta - \varphi)}{1 - a\cos(\theta - \varphi)} \right.$$
$$\left. - \frac{1}{2} - \frac{\cos(\theta - \varphi)}{2a} \ln|1 - 2a\cos(\theta - \varphi) + a^2| \right] d\theta \ .$$

Now

$$v\left(\varrho e^{i\theta}\right) = -\frac{1}{4\pi} \int_0^{2\pi} g_1\left(\varrho e^{i\theta}\right) \left[\frac{r\cos^2(\theta - \varphi)}{\varrho\sin(\theta - \varphi)} \arctan\frac{\varrho\sin(\theta - \varphi)}{r - a\cos(\theta - \varphi)} \right.$$
$$\left. - \frac{1}{2} - \frac{r\cos(\theta - \varphi)}{2\varrho} \ln\frac{r^2 - 2r\varrho\cos(\theta - \varphi) + \varrho^2}{r^2} \right] d\theta \ ,$$

since $a = \varrho/r$. Replacing $\theta - \varphi$ by α and using (4.24, 25) and the relations

$$|x - y|^2 = r^2 + \varrho^2 - 2r\varrho\cos(\theta - \varphi) , \quad \varrho = |x - x_0| ,$$

$$\frac{1}{2\pi}\int_0^{2\pi} \Delta u\left(\varrho e^{i\theta}\right) d\theta = \int_{S(x_0,r)} \Delta u(y)dS_y ,$$

we obtain the desired relation. □

It is also possible to obtain the mean value relations generalizing the relations (4.15, 16) for an arbitrary point $x \in K(x_0, r)$.

Theorem 4.4. A biharmonic function satisfies the mean value relations

$$u(x) = -\frac{(r^2 - |x - x_0|^2)^2}{2r^2\omega_2}\int_{S(x_0,r)} \frac{1}{|x - y|^2}\frac{\partial u}{\partial r}(y)dS_y$$

$$+ \frac{(r^2 - |x - x_0|^2)^2}{2r^3\omega_2}\int_{S(x_0,r)} \frac{[2r^2 - 2r|x - x_0|\cos(\alpha)]u(y)dS_y}{|x - y|^4} ,$$

where $\omega_2 = 2\pi$, α is the angle between the vectors $\overrightarrow{x_0 x}$, $\overrightarrow{x_0 y}$.
By differentiating this relation we get

$$\Delta u(x) = \frac{4}{r\omega_2}(r^2 - t^2)\int_{S(x_0,r)} \frac{1}{|x - y|^2}\frac{\partial u}{\partial r}(y)dS_y$$

$$- \frac{2(r^2 - t^2)^2}{r^2\omega_2}\int_{S(x_0,r)} \frac{1}{|x - y|^2}\frac{\partial u}{\partial r}(y)dS_y$$

$$+ K(x_1, x_1^0)\int_{S(x_0,r)} A(x_1, x_1^0, y_1)\frac{\partial u}{\partial r}dS_y$$

$$+ K(x_2, x_2^0)\int_{S(x_0,r)} A(x_2, x_2^0, y_2)\frac{\partial u}{\partial r}dS_y$$

$$- \frac{8}{r^2\omega_2}(r^2 - 2t^2)\int_{S(x_0,r)} \frac{r - t\cos(\alpha)}{r^4}u(y)dS_y$$

$$+ \frac{(r^2 - t^2)}{r^3\omega_2}\int_{S(x_0,r)} B(y)u(y)dS_y$$

$$- \frac{K(x_1, x_1^0)}{r}\int_{S(x_0,r)} A(x_1, x_1^0, y_1)dS_y$$

$$- \frac{K(x_2, x_2^0)}{r}\int_{S(x_0,r)} A(x_2, x_2^0, y_2)dS_y ,$$

where

$$t = |x - x_0| , \quad x = (x_1, x_2) , \quad x_0 = (x_1^0, x_2^0) ,$$

155

$$K(z_1, z_2) = \frac{2}{r\omega_2} (r^2 - t^2)(z_1 - z_2) \, ,$$

$$A(p, q, z) = [2(q - p)r^2 - 8(r^2 - rt\cos(\alpha))(p - z)]/|x - y|^6 \, ,$$

$$B(y) = \left[16r(r - t\cos(\alpha)) - 8r^2 \right.$$
$$\left. + 5[(x_1 - y_1)(x_1 - x_1^0) + (x_2 - y_2)(x_2 - x_2^0)] \right] |x - y|^{-6} \, .$$

We present now the mean value relations for the equation

$$\Delta\Delta u(x) + a_1 \Delta u(x) + a_2 u(x) = 0 \quad (x \in \mathbf{R}^2) \, . \tag{4.27}$$

We introduce the notation

$$\delta = (k_2^2 - k_1^2)I_0(k_1 r)I_0(k_2 r) \, ,$$
$$d_{11}(I_n) = k_1^2 I_n(k_1 r) - k_2^2 I_n(k_2 r) \, ,$$
$$d_{12}(I_n) = -(I_n(k_1 r) - I_n(k_2 r)) \, ,$$
$$d_{21}(I_n) = -[I_n(k_1 r) - I_n(k_2 r)] \, ,$$
$$d_{22}(I_N) = -\left[k_2^2 I_n(k_1 r) - k_1^2 I_n(k_2 r)\right] \, ,$$

where $k_1^2 \neq k_2^2$ are the roots of the characteristic equation $\lambda^2 + a_1\lambda + a_2 = 0$; $I_n (n \geq 0)$ is the Bessel function of order n. We put

$$c_{11}(r) = \frac{d_{11}(I_0)}{\delta} \, , \quad c_{12}(r) = \frac{d_{12}(I_0)}{\delta} \, ,$$
$$c_{21}(r) = \frac{d_{21}(I_0)}{\delta} \, , \quad c_{22}(r) = \frac{d_{22}(I_0)}{\delta} \, .$$

For any circle $K(x_0, r)$ the following statement is true [4.3].

Theorem 4.5. An arbitrary solution to (4.27) satisfies the mean value relations

$$u(x_0) = \frac{c_{11}(r)}{2\pi} \int_0^{2\pi} u\left(x_0 + re^{i\theta}\right) d\theta + \frac{c_{12}(r)}{2\pi} \int_0^{2\pi} \Delta u\left(x_0 + re^{i\theta}\right) d\theta \, ,$$

$$\Delta u(x_0) = \frac{c_{21}(r)}{2\pi} \int_0^{2\pi} u\left(x_0 + re^{i\theta}\right) d\theta + \frac{c_{22}(r)}{2\pi} \int_0^{2\pi} \Delta u\left(x_0 + re^{i\theta}\right) d\theta \, .$$

We can generalize this relation for the vector $(u(x), \partial u/\partial r)$.

Theorem 4.6. Any solution to (4.27) satisfies the relations

$$u(x_0) = \frac{b_{11}(r)}{2\pi r} \int_{S(x_0, r)} u(y)dS_y + \frac{b_{12}(r)}{2\pi r} \int_{S(x_0, r)} \frac{\partial u}{\partial r}(y)dS_y \, ,$$

$$\Delta u(x_0) = \frac{b_{21}(r)}{2\pi r} \int_{S(x_0, r)} u(y)dS_y + \frac{b_{22}(r)}{2\pi r} \int_{S(x_0, r)} \frac{\partial u}{\partial r}(y)dS_y \, , \tag{4.28}$$

where

$$b_{11}(r) = \frac{d_{11}(I_0) - d_{11}(I_2)}{\delta_1} , \quad b_{12}(r) = \frac{d_{12}(I_0)}{\delta_1} ,$$

$$b_{21}(r) = \frac{d_{21}(I_0) - d_{21}(I_2)}{\delta_1} , \quad b_{22}(r) = \frac{d_{22}(I_0)}{\delta_1} , \tag{4.29}$$

$$\delta_1 = \delta + s , \qquad\qquad s = k_1^2 I_0(k_2 r) I_2(k_1 r) - k_2^2 I_0(k_1 r) I_2(k_2 r)$$

and the quantities δ and $d_{ij}(I_n)$ are defined in Theorem 4.5.

Proof. The general solution to (4.27) at a point $x \in K(x_0, r)$ has the form [in polar coordinates $x = (r, \theta)$]:

$$u(r, \theta) = A_0 I_0(k_1 r) + B_0 I_0(k_2 r) + \sum_{n=1}^{\infty} (A_n I_n(k_1 r) + B_n I_n(k_2 r)) \sin(n\theta)$$

$$+ \sum_{n=1}^{\infty} (C_n I_n(k_1 r) + D_n I_n(k_2 r)) \cos(n\theta) ,$$

where A_n, B_n, C_n and D_n ($n \geq 0$) are arbitrary constants. By differentiation with respect to r we obtain (keeping in mind that $2I_n' = I_{n-1} + I_{n+1}$, $I_0' = I_1$)

$$\frac{\partial u}{\partial r}(r\theta) = A_0 k_1 I_1(k_1 r) + B_0 k_2 I_2(k_2 r) + \sum_{n=1}^{\infty} \left[A_n \frac{k_1}{2} \left(I_{n-1}(k_1 r) + I_{n+1}(k_1 r) \right) \right.$$

$$\left. + B_n \frac{k_2}{2} \left(I_{n-1}(k_2 r) + I_{n+1}(k_2 r) \right) \right] \sin(n\theta)$$

$$+ \sum_{n=1}^{\infty} \left[C_n \frac{k_1}{2} \left(I_{n-1}(k_1 r) + I_{n+1}(k_1 r) \right) \right.$$

$$\left. + D_n \frac{k_2}{2} \left(I_{n-1}(k_2 r) + I_{n+1}(k_2 r) \right) \right] \cos(n\theta) .$$

After integration we get

$$\frac{1}{2\pi} \int_0^{2\pi} u(r, \theta) d\theta = A_0 I_0(k_1 r) + B_0 I_0(k_2 r) ,$$

$$\frac{1}{2\pi} \int_0^{2\pi} \frac{\partial u}{\partial r}(r, \theta) d\theta = A_0 k_1, I_1(k_1 r) + B_0 k_2 I_1(k_2 r) . \tag{4.30}$$

Multiplying the last equality by $2/r$ we get [keeping in mind that $2I_1 = x(I_0 - I_2)$]:

$$\Delta u(x_0) = A_0 k_1^2 \left(I_0(k_1 r) - I_2(k_1 r) \right) + B_0 k_2^2 \left(I_0(k_2 r) - I_2(k_2 r) \right) . \tag{4.31}$$

Here we used the equality

$$\Delta u(x_0) = \frac{1}{r\pi} \int_0^{2\pi} \frac{\partial u}{\partial r}(r, \theta) d\theta .$$

Consider the equalities (4.30, 31) at the center of the circle $K(x_0, r)$. Since $I_0(0) = 1$, $I_2(0) = 0$, we get

$$u(x_0) = A_0 + B_0 ,$$
$$\Delta u(x) = k_1^2 A_0 + k_2^2 B_0 . \tag{4.32}$$

Now we find A_0 and B_0 from (4.30, 31). Let

$$S_1 = \frac{1}{2\pi} \int_0^{2\pi} u(r, \theta) d\theta , \quad S_2 = \frac{1}{r\pi} \int_0^{2\pi} \frac{\partial u}{\partial r}(r, \theta) d\theta .$$

Then

$$\delta_1 = \begin{vmatrix} I_0(k_1 r) & I_0(k_2 r) \\ k_1^2 \left(I_0(k_1 r) - I_2(k_1 r) \right) & k_2^2 \left(I_0(k_2 r) - I_2(k_2 r) \right) \end{vmatrix}$$

$$= \delta - k_2^2 I_0(k_1 r) I_2(k_2 r) + k_1^2 I_0(k_2 r) I_2(k_1 r) = \delta + s ,$$

$$A_0 = \frac{1}{\delta_1} \begin{vmatrix} S_1 & I_0(k_2 r) \\ S_2 & k_2^2 \left(I_0(k_2 r) - I_2(k_2 r) \right) \end{vmatrix}$$

$$= \left[S_1 k_2^2 \left(I_0(k_2 r) - I_2(k_2 r) \right) - S_2 I_0(k_2 r) \right] \delta_1^{-1} ,$$

$$B_0 = \frac{1}{\delta_1} \begin{vmatrix} I_0(k_1 r) & S_1 \\ k_1^2 \left(I_0(k_1 r) - I_2(k_1 r) \right) & S_2 \end{vmatrix}$$

$$= \left[S_2 I_0(k_1 r) - k_1^2 S_1 \left(I_0(k_1 r) - I_2(k_1 r) \right) \right] \delta_1^{-1} .$$

Substituting these coefficients into (4.32) we obtain (4.28). □

The relations of the type (4.20, 21, 28) could be applied in the following scheme. Inside the domain $G \backslash \Gamma_\varepsilon$ on each step of the walk on circles one uses the relations for the vector $(u(x), \Delta u(x))^{\mathsf{T}}$. In the last circle one uses the relation of the type (4.20, 21) or (4.28). We now present also the similar mean value relation of the type (4.20, 21) for (4.27).

Theorem 4.7. Any solution to (4.27) satisfies the mean value relations

$$u(x_0) = \frac{l_{11}(r)}{2\pi r} \int_{S(x_0, r)} u(y) dS_y + \frac{l_{12}(r)}{2\pi r} \int_{S(x_0, r)} \left(\Delta u(y) - \alpha \frac{\partial u}{\partial r} \right) dS_y$$

$$\Delta u(x_0) = \frac{l_{21}(r)}{2\pi r} \int_{S(x_0, r)} u(y) dS_y + \frac{l_{22}(r)}{2\pi r} \int_{S(x_0, r)} \left(\Delta u(y) - \alpha \frac{\partial u}{\partial r} \right) dS_y ,$$

where

$$l_{11}(r) = \left(1 - \frac{\alpha r}{2} \right) d_{11}(I_0) - \frac{\alpha r}{2} \frac{d_{11}(I_2)}{\delta_2} ,$$

$$l_{12}(r) = \frac{d_{12}(I_0)}{\delta_2} ,$$

$$l_{21}(r) = \left(1 - \frac{\alpha r}{2}\right) d_{21}(I_0) - \frac{\alpha r}{2} \frac{d_{21}(I_2)}{\delta_2} \,,$$

$$l_{22}(r) = \frac{d_{22}(I_0)}{\delta_2} \,,$$

$$\delta_2 = \left(1 - \frac{\alpha r}{2}\right) \delta - \frac{\alpha r}{2} S \,,$$

and α, δ, S, d_{ij} are determined above.

The proof follows immediately from the relations of Theorems 4.5 and 4.6.

4.1.3 Direct and Adjoint Algorithms for Calculating the Fields of Solution and Derivatives

In Sects. 4.1.1 and 4.1.2 we constructed ε-biased estimates for calculating the solution at a fixed point x_0. This means that we have to construct the ensemble of the walk on circles trajectories for each point x_0. However, using the homogeneity of the boundary conditions (4.2, 3), it is possible to construct an adjoint method (like in Sect. 3.2.2) for calculating the solution (and the derivatives) at a set of points simultaneously. This means that one ensemble of trajectories is used to calculate the solutions (and the derivatives) at a set of points. In addition, the adjoint method enables one to take into account the right-hand side of equations in a simple manner.

Consider the boundary value problem

$$\Delta^2 u = f \,, \quad (x \in G) \,, \quad u|_\Gamma = 0 \,, \quad Bu|_\Gamma = 0 \,, \tag{4.33}$$

where B is the corresponding boundary operator. We seek the solution to (4.33) in the form

$$u(x) = \int_G u_\delta(x, y) f(y) d\sigma(y) \,, \tag{4.34}$$

where u_δ is the Green's function

$$\Delta^2 u_\delta = \delta(x - y) \,, \quad u_\delta|_\Gamma = Bu|_\Gamma = 0 \,. \tag{4.35}$$

We put

$$u_\delta(x, y) = \mathcal{E}(x, y) + W(x, y) \,,$$

where

$$\mathcal{E}(x, y) = \frac{|y - x|^2}{8\pi} \ln|y - x|$$

is the fundamental solution of the biharmonic equation, i.e., $\Delta^2 \mathcal{E} = \delta(x - y)$ in $\mathbb{R}^2$. Then $W(x, y)$ satisfies the conditions:

$$\Delta^2 W = 0 , \quad W|_{\partial G} = -\mathcal{E}|_{\partial G} , \quad BW|_{\partial G} = -B\mathcal{E}|_{\partial G} . \qquad (4.36)$$

Thus we can use an ε-biased estimate of the type (4.18) to solve (4.34). Therefore, using the double randomization principle we can construct on the basis of (4.34) an ε-biased estimate for $u(x)$. We denote by $\pi(x)$ some appropriate probability density of the point x in the domain G [i.e., $\pi(x) \neq 0$ if $f(x) \neq 0$]. Let $\xi(x, x_0)$ be an ε-biased estimate of $W(x, y)$ at the point x_0. Denote by E_{x_0} the averaging over the trajectories $\{\xi_k\}_{k=0}^{N_\varepsilon}$ starting from x_0. We shall write $J = M_\pi F$ for

$$J = \int_G F(x)\pi(x)dx .$$

From (4.34) we get

$$u(x_0) \cong M_\pi E_{x_0} \left[\eta_0(x_0, x)|x \right] , \qquad (4.37)$$

where

$$\eta_0(x_0, x) = \frac{f(x)}{\pi(x)} \{ \mathcal{E}(x_0, x) + \xi_0(x_0, x) \} . \qquad (4.38)$$

Thus (4.37, 38) present an analog of the algorithm described in Chap. 3 (Theorem 3.4).

We now describe the adjoint scheme. Suppose that we need to calculate the solution $u(x_i)$ to (4.33) and a differential expression

$$Lu(x_i) , \quad i = 0, 1, \ldots, m .$$

The Green's function $u_\delta(x, y)$ is symmetric, therefore

$$u(x_i) \cong M_\pi E_x \left[\eta(x_i, x)|x \right] ,$$

where

$$\eta(x_i, x) = \frac{f(x)}{\pi(x)} \{ \mathcal{E}(x_i, x) + \xi_1(x_i, x) \} , \qquad (4.39)$$
$$i = 0, 1, \ldots, m ,$$

where $\xi_1(x, x_0)$ is the ε-biased estimate of $W(x_0)$. Analogously,

$$Lu(x_i) \cong M_\pi E_x[\zeta(x_i, x)|x] ,$$

where

$$\zeta(x_i, x) = \frac{f(x)}{\pi(x)} \{ L\mathcal{E}(x_i, x) + \xi_1(x_i, x) \} , \quad i = 0, 1, \ldots, m , \qquad (4.40)$$

where $\xi_1(x_i, x)$ is the ε-biased estimate of the solution to (4.36), where $\mathcal{E}$ is replaced by $L\mathcal{E}$.

Thus the scheme of numerical realization of the adjoint algorithm completely coincides with the scheme described in Sect. 3.2.2.

To conclude this section we present in Table 4.1 the differential expressions Lu of a biharmonic function u applied in the plate theory, and expressions $L\mathcal{E}$ and $L\Delta\mathcal{E}$ used in the adjoint algorithm.

Table 4.1

$Lu(x_1, x_2)$	$L\mathcal{E}(x_1, x_2, y_1, y_2)$	$L\Delta\mathcal{E}(x_1, x_2, y_1, y_2)$
$u(x_1, x_2)$	$|x-y|^2 \ln|x-y|/8\pi$	$(1 + \ln|x-y|)/2\pi$
$\Delta u(x_1, x_2)$	$(1 + \ln|x-y|)/2\pi$	0
$M_{x_i} u(x_1, x_2)$ $(i = 1, 2)$	$-D\left[(1+\nu)\dfrac{2\ln(|x-y|)+1}{8\pi} + \dfrac{(x_i - y_i) + \nu(x_2 - y_2)^2}{4\pi|x-y|^2}\right]$	$-D\left[\dfrac{1+\nu}{2\pi|x-y|^2} - \dfrac{(x_i - y_i) + \nu(x_2 - y_2)^2}{\pi|x-y|^4}\right]$
$Q_{x_i} u(x_1, x_2)$ $(i = 1, 2)$	$-\dfrac{D(x_i - y_i)^2}{2\pi|x-y|^2}$	0
$Hu(x_1, x_2)$	$-\dfrac{D(x_1 - y_1)(x_2 - y_2)}{\pi|x-y|^2}$	0

4.2 Metaharmonic Equations

In this section we treat the general case of metaharmonic equations. These equations are applied to many problems of mathematical physics, for example, in the plate and envelope theories, in the construction of optimal cubic formulas [4.4], in eigen-value problems, in transfer theory, iterative processes, etc.

In Sect. 4.1.2 we directly constructed the mean value relations for the equation

$$\Delta^2 u + a_1 \Delta u + a_2 u = 0 .$$

However, it is more convenient to split the metaharmonic operator.

4.2.1 Mean Value Theorems for Metaharmonic Equations

We introduce the metaharmonic operator in $\mathbf{R}^m$,

$$L(\Delta) = (\Delta - \lambda_n)^{p_n}(\Delta - \lambda_{n-1})^{p_{n-1}} \ldots (\Delta - \lambda_1)^{p_1} ,$$

where $p_1 \leq p_2 \leq \ldots \leq p_n$ are positive integers, $\lambda_1, \lambda_2, \ldots, \lambda_n$ are real or complex numbers; $\nu = p_1 + p_2 + \ldots + p_n$ is the order of the metaharmonic operator.

Theorem 4.8. The solution of the equation

$$(\Delta - \lambda)^p u = 0$$

satisfies the mean value relation

$$v(x_0) = N_{x_0}^r(Av) . \tag{4.41}$$

Here

$$v = \left(u, (\Delta - \lambda)u, \ldots, (\Delta - \lambda)^{p-1}u\right)^{\mathrm{T}},$$
$$A = \left(I + \beta_1 H + \ldots + \beta_{p-1} H^{p-1}\right) w_m^{-1}(r, \lambda),$$

where the entries $\beta_1, \ldots, \beta_{p-1}$ are determined from the recursion relation

$$\beta_1 = -\alpha_1, \quad \beta_2 = -\alpha_2 - \beta_1\alpha_1, \ldots, \beta_k = -\alpha_k - \alpha_{k-1}\beta_1 - \ldots - \beta_{k-1}, \quad (4.42)$$

I is the unit $p \times p$ matrix, H is a matrix whose non-zero entries have the form $h_{i,i+1} = 1$.

In (4.42) $\alpha_i = \gamma_i / w_m(r, \lambda)$, $i = 1, \ldots, p - 1$ where

$$\gamma_k = \frac{1}{k!} \frac{\partial^k w_m}{\partial \lambda^k} (r, \lambda) = \frac{r^{2k}}{k!} \frac{w_{m+2k}(r, \lambda)}{2^k m(m+2)\ldots(m+2k-2)}.$$

Proof. Expanding in (3.72) the function $W_m(ir\lambda^{1/2})$ in a power series in $(\Delta - \lambda)$, we obtain, using the relation $(\Delta - \lambda)^p u = 0$, [4.5]:

$$N_{x_0}^r(u) = \sum_{k=1}^{p} \frac{1}{(k-1)!} \frac{\partial^{k-1} w_m(r, \lambda)}{\partial \lambda^{k-1}} (\Delta - \lambda)^{k-1} u(x_0). \qquad (4.43)$$

Using relations similar to (4.43) for the functions $(\Delta - \lambda)u$, $(\Delta - \lambda)^2 u, \ldots, (\Delta - \lambda)^{p-1}u$, we obtain a system of equations which can be written in matrix form,

$$N_{x_0}^r(v) = Rv(x_0), \qquad (4.44)$$

where $R = \{r_{ij}\}_{i,j=1}^{p}$ is a matrix with diagonal entries

$$\gamma_0 = r_{ii} = w_m(r, \lambda), \quad i = 1, 2, \ldots, p$$

and non-zero entries

$$\gamma_1 = r_{i,i+1} = \frac{1}{1!} \frac{\partial^2 w_m}{\partial \lambda} (r, \lambda) + \frac{1}{1!} \frac{r^2}{2m} w_{m+2}$$

$$\gamma_2 = r_{i,i+2} = \frac{1}{1!} \frac{\partial^2 w_m}{\partial \lambda^2} (r, \lambda) = \frac{1}{2!} \frac{r^2 w_{m+4}}{2^2 m(m+2)}$$

$$\cdots \cdots \cdots \cdots$$

$$\gamma_k = r_{i,i+k} = \frac{1}{k!} \frac{\partial^k w_m}{\partial \lambda^k} (r, \lambda) = \frac{1}{k!} \frac{r^{2k} w_{m+2k}(r, \lambda)}{2^k m(m+2)\ldots(m+2k-2)},$$
$$i, k = 1, 2, \ldots, p - 1.$$

Consequently, using the specific structure of R, this matrix can be written in the form

$$R = \left(I + \alpha_1 H + \alpha_2 H^2 + \ldots + \alpha_{p-1} H^{p-1}\right) w_m,$$
$$\alpha_i = \frac{\gamma_i}{w_m} \quad i = 1, \ldots, p - 1,$$

162

where the matrix H is defined in the theorem. It is clear that the matrix $A = R^{-1}$ has the same structure, i.e.,

$$R^{-1} = \left(I + \beta_1 H + \beta_2 H^2 + \ldots + \beta_{p-1} H^{p-1}\right) w_m^{-1} .$$

From $R^{-1} R = I$ we obtain (using the condition $H^k = 0$ for $k \geq p$) the recurrent relations (4.42) for the entries $\beta_1, \ldots, \beta_{p-1}$. Hence, the matrix $A = R^{-1}$ is easily constructed explicitly, and we have from (4.44)

$$v(x_0) = N_{x_0}^r (Av)$$

which coincides with (4.44). $\qquad\qquad\qquad\qquad\qquad\qquad\qquad\qquad\qquad\qquad$ □

4.2.2 Vector Walk on Spheres Algorithm

Consider the Riqué boundary value problem for the metaharmonic equation in a domain $G \subset \mathbf{R}^m$,

$$Lu(x) = 0 , \quad x \in G \qquad\qquad\qquad\qquad\qquad\qquad (4.45)$$

$$u|_{\partial G} = \varphi_1, \ldots \Delta u|_{\partial G} = \varphi_2, \ldots, \Delta^{\nu-1} u|_{\partial G} = \varphi_\nu , \qquad\qquad (4.46)$$

where ν is the order of the metaharmonic operator L. Here and in the following discussion $v|_\Gamma = \varphi$ means that

$$v(x) \rightarrow \varphi(\xi) , \quad \text{as} \quad x \rightarrow \xi \in \Gamma (x \in G) .$$

It is not difficult to pass from (4.45, 46) to the following boundary value problem:

$$Lu(x) = 0 , \quad x \in G ,$$

$$u|_{\partial G} = \psi_1 , \quad (\Delta - \lambda_1) u|_{\partial G} = \psi_2, \ldots, (\Delta - \lambda_1)^{p_1-1} u = \psi_{p1}, \ldots$$

$$(\Delta - \lambda_n)^{p_n-1} (\Delta - \lambda_{n-1})^{p_n-1} \ldots (\Delta - \lambda_1)^{p_1} u = \psi_\nu$$

where Ψ are linearly dependent on $\varphi_1, \varphi_2, \ldots, \varphi_\nu$. We shall seek the solution of (4.45, 46) in the form

$$u = u_1 + u_2 + \ldots + u_n , \qquad\qquad\qquad\qquad\qquad\qquad (4.47)$$

where u_i satisfies the equation

$$(\Delta - \lambda_i)^{p_i} u = 0 , \quad i = 1, 2, \ldots, n . \qquad\qquad\qquad\qquad (4.48)$$

From (4.47, 48) it is obvious that $Lu = 0$. To obtain a unique representation of the solution $u(x)$ in the form of (4.47), it is necessary to derive the appropriate boundary conditions for (4.48).

We introduce the notation

$$(\Delta - \lambda_i)^{l-1} u_i = v_i^{(l)}$$

$$v_i^{(l)}|_{\partial G} = \psi_{il} , \quad i = 1, 2, \ldots, n , \quad l = 1, 2, \ldots, p_i$$

$$\qquad\qquad\qquad\qquad\qquad\qquad\qquad\qquad\qquad\qquad\qquad\qquad (4.49)$$

and consider vectors of dimension ν:

$$v = \left(v_1^{(1)}, v_1^{(2)}, \ldots, v_1^{(p_1)}, v_2^{(1)}, v_2^{(2)}, \ldots, v_2^{(p_2)}, \ldots, v_n^{(1)}, v_n^{(2)}, \ldots, v_n^{(p_n)}\right)$$

$$\psi = \left(\psi_{11}, \psi_{12}, \ldots, \psi_{1p_1}, \psi_{21}, \psi_{22}, \ldots, \psi_{2p_2}, \ldots, \psi_{n1}, \psi_{n2}, \ldots \psi_{np_n}\right)$$

$$\Psi = (\Psi_1, \ldots, \Psi_\nu) .$$

Let

$$\mu_1 = \lambda_1 , \quad \mu_2 = \lambda_1, \ldots, \mu_{p_1} = \lambda_1 ; \quad \mu_{p_1+1} = \lambda_2, \ldots$$

$$\mu_{p_1+p_2} = \lambda_2, \ldots, \quad \mu_{p_1+p_2+\ldots+p_{n-1}+1} = \lambda_n, \ldots, \quad \mu_{\nu-1} = \lambda_n .$$

We derive a linear relation between the vector v and

$$U = \left(u, (\Delta - \mu_1)u, \ldots, (\Delta - \mu_i)(\Delta - \mu_{i-1})\ldots(\Delta - \mu_1)u, \ldots, \prod_{i=1}^{\nu-1}(\Delta - \mu_i)u\right)$$

$$= (U_1, U_2, \ldots, U_\nu)$$

in the form

$$BV = U , \tag{4.50}$$

i.e., the problem is to find the matrix $B = \{b_{ij}\}$. To do this, we apply to (4.47) the operator $\prod_{i=1}^{k}(\Delta - \mu_i)$ for $k = 0, 1, \ldots, \nu - 1$ ($k = 0$ corresponds to the identity operator). Hence,

$$U_1 = v_1^{(1)} + v_2^{(1)} + \ldots + v_n^{(1)} \quad (k = 0)$$

and from (4.50),

$$U_l = b_{11}^{(l)} v_1^{(1)} + b_{12}^{(l)} v_1^{(2)} + b_{1p_1}^{(l)} v_1^{(p_1)} + b_{21}^{(l)} v_2^{(1)} + b_{22}^{(l)} v_2^{(2)} + \ldots + b_{2p_2}^{(l)} v_2^{(p_2)} + \ldots$$

$$+ b_{n1}^{(l)} v_n^{(l)} + b_{n2}^{(l)} v_n^{(2)} + \ldots + b_{np_n}^{(1)} v_n^{(p_n)}$$

$l = 1, 2, \ldots, \nu$; the l-th row of the matrix B has the form

$$\left(b_{11}^{(l)}, b_{12}^{(l)}, \ldots, b_{1p_1}^{(l)}, b_{21}^{(l)}, b_{22}^{(l)}, \ldots, b_{2p_2}^{(l)}, \ldots, b_{n1}^{(l)}, b_{n2}^{(l)}, \ldots, b_{np_n}^{(l)}\right) .$$

Applying the operator

$$(\Delta - \mu_i) = [\Delta - \lambda_i) + (\lambda_i - \mu_i)]$$

in (4.50) to the i-th row we obtain the following recurrence formula:

$$b_{i1}^{(l+1)} = (\lambda_i - \mu_l) b_{i1}^{(l)} , \quad b_{ik}^{(i+1)} = (\lambda_i - \mu_l) b_{ik}^{(l)} + b_{i,k-1}^{(l)} ,$$

$$i = 1, \ldots, n ; \quad k = 2, 3, \ldots, p_i . \tag{4.51}$$

As follows from (4.51), the matrix B is triangular with non-zero diagonal entires $b_{ii} = 1$, $i = 1, 2, \ldots, p_1$,

$$b_{ii} = \prod_{j=1}^{k-1} (\lambda_k - \lambda_j)^{p_j} , \quad p_1 + p_2 + \ldots + p_{k-1} \le i \le p_1 + p_2 + \ldots + p_k .$$

To obtain the boundary values ψ_{il}, we take the trace of (4.50) on the boundary and then take the inverse of the matrix B (this procedure is not difficult because B is triangular: $\psi = B^{-1}\Psi$). The original problem is thus reduced to a boundary value problem of the type

$$(\Delta - \lambda)^p u = 0 , \quad x \in G$$
$$u|_{\partial G} = \psi_1 , \quad (\Delta - \lambda)u|_{\partial G} = \psi_2, \ldots, (\Delta - \lambda)^{p-1}u|_{\partial G} = \psi_p \tag{4.52}$$

for which the mean value relation (4.31) was obtained in Sect. 4.2.1. The unbiased vector estimate

$$\xi(x_0) = \left(\xi_1(x_0), \ldots, \xi_p(x_0)\right)^{\mathrm{T}}$$

for the solution of (4.52) is constructed according to the scheme described in Chap. 1.

Let $\{x_n\}_{n=0}^{\infty}$ be the ε-spherical process starting from the point x_0, and let $N = N_\varepsilon$ be the random number of steps of this process inside the domain $G\backslash\partial G_\varepsilon$ until $\xi_n \in \Gamma$. Let

$$C^{(1)}(x_0, 0) = C(x_0, x_1), \ldots, C(x_0, \ldots, x_n)$$
$$= C^{(n-1)}(x_0, \ldots, x_{n-1})C(x_{n-1}, x_{n-1}) ,$$

where

$$C(x, x') = R^{-1} = w_m^{-1}(r, \lambda) \left(I + \alpha^{(1)}r^2 H + \alpha^{(2)}r^4 H^2 + \ldots \right.$$
$$\left. + \alpha^{(p-1)}r^{2(p-1)}H^{p-1}\right) ,$$

$\alpha^{(i)} = \beta_i/r^{2i}$, and $r = d(x)$ is the distance from the point x to the boundary Γ. Then the random vector variable

$$\xi(x_0) = C^{(N)}(x_0, \ldots, x_n)g(x_N)$$

where

$$g_l(x) = \begin{cases} 0 , & x \in G\backslash\partial G_\varepsilon \\ v_l(x) , & x \in \partial G_\varepsilon , \end{cases} \quad l = 1, \ldots, p$$

is an unbiased estimate for $v(x_0)$. This conclusion follows from Theorem 4.10 proved below.

Let us now introduce an ε-biased random estimate

$$\xi^{(\varepsilon)}(x_0) = C^{(N)}(x_0, x_1, \ldots, x_N)\psi(x_N^*)$$

where x_N^* is a point of Γ nearest to the point $x_N \in \Gamma$. Consider a Wiener process $w_x(t)$ in $\mathbf{R}^m$ starting at $x : w_x(0) = x$. Then $x_n = w_{x_{n-1}}(\tau_n)$, where τ_n is the first passage time of the Wiener process $w_{x_{n-1}}(t)$, i.e., the time when the process $w_{x_{n-1}}(t)$ first reaches the surface of the sphere $S(x_{n-1}, d(x_{n-1}))$. Hence,

$$\tau = \tau_{x_0} = \sum_{n=1}^{\infty} \tau_n$$

is the time at which the process $w_{x_0}(t)$ reaches the boundary ∂G for the first time.

Let $\mathcal{F}$ be a σ-algebra generated by the sequence $\{x_n\}_{n=0}^{\infty}$. Denote by $M^{\mathcal{F}} \xi$ the conditional expectation of a random variable ξ with respect to the σ-algebra $\mathcal{F}$, and denote by M the mean over all Wiener trajectories starting from point x_0.

Let

$$\alpha = \sup_{r,i} \left\{ 1, \alpha^{(i)} \right\}, \quad \|C^{(n)}\| = \max_{1 \le i \le p} \sum_{j=1}^{p} |C_{ij}^{(n)}| .$$

The proof of convergence of the Neumann series and the finiteness of variance in Theorem 4.10 follows from the inequality of Theorem 4.9.

Theorem 4.9.   For the random vector $\zeta^{(\varepsilon)}(x_0)$, the following inequality holds:

$$\|\zeta^{(\varepsilon)}(x_0)\| \le c M^{\mathcal{F}} e^{-\lambda \tau / 2} M^{\mathcal{F}} (1 + \tau)^{p-1} \sup_{x \in \partial G} \|\psi(x)\| . \tag{4.53}$$

The proof of (4.53) follows from Lemmas 4.1 and 4.2.

Lemma 4.1.   The following inequality holds:

$$\|C^{(n)}\| \le \alpha^p \frac{\left(1 - \sum_{k=1}^{n} r_k^2\right)^{p-1}}{|w_m(r_1, \lambda) \ldots w(r_m, \lambda)|} \tag{4.54}$$

where $r_k = d(x_{k-1})$, $k = 1, 2, \ldots, n$.

Lemma 4.2.   The following inequality holds:

$$|w_m^{-1}(r_1, \lambda) \ldots w_m^{-1}(r_n, \lambda)| \left(1 + \sum_{k=1}^{n} r_k^2\right)^{p-1}$$
$$\le m^{p-1} M^{\mathcal{F}} e^{-\lambda \tau / 2} M^{\mathcal{F}} (1 + \tau)^{p-1} . \tag{4.55}$$

Proof of Lemma 4.1. Let

$$w^{(n)} = \prod_{i=1}^{n} |w_m(r_i \lambda)| .$$

It is clear that

$$\|C^{(n)}\| \leq \left\| \frac{1}{w^{(n)}} \prod_{i=1}^{n} \left(I + \alpha r_i^2 H + \ldots + \alpha r_i^{2(p-1)} H^{p-1} \right) \right\|$$

$$= \left\| \frac{1}{w^{(n)}} \left(I + q^{(1)} H^2 + q^{(2)} H^2 + \ldots + q_n^{(p-1)} H^{p-1} \right) \right\| ,$$

where

$$q_n^{(l)} = \alpha^l \sum_{k_1+k_2+\ldots+k_n=l} r_1^{2k_1} r_2^{2k_2} \ldots r_n^{2k_n} , \quad l = 1, 2, \ldots, p-1 .$$

Therefore,

$$\|C^{(n)}\| = \frac{1}{w^{(n)}} \left[1 + \alpha^p \sum_{0<k_1+\ldots+k_n<p} r_1^{2k_1} \ldots r_n^{2k_n} \right]$$

$$\leq \frac{\alpha^p}{w^{(n)}} \left[1 + \sum_{0<k_1+\ldots+k_n<p} r_1^{2k_1} \ldots r_n^{2k_n} \right] \tag{4.56}$$

and hence, inequality (4.54) follows from the obvious inequality

$$1 + \sum_{0<k_1+\ldots+k_n<p} y_1^{k_1} \ldots y_n^{k_n} \leq \left(1 + \sum_{k=1}^{n} y_k \right)^{p-1}$$

for $y_i = r_i^2$, $i = 1, 2, \ldots, n$.

Proof of Lemma 4.2. This follows from the equality

$$w_m^{-1}(r, \lambda) = M^{\mathcal{F}} \exp \left(-\frac{\lambda \tau_i}{2} \right)$$

and the inequality

$$\left(1 + \sum_{k=1}^{n} r_k^2 \right)^{p-1} \leq m^{p-1} M^{\mathcal{F}} (1 + \tau)^{p-1} . \tag{4.57}$$

To prove inequality (4.57), note that the strict Markov property of the Wiener process yields

$$M^{\mathcal{F}} \left(\tau_1^{k_1} \tau_2^{k_2} \ldots \tau_n^{k_n} \right) = M^{\mathcal{F}} \tau_1^{k_1} M^{\mathcal{F}} \tau_2^{k_2} \ldots M^{\mathcal{F}} \tau_n^{k_n} . \tag{4.58}$$

Next we obtain from the well known property of the Wiener process that

$$M^{\mathcal{F}} \tau_i^{k_i} = r^{2k_i} \mu_{k_i} , \quad i = 1, 2, \ldots \tag{4.59}$$

where $\mu_{k_i} = M(\tau_i^*)$, τ_i^* is the first passage time of the Wiener process $w_0(t)$ [starting at the origin and reaching the surface $S(0,1)$ at the time τ_i^*]. Now we shall prove (4.57). Let $C_A^{(s)} \equiv C_{k_1, k_2, \ldots k_n}^{(s)}$ be the coefficients in the expansion

$$(y_1 + y_2 + \ldots + y_n)^s = \sum_A C^{(s)}_{k_1,k_2,\ldots,k_n} y_1^{k_1} \ldots y_n^{k_n} , \qquad (4.60)$$

where the set of indices is defined as $A = \{k_1 + k_2 + \ldots + k_n = s\}$. From (4.58–60) we get

$$M^{\mathcal{F}}(1 + \tau_1 + \ldots + \tau_n)^s = M^{\mathcal{F}}\left(\sum_A C^{(s)}_A \tau_1^{k_1} \tau_2^{k_2} \ldots \tau_n^{k_n}\right)$$

$$= \sum_A C^{(s)}_A \mu_{k_1} \mu_{k_2} \ldots \mu_{k_n} r_1^{2k_1} \ldots r_n^{2k_n} . \qquad (4.61)$$

From (4.61) we obtain, using the inequality

$$\mu_k = M(\tau_1^*)^k \ge (M\tau_1^*)^k = \mu_1^k = \left(\frac{1}{m}\right)^k ,$$

that

$$M^{\mathcal{F}}(1 + \tau_1 + \ldots + \tau_n)^s \ge \sum_A C^{(s)}_A \left(\frac{1}{m}\right)^{k_0+k_1+\ldots+k_n} r_1^{2k_1} \ldots r_n^{2k_n}$$

$$\ge \frac{1}{m^s}\left(1 + \sum_{k=1}^n r_k^2\right)^s .$$

From the last inequality (for $s = p - 1$) taken together with the inequality

$$\tau_{x_0} = \sum_{k=1}^\infty \tau_k \ge \sum_{k=1}^n \tau_k$$

we obtain (4.57). Lemma 4.2 and Theorem 4.9 are thus proved. $\qquad \square$

We shall now formulate the main result.

Theorem 4.10. Let λ_1 be the first eigen-value of the Laplace operator Δ. The Neumann series representing $M\zeta^{(\varepsilon)}(x_0)$ for (4.52) converges if

$$M(1 + \tau)^{t(p-1)} < \infty$$

where $\operatorname{Re}\{\lambda\} > \lambda_1$, $1/q + 1/t = 1$, $q < |\lambda_1|/|\lambda|$ and variance of $\zeta^{(\varepsilon)}(x_0)$ is uniformly (with respect to ε) bounded if

$$\operatorname{Re}\{\lambda\} > \frac{\lambda_1}{2} , \quad M(1 + \tau)^{2t(p-1)} < \infty$$

where

$$\frac{1}{q} + \frac{1}{t} = 1 , \quad q < \frac{|\lambda_1|}{2|\lambda|} .$$

Proof. Recall that $Me^{-c\tau} < \infty$ if $\operatorname{Re}\{c\} > \lambda_1/2$. Assume that $\operatorname{Re}\{\lambda\} \le 0$;

if Re $\{\lambda\} > 0$, the proof is simple. Without loss of generality of the proof, we assume that Im $\{\lambda\} = 0$. From $|\lambda| < |\lambda_1|$ it follows that there exists some $q > 1$ such that $q|\lambda| < |\lambda_1|$. Using Hölder's inequality, we then obtain

$$M\left\{M^{\mathcal{F}}e^{-\lambda\tau/2}M^{\mathcal{F}}(1+\tau)^{p-1}\right\}$$

$$\leq \left[M\left\{M^{\mathcal{F}}e^{-\lambda\tau/2}\right\}^q\right]^{1/q}\left[M\left\{M^{\mathcal{F}}(1+\tau)^{p-1}\right\}^t\right]^{1/t}$$

where $q^{-1} + t^{-1} = 1$. Now

$$M\left\{M^{\mathcal{F}}e^{-\lambda\tau/2}\right\}^q \leq MM^{\mathcal{F}}e^{-\lambda\tau/2} = Me^{-\lambda\tau/2} ,$$

$$M\left\{M^{\mathcal{F}}(1+\tau)^{p-1}\right\}^t \leq MM^{\mathcal{F}}(1+\tau)^{p-1} = M(1+\tau)^{t(p-1)} .$$

Since $q|\lambda| < |\lambda_1|$, we conclude that $Me^{-\lambda\tau/2} < \infty$ and the convergence of the Neumann series is proved. The finiteness of the variance can be proved in a similar way since the finiteness of the second moment of $\|\zeta^{(\varepsilon)}\|$ follows from (4.53) provided that Re $\{\lambda\} > \lambda_1/2$. □

Remark. The condition $M(1+\tau)^{t(p-1)} < \infty$ holds, for example, for arbitrary bounded domains. It also holds for domains bounded in one direction, i.e., when the domain lies between two parallel planes. This condition is unnecessary if Re $\{\lambda\} > 0$. Hence, Theorem 4.10 holds in this case for arbitrary domains.

4.2.3 Scalar Algorithms

As we mentioned in Sect. 4.2.2, (4.45, 46) can be reduced to

$$\begin{aligned}(\Delta - \lambda_0)^p u(x) &= 0 , \quad x \in G , \\ (\Delta - \lambda_0)^k u|_{\partial G} &= \varphi_k , \quad \lambda_0 > \lambda_1 ,\end{aligned} \tag{4.62}$$

where $k = 0, 1, \ldots, p - 1$; $(\Delta - \lambda_0)^0 u(x) \equiv u$; λ_1 is the first eigen-value of the Laplace equation. We now show that the solution to (4.62) can be obtained by the usual scalar algorithm for

$$\Delta v(x, \lambda) - \lambda v = 0 , \quad x \in G ; \quad v|_\Gamma = \varphi . \tag{4.63}$$

Indeed, let us consider an auxiliary boundary value problem

$$\begin{aligned}\Delta v(x, \lambda) - \lambda v &= 0 , \quad x \in G , \\ v|_\Gamma &= \varphi_{p-1} + (\lambda - \lambda_0)\varphi_{p-2} + \ldots + (\lambda - \lambda_0)^{p-1}\varphi_0 ,\end{aligned} \tag{4.64}$$

where λ is a complex parameter, such that Re $\{\lambda\} > \lambda_1$. We seek the solution to (4.64) in the form of a power series

$$v(x, \lambda) = \sum_{j=0}^{\infty} (\lambda - \lambda_0)^j v_j(x) \,, \qquad\qquad (4.65)$$

since $v(x, \lambda)$ is analytic for $\lambda : \mathrm{Re}\, \{\lambda\} > \lambda_1$. Substituting this representation into (4.64) yields

$$\Delta v_j - \lambda_0 v_j = v_{j-1} \,, \quad j = 0, 1, \dots \,.$$

Consequently,

$$(\Delta - \lambda_0)^p v_{p-1}(x) = 0 \,, \quad x \in G \,.$$

In addition,

$$
\begin{aligned}
&v_{p-1}\big|_{\Gamma} = \varphi_0 \,, \\
&(\Delta - \lambda_0) v_{p-1} = v_{p-2}(x) \xrightarrow[x \to \xi]{} \varphi_1(\xi) \\
&\cdots\cdots\cdots\cdots \\
&(\Delta - \lambda_0)^p v_{p-1} = v_0(x) \xrightarrow[x \to \xi]{} \varphi_{p-1}(\xi) \,.
\end{aligned}
\qquad\qquad (4.66)
$$

Thus, $v_{p-1}(x)$ solves (4.62). From (4.65) we get

$$u(x) = v_{p-1}(x) = \frac{1}{2\pi i} \int_{C_{\lambda_0}} \frac{v(x, \lambda)d\lambda}{(\lambda - \lambda_0)^p} \qquad\qquad (4.67)$$

where C_{λ_0} is a circle $S(\lambda_0, R)$ on a complex plane such that $R < |\lambda_1 + \lambda_0|$.

Using (4.67), it is possible to construct an ε-biased estimate of $u(x)$ on the trajectory $\{\xi_k\}_{k=0}^{N_\varepsilon}$ of the ε-spherical process starting from the point x:

$$\eta(x) = \frac{1}{2\pi i} \sum_{l=1}^{L} \mu_l \frac{\zeta_\varepsilon(x, \lambda_l)}{(\lambda_l - \lambda_0)^p} \qquad\qquad (4.68)$$

where $\{\lambda_l\}_{l=1}^{L} \in C_{\lambda_0}$ are the nodes, and $\{\mu_l\}_{l=1}^{L}$ are the weights of a cubic formula; $\zeta_\varepsilon(x, \lambda_l)$ is an ε-biased estimate for (4.64) constructed on one trajectory $\{\xi_k\}_{k=0}^{N_\varepsilon}$ simultaneously for all $\{\lambda_l\}_{l=1}^{L}$. It is clear that in (4.68) the random estimate $\zeta_\varepsilon(x, \lambda_l)$ can be constructed on the trajectories of the walk on boundary process. This permits one to calculate the solution to (4.62) simultaneously at a set of points.

4.3 Spatial Problems of the Elasticity Theory

In this section we consider the Lamé equation. It should be noted that the direct generalization of the walk on spheres algorithm to the Dirichlet problem for Lamé's equation is impossible, because the corresponding Neumann series for the local integral equation diverges. (Note that in [4.6, 7] incorrect schemes were proposed.) We overcome this difficulty by introducing some regularization procedure. The difficulties in constructing the walk on boundary algorithms lie in the necessity to evaluate the principal-value integrals.

4.3.1 Walk on Boundary Algorithms for the Lamé Equation

Consider the interior Dirchlet problem for the Lamé system of equations

$$\mu \Delta u(x) + (\lambda + \mu) \text{grad div } u(x) = 0, \quad x \in \Gamma \subset \mathbf{R}^3 , \tag{4.69}$$

$$u|_\Gamma = \varphi ,$$

where λ, μ are the Lamé constants of elasticity, $u = (u_1(x), u_2(x), u_3(x))^{\mathrm{T}}$, $\varphi(y) = (\varphi_1(y), \varphi_2(y), \varphi_3(y))^{\mathrm{T}}$. As in the problems of potential theory (Chap. 3), we assume that ∂G consists of a finite number of parts which are Ljapunov surfaces, and the boundary points are considered as $\Gamma = \partial G \backslash \Gamma_0$ where $\Gamma_0 \subset \partial G$ is a set of zero surface measure σ consisting of corner points of ∂G and the points of discontinuity of the function φ.

Let

$$U(x) = \int_\Gamma P[x,y]\Phi(y)d\sigma(y) \tag{4.70}$$

be the double layer vector potential [4.8], where Φ is an unknown vector function from the Hölder functional space $C^{0,\alpha}(\Gamma)$, $\alpha > 0$,

$$P[x,y] = K[x,y] , \quad x \in G ,$$

$$K[x,y] = \{k_{ij}\} , \quad i,j = 1,2,3 ;$$

$$k_{ij}(x,y) = \frac{\delta_{ij}}{2\pi} \frac{\partial}{\partial n(y)} \frac{1}{|x-y|}$$

$$+ \mu \sum_{k=1}^{3} M_{jk} \left[(\lambda' - \mu') \frac{\delta_{ki}}{|y-x|} + 2\mu \frac{(y_k - x_k)(y_i - x_i)}{|y-x|^3} \right] ,$$

$$\lambda' = (\lambda + 3\mu)[4\pi\mu(\lambda + 2\mu)]^{-1} , \quad \mu' = (\lambda + \mu)[4\pi\mu(\lambda + 2\mu)]^{-1} ,$$

$$M_{jk} = n_k(y) \frac{\partial}{\partial y_j} - n_j(y) \frac{\partial}{\partial y_k} , \quad k,j = 1,2,3 .$$

After some transformations we get

$$k_{ii} = \frac{1}{2\pi} \frac{\cos(\varphi_{yx})}{|y-x|^2} \frac{1}{\lambda + 2\mu} \left[\mu + 3(\lambda + \mu) \frac{(y_i - x_i)^2}{|y-x|^2} \right] ,$$

$$k_{ij} = \frac{3}{2\pi} \frac{\cos(\varphi_{yx})}{|y-x|^4} \frac{\lambda + \mu}{\lambda + 2\mu} (y_i - x_i)(y_j - x_j) \tag{4.71}$$

$$- \frac{\mu}{\pi(\lambda + 2\mu)} \left[n_i(y) \frac{y_j - x_j}{|y-x|^3} - n_j(y) \frac{y_i - x_i}{|y-x|^3} \right] , \quad i \neq j .$$

It is known [4.8] that the potential (4.69) solves the problem (4.69) if Φ satisfies the vector singular boundary integral equation

$$\Phi(y) = -\kappa \int_{\partial G} K[y, y']\Phi(y')d\sigma(y') + \varphi(y) , \quad (\kappa = 1) , \qquad (4.72)$$

or in the operator form

$$\Phi = -\kappa K[\,]\Phi + \varphi .$$

It is also known that the spectral properties of (4.72) coincide with those of the boundary integral equation of the classical potential theory [4.8]. Although the solution to (4.69) is represented as a linear functional (4.70) of the solution of (4.72), the standard estimate of the Monte Carlo method is not applicable here since (a) $\lambda = -1$ is the eigen-value of (4.72), and (b) it is necessary to evaluate the principal-value integrals.

The latter difficulty is discussed in detail in Chaps. 2, 3 so we can use an appropriate analytical continuation; for example, using the conformal mapping $\lambda = 2\eta/(1 - \eta)$ we obtain

$$\Phi(y) = \varphi(y) + \sum_{i=0}^{\infty} \frac{2}{3} \left[\frac{I + 2K[\,]}{3} \right]^i K[\,]\varphi(y) .$$

Generally, we shall have

$$u(x) = \sum_{i=0}^{n} l_i^{(n)} \int_{\Gamma} P[x, y]K^i[\,]\varphi(y)d\sigma(y) + \varepsilon(n) , \qquad (4.73)$$

where the coefficients are determined by the method used (Chap. 2).

Now let us turn to the difficulty (b). The kernel includes singular components

$$k_{ij}(y, y') = k_{ij}^{(r)}(y, y') + k_{ij}^{(s)}(y, y') ,$$

where $k_{ij}^{(s)}(y, y') = 0(|y - y'|^{-2})$ as $y' \to y$ on the boundary, $i \neq j$. Therefore, any standard non-biased estimate ξ of the type (1.16) for $(P, K[\,]\varphi)$ will have infinite moments $M|\xi|^m = \infty$, $m \geq 1$, since the increase of the density $p(y, y')$, as $y' \to y$, cannot be faster than $0(|y - y'|^{-2+\varepsilon})$, $\varepsilon > 0$. Here

$$k_{ij}^{(r)}(y, y') = 0(|y - y'|^{-2}) \quad \text{as} \quad y' \to y \quad \text{on} \quad \partial G .$$

Now we construct an estimate with finite variance. Consider the singular part of the kernel. Let

$$\varphi(y) = \int_{\Gamma} k(y, y')\varphi(y')d\sigma(y') + f(y) \qquad (4.74)$$

be a 3D singular equation where Γ is a Ljapunov surface with exponent γ. The kernel of (4.74) behaves like

$$k(y,y') = 0(|y - y'|^{-2}) \quad \text{as} \quad y' \to y \,,$$

therefore the integral in (4.74) is defined as follows:

$$\int_\Gamma k(y,y')\varphi(y')d\sigma(y') = \lim_{\varepsilon \to 0} \int_{\Gamma(\varepsilon)} k(y,y')\varphi(y')d\sigma(y') \,, \tag{4.75}$$

where

$$\Gamma^{(\varepsilon)} = \Gamma \backslash \{y' : |y - y'| \le \varepsilon\} \,.$$

An integral of the type of (4.75) exists iff [4.8]

$$\int_S r(y,\theta)dS = 0 \,,$$

where

$$r(y,\theta = k(y,y')|y - y'|^2 \,, \quad \theta = \frac{\pi(y') - y}{|\pi(y') - y|} \,.$$

$S = S(y,1)$ is a unit circle lying in the plane tangent to the boundary at $y \in \Gamma$. Here $\pi(y')$ is the projection of y' on the tangent plane.

This condition can be used to construct an effective Monte Carlo estimate for calculating principal-value integrals. It should be noted that a Monte Carlo method for evaluating principal-value integrals has been proposed which is applicable if it is possible to find domains where the integrand preserves its sign [4.9]. In our case there is no such information, but we can utilize the symmetry properties of the kernel $k(y,y')$.

We now investigate the behavior of the regular and singular parts of the kernel $k_{ij}(y,y')$ as $d \equiv |y - y'| \to 0$. To this end we define a symmetric transformation $\alpha_y(y')$ of a point $y' \in \Gamma$ as follows:

$$\alpha_y(y') \in \Gamma : |y - \alpha_y(y')| = |y - y'| \,,$$

where the angle between the vector $\pi(\alpha_y(y') - y)$ and an arbitrary vector τ lying in the tangent plane differs from the angle between τ and $\pi(y' - y)$ exactly by the angle π. We introduce a spherical coordinate system whose origin coincides with the point y and the axis z goes through $n(y)$. Then

$$y' = (d,\theta,\varphi) \,, \quad \alpha_y(y') = (d,\bar\theta,\varphi + \pi) \,,$$

where $\bar\theta = \theta(1 + 0(1))$.

Let us also define a transformation based on the symmetry:

$$\chi_y(y') = (\bar d,\theta,\varphi + \pi) \in \Gamma \,,$$

where $\bar d = d(1 + 0(1))$. Note that all the arguments considered below for α_y remain valid for χ_y.

Any boundary point is either an elliptic point or an interior point of the planar part of the boundary. In the first case

$$k_{ij}^{(r)}(y, y') = \frac{3\cos(\varphi_{yy'})}{2\pi|y - y'|^4} \frac{\lambda + \mu}{\lambda + 2\mu} (y_i' - y_i)(y_j' - x_j)$$

$$= k_{ij}^{(r)}(y, \alpha_y(y')) (1 + 0(d^\gamma)) = 0 \left(d^{-2+\gamma}\right) , .$$

$$k_{ij}^{(s)}(y, y') = -\frac{\mu}{\pi(\lambda + 2\mu)} \left[n_i(y') \frac{y_j' - y_j}{|y - y'|^3} - n_j(y') \frac{y_i' - y_i}{|y - y'|^3} \right]$$

$$= -k_{ij}^{(s)}(y, \alpha_y(y')) (1 + 0(d^\gamma)) .$$

Consequently,

$$k_{ij}(y, \alpha_y(y')) = -k_{ij}(y, y') (1 + 0(d^\gamma)) . \tag{4.76}$$

Recall that γ is the Ljapunov exponent of the boundary Γ. If y, y' and $\alpha_y(y')$ all lie on a plane, then

$$k_{ij}(y, \alpha_y(y')) = -k_{ij}(y, y') (1 + 0(d^\gamma)) .$$

Now, fix a point $y \in \Gamma$ and construct a Monte Carlo estimate for $K[\,]\varphi(y)$. First, let y be an elliptic point. Then the density of distribution of the point y' over the solid angle, i.e., with respect to the measure

$$d\Omega(y') = \frac{\cos(\varphi_{yy'})}{|y - y'|^2} d\sigma(y')$$

is taken as

$$\bar{p}^{(\beta)}(\theta, \varphi) = c[\sin(\theta)]^{-\beta} , \quad \beta \geq 0 .$$

This means that the angles determining the direction $\overrightarrow{y, y'}$ are sampled as follows: the angle φ is sampled according to uniform distribution on $(0, 2\pi)$, and the angle θ is sampled from the density $c[\sin(\theta)]^{-\beta}$. In particular, if $\beta = 0$, the last distribution is also uniform, i.e.,

$$p^{(\beta)}(y, y') = \bar{p}^{(\beta)} \frac{d\Omega}{d\sigma} (y, y')$$

coincides in this case with the density $p(y, y')$ of the isotropic walk on boundary process [see (1.49)]. Here

$$\frac{d\Omega}{d\sigma}(y, y') = \frac{\cos(\varphi_{yy'})}{|y - y'|^2} .$$

Hence,

$$p^{(\beta)}(y, y') = 0(|y - y'|^{2+\gamma-\beta})$$

as $y' \to y$, and we require that $\beta < \gamma$. The constant c is chosen from the condition

$$\int_{S_y} p^{(\beta)}(y, y')d\sigma(y') = 1 ,$$

where S_y is a neighborhood of the point y.

Let $p_1(y, y')$ be a probability density defined on $\Gamma \backslash S_y$. Then the distribution density of the point y' on Γ is defined as

$$p(y, y') = \omega p^{(\beta)}(y, y')I_S(y') + (1 - \omega)p_1(y, y')I_{\Gamma \backslash S}(y') \, ,$$

where I_A is the indicator function of a set A. According to this density, the point y' is sampled with probability ω in S using the density $p^{(\beta)}$, and with probability $1 - \omega$ this point is sampled in $\Gamma \backslash S$ according to p_1.

Now consider the case when y belongs to a planar part of the boundary, i.e., there exists an ε_0 such that the circle $C(y, \varepsilon_0) = \{y \in \Gamma : |y - y'| < \varepsilon_0\}$ also belongs to this plane. In this case the point y' is sampled in $C(y, \varepsilon_0)$ according to the density $p^{(\beta)}(y, y') = c|y - y'|^{-1-\beta}$ with probability ω (i.e., the angle φ is uniformly distributed), and it is sampled in $\Gamma \backslash S$ with probability $1 - \omega$ according to the density p_1. Let $y_0 = y$, $y_{11} = y'$, $y_{12} = \alpha_y(y')$ in S [or $y_{12} = \chi_y(y')$]. The point y_{12} is defined by the density p_1 in $\Gamma \backslash S$. Consider the following estimate:

$$\zeta_1^*(y_0) = -\frac{1}{2} \left\{ \frac{K[y_0, y_{11}]\varphi(y_{11})}{p^{(\beta)}(y_0, y_{11})} + \frac{K[y_0, y_{12}]\varphi(y_{12})}{p_1^{(\beta)}(y_0, y_{12})} \right\} \, . \tag{4.77}$$

Here $p_1^{(\beta)}(y_0, y_{12})$ is the density of the point y_{12}; it equals $p^{(\beta)}(y_0, y_{12})$ if $y_{12} = \chi_y(y')$. If $y_{12} = \alpha_y(y')$ then it is equal to $p^{(\beta)}(y_0, y_{12}) = 0(1 + 0(1))$ in S (in the general case; in the case of a sphere or a plane the term $0(1)$ can be omitted). Note that in any case

$$p_1^{(\beta)}(y_0, y_{12}) = p^{(\beta)}(y_0, y_{12})(1 + 0(1)) \quad \text{as} \quad y_{11} \to y_0 \, .$$

By the construction, obviously,

$$M\zeta_1^*(y_0) = K[\,]\varphi(y_0) \, .$$

The variance of the estimate (4.77) is finite. Indeed, it is sufficient to consider the singular part of a non-diagonal element of the matrix $K[y_0, y_{11}]$ and show that the estimate

$$\xi(y_0) = \frac{1}{2} \left\{ \frac{k_{ij}^{(s)}(y_0, y_{11})\varphi(y_{11})}{p^{(\beta)}(y_0, y_{11})} + \frac{k_{ij}^{(s)}(y_0, y_{12})\varphi(y_{12})}{p^{(\beta)}(y_0, y_{12})} \right\}$$

for the integral

$$\int_\Gamma k_{ij}^{(s)}(y_0, y_1)\varphi_j(y_1)d\sigma(y_1)$$

has a finite second moment.

Assume that the function φ belongs to the Hölder class $C^{0,\delta}$. From (4.76) we get

$$\zeta = \frac{1}{4} \frac{[k_{ij}^{(s)}(y_0, y_{11})]^2}{p^{(\beta)}(y_0, y_{11})} \{\varphi_j(y_{11}) - \varphi_j(y_{12})(1 + 0(d^{\gamma}))\}^2 (1 + 0(1))$$
$$= 0 \left(d^{-3+\beta+2\min(\gamma,\delta)} \right) .$$

If there are no plane parts of Γ, then

$$\zeta = 0 \left(d^{-2-\gamma+\beta+2\min(\gamma,\delta)} \right) .$$

Thus

$$M\zeta^2 = \int_{\Gamma} \zeta(y_0, y_{11}) d\sigma(y_{11}) < \infty$$

if $\beta > 1 - 2\min(\gamma, \delta)$. If there are no plane parts, then $M\zeta^2 < \infty$ if $\beta > \gamma - 2\min(\gamma, \delta)$. In the last case the density $p^{(\beta)}$ can be taken as the density of the isotropic walk on boundary process ($\beta = 0$) provided that $\delta \geq \gamma/2$, which is certainly possible if $\delta \geq 1/2$. If, besides, Γ is convex, then the standard density (1.49) can be used for all boundary points.

Thus, assume that y_0 is sampled from the density p_0 [see (1.48)]. Then according to $p^{(\beta)}(y_0, y_{11})$ the point y_{11} is sampled. Symmetric transformation yields $y_{12} = \alpha_{y_0}(y_{11})$ or $y_{12} = \chi_{y_0}(y_{11})$. Then

$$M\xi_i^* = (P, K^i[]\varphi) ,$$

where

$$\xi_i^* = \frac{P[x, y_0]}{p_0(y_0)} \zeta_i^*(y_0)$$

has a finite variance. Here

$$\zeta_i^*(y_0) = -\frac{1}{2} \left\{ \frac{K[y_0, y_{11}]}{p^{(\beta)}(y_0, y_{11})} \zeta_{i-1}^*(y_{11}) + \frac{K[y_0, y_{11}]}{p^{(\beta)}(y_0, y_{11})} \zeta_{i-1}^*(y_{12}) \right\} \qquad (4.78)$$

and ζ_i^* is defined by (4.77). Substituting ξ_i^* into (4.73) we obtain an $\varepsilon(n)$-biased estimate for $u(x)$. The decrease of $\varepsilon(n)$, as n increases, behaves like the error in the scalar case described in Chap. 3.

4.3.2 Walk on Spheres Algorithm for the First Boundary Value Problem

Consider the boundary value problem (4.69) for a bounded domain $G \subset \mathbb{R}^n$. Let $B(x, r)$ and $S(x, r) = \partial B(x, r)$ be a ball and a sphere, respectively, centered at x with radius r such that $B(x, r) \subseteq G$. Then the following mean value relation holds [4.10]:

$$u_i(x) = \frac{n}{2(n + \alpha)\sigma_n} \int_{S(0,1)} [(2 - \alpha)\delta_{ij} + (n + 2)\alpha w_i w_j] u_j d\Omega , \qquad (4.79)$$

$i, j = 1, \ldots, n$. We rewrite it in vector form:

$$u = \int_{S(0,1)} Au\, d\Omega \; .$$

Here $w_i = (x_i - y_i)/r$ is the direction cosine of the vector $w \in S(0,1)$, δ_{ij} is the Kronecker symbol, $\alpha = 1 + \lambda/\mu$. The integral of the vector in (4.79) is understood as a vector whose components are integrals of the corresponding components of the vector Au. Using the mean value relation (4.79) in the matrix-integral form (Sects. 1.1.1, 2)

$$u(x) = K[\,]u + \tilde{\varphi} \tag{4.80}$$

where $K[\,]$ is the corresponding matrix-integral operator with a generalized kernel of type (1.25) and the function $\tilde{\varphi}$ is defined according to the general scheme described in Sect. 1.1.2 [see (1.26)].

Note now that the Neumann series for (4.80) diverges. Hence it is not possible to use direct generalization of the walk on spheres algorithm in this case. We suggest here a regularization procedure utilizing the fact that the properties of the spectrum of (4.69) are completely analogous to those of the Laplace operator. Consequently, here we can apply the approach described in Sect. 3.1.4. Indeed, consider the problem

$$\Delta^* u(x, \nu) - \nu u(x, \nu) = 0 \; , \quad x \in G \; , \quad u|_\Gamma = \varphi \; , \tag{4.81}$$

where ν is a complex parameter; the operator Δ^* is defined by

$$\Delta^* u = \mu \Delta u + (\lambda + \mu)\mathrm{grad\, div}\, u \; .$$

Now we obtain for (4.81) a mean value relation of type (4.79) where, obviously, the entries of the matrix A must depend on ν : $A = A_\nu$.

Let η be an arbitrary measure [an averaging measure over a sphere $S(x, r)$] and let N_η be the corresponding averaging operator.

$$N_\eta u(x) = \int_{S(x,r)} u(x + re)d\eta(e) \; .$$

Note that N_η may appear as a vector operator. We introduce a measure

$$d\eta_1(s) = \left\{ \frac{1}{\omega_n} \left[a\delta_{ij} + bs_i s_j \right] d\Omega(s) \right\}_{i,j=1}^n \; , \tag{4.82}$$

where a and b are some constants, s_i is the i-th direction cosine of the vector $s \in S(0,1)$. The corresponding operator averaging the vector function $u(x) = (u_1(x), \ldots, u_n(x))^\mathrm{T}$ over the sphere $S(x, r)$ with respect to measure (4.82) will be denoted by $N^1 u(x)$:

$$N^1 u(x) = \int_{S(x,r)} u(x + rs)d\eta_1(s) \tag{4.83}$$

177

where the integral of the vector $u(x + rs)$ is understood to be a vector whose components are integrals of the corresponding elements. The integrand in (4.83) is regarded as a product of the matrix $d\eta_1(s)$ by the vector $u(x + rs)$.

Now we present some auxiliary results which will be used in obtaining the main result, namely, Theorem 4.16 on the mean value relation for (4.81) [4.11].

Theorem 4.11. For real vector functions $u(x) = (u_1(x), \ldots, u_n(x))^T$, $x \in G \subset R^n$, analytic in G, the following expansion

$$N^1 u(x) - u(x) = \left(a - 1 + \frac{b}{n}\right) u(x) + \sum_{k=1}^{\infty} \frac{r^{2k}(2k-1)!!}{(2k)!n(n+2)\ldots(n+2k)}$$

$$\times \left\{(n+2k)a\Delta^k u(x) + b\Delta^k u(x) + 2kb\Delta^{k-1}\text{grad div } u(x)\right\} \tag{4.84}$$

holds if the series on the right-hand side of (4.84) converges.

Proof. For simplicity, consider the case $n = 3$. The proof for arbitrary dimensions is analogous.

Consider the first component of the vector on the left-hand side of (4.84):

$$\frac{1}{4\pi} \int \left(a + bs_1^2\right) u_1(x + rs) d\Omega(s) - u_1(x)$$

$$+ \frac{b}{4\pi} \int s_1 s_2 u_2(x + rs) d\Omega(s) + \frac{b}{4\pi} \int s_1 s_3 u_3(x + rs) d\Omega(s) . \tag{4.85}$$

Obviously,

$$u_1(x + rs) = u_1(x) + r \sum_{k=1}^{3} \frac{\partial u_1(x)}{\partial x_k} s_k + \frac{r^2}{2!} \sum_{j,k=1}^{3} \frac{\partial^2 u_1(x)}{\partial x_k \partial x_j} s_k s_j + \ldots . \tag{4.86}$$

Consider the k-th sum, i.e., the factor of the term $r^{2k}/(2k)!$ (all odd terms are, obviously, equal to zero):

$$\sum_{A_k^\alpha} C(2\alpha_1, 2\alpha_2, 2\alpha_3) \frac{\partial^{2k}}{\partial x_1^{2\alpha_1} \partial x_2^{2\alpha_2} \partial x_3^{2\alpha_3}} \int s_1^{2\alpha_1} s_2^{2\alpha_2} s_3^{2\alpha_3} \frac{d\Omega(s)}{4\pi} , \tag{4.87}$$

where the set of indices is defined as $A_k^\alpha = \{\alpha : \alpha_1 + \alpha_2 + \alpha_3 = k\}$. Taking into account the relation

$$\int s_1^{2\alpha_1} s_2^{2\alpha_2} s_3^{2\alpha_3} d\Omega(s) = 4\pi \frac{(2\alpha_1 - 1)!!(2\alpha_2 - 1)!!(2\alpha_3 - 1)!!}{[2(\alpha_1 + \alpha_2 + \alpha_3) + 1]!!} , \tag{4.88}$$

$$\Delta^k = \sum_{A_k^\alpha} C(\alpha_1, \alpha_2, \alpha_3) \frac{\partial^{2k}}{\partial x_1^{\alpha_1} \partial x_2^{2\alpha_2} \partial x_3^{2\alpha_3}} , \qquad \alpha_i = 0, 1, \ldots ; \quad i = 1, 2, 3,$$

we obtain

$$\frac{1}{4\pi} \int \left(a + bs_1^2\right) u_1(x + rs) d\Omega(s) - u_1(x) = \sum_{k=1}^{\infty} \frac{r^{2k}}{(2k + 1)!(2k + 3)}$$

$$\times \left\{ a(2k + 3)\Delta^k u_1(x) + b\Delta^k u_1(x) + 2kb\Delta^{k-1} \frac{\partial^2 u_1(x)}{\partial x_1^2} \right\}$$

$$+ \left(a - 1 + \frac{b}{3}\right) u_1(x) . \tag{4.89}$$

Similar expansions for the second and third terms in (4.85) can also be derived on the basis of (4.86–89):

$$\frac{1}{4\pi} \int bs_1 s_2 u_2(x + rs) d\Omega(s) = b \sum_{k=1}^{\infty} M_k'(r)$$

$$\times \left\{ 2k \frac{\partial^2}{\partial x_1 \partial x_2} \Delta^{k-1} u_2(x) \right\} ,$$

$$\frac{1}{4\pi} \int bs_1 s_3 u_3(x + rs) d\Omega(s) = b \sum_{k=1}^{\infty} M_k(r) \tag{4.90}$$

$$\times \left\{ 2k \frac{\partial^2}{\partial x_1 \partial x_3} \Delta^{k-1} u_3(x) \right\} ,$$

where $M_k(r) = r^{2k}/(2k+1)!(2k+3)$. It follows from (4.89, 90) that the expression (4.85) transforms to

$$\left(a - 1 + \frac{b}{3}\right) u_1(x) + \sum_{k=1}^{\infty} \frac{r^{2k}}{(2k + 1)!(2k + 3)}$$

$$\times \left\{ a(2k + 3)\Delta^k u_1(x) + b\Delta^k u_1(x) + 2kb\Delta^{k-1} \frac{\partial}{\partial x_1} \operatorname{div} u(x) \right\} . \tag{4.91}$$

Repeating calculations (4.85–91) for the second and third components of the vector on the left-hand side of (4.84), we obtain

$$N^1 u(x) - u(x) = \sum_{k=1}^{\infty} \frac{r^{2k}}{(2k + 1)!(2k + 3)} \left\{ a(2k + 3)\Delta^k u(x) \right.$$

$$+ b\Delta^k u(x) + 2kb\Delta^{k-1} \operatorname{grad} \operatorname{div} u(x)$$

$$\left. + \left(a - 1 + \frac{b}{3}\right) u(x) \right\} .$$

$\square$

Using (4.84), we obtain an expansion of the solution to the equation (4.81).

Theorem 4.12. For the solution to equation (4.81) the following expansion holds:

$$N^1 u(x) - u(x) = \sum_{k=1}^{\infty} \frac{r^{2k}(2k-1)!!}{(2k)!\,n(n+2)\ldots(n+2k)}$$
$$\times \left\{ \frac{kn(n+2)\nu\Delta^{k-1}u(x)}{\lambda + \mu(n+1)} - n(k-1)\Delta^k u(x) \right\} \qquad (4.92)$$

Proof. Assume that $\beta = (n+2)(\lambda+\mu)/2[\lambda+\mu(n+1)]$. Setting $a = 1 - \beta$ and $b = n\beta$ in (4.84), we obtain (4.92). $\qquad \square$

To prove the next theorem, we use the following properties of the solution to (4.92):

$$\Delta^k \mathrm{div}\, u(x) = \left(\frac{\nu}{\lambda + 2\mu} \right)^k \mathrm{div}\, u(x) \qquad (4.93)$$

$$\nu^k u = \Delta^{*k} u = \mu^k \Delta^k u + \left(\frac{\nu}{\lambda + 2\mu} \right)^{k-1} \frac{\nu u - \mu \Delta u}{\lambda + \mu} \left\{ (\lambda + 2\mu)^k - \mu^k \right\} \quad (4.94)$$

which are directly implied by (4.81).

Theorem 4.13. The solution to (4.81) satisfies the mean value relation

$$u(x) = N^1 u(x) - \frac{(W_1 - 1)\mu}{(\lambda + \mu)\nu} \left\{ (\lambda + 2\mu)\Delta u - \nu u \right\}$$
$$+ \frac{(W_2 - 1)(\lambda + 2\mu)}{(\lambda + \mu)\nu} \left\{ \mu \Delta u - \nu u \right\}, \qquad (4.95)$$

where

$$W_1 = W_1(\lambda, \mu, \nu, r) = (1 - \beta) \frac{\sinh(rk_1)}{rk_1} + 3\beta \left[\frac{\cosh(rk_1)}{r^2 k_1^2} - \frac{\sinh(rk_1)}{r^3 k_1^3} \right],$$

$$W_2 = W_2(\lambda, \mu, \nu, r) = (1 + 2\beta) \frac{\sinh(rk_2)}{rk_2} - 6\beta \left[\frac{\cosh(rk_2)}{r^2 k_2^2} - \frac{\sinh(rk_2)}{r^3 k_2^3} \right],$$

$$k_1 = \left[\frac{\nu}{\mu} \right]^{1/2}, \qquad k_2 = \left[\frac{\nu}{\lambda + 2\mu} \right]^{1/2}.$$

Proof. Using expansion (4.92) of Theorem 4.12 and the properties (4.93, 94) (in $\mathbf{R}^3$), we obtain

$$N^1 u(x) - u(x) = \sum_{k=1}^{\infty} \frac{3r^{2k}}{(2k+1)!(2k+3)}$$
$$\times \left[\left\{ \left(\frac{\nu}{\mu} \right)^{k-1} \frac{(\lambda + 2\mu)\Delta u - \nu u}{\lambda + \mu} \right\} \left\{ \frac{(\mu - \lambda)k}{\lambda + 4\mu} + 1 \right\} \right.$$
$$\left. - \left(\frac{\nu}{\lambda + 2\mu} \right)^{k-1} \frac{\mu \Delta u - \nu u}{\lambda + \mu} \left\{ \frac{2(2\lambda + 3\mu)k}{\lambda + 4\mu} + 1 \right\} \right].$$

Summing up the series on the right-hand side of this equality yields (4.95). $\qquad \square$

Remark. The mean value relation for the Lamé equation $\Delta^* u = 0$ follows immediately from (4.92). Indeed, using (4.94), we obtain from (4.92):

$$u(x) = N^1 u(x) . \tag{4.96}$$

The following result is directly implied [4.8, Theorem III.2.4].

Theorem 4.14. The general solution to (4.81) lying in $C^2(G)$ can be represented in the form $u(x) = u_p(x) + u_s(x)$, where u_p and u_s are regular vector functions determined by the equations

$$
\begin{aligned}
(\Delta - k_p^2) u_p &= 0 , & \operatorname{rot} u_p &= 0 , \\
(\Delta - k_s^2) u_s &= 0 , & \operatorname{div} u_s &= 0 , \\
k_p = \left[\frac{\nu}{\lambda + 2\mu} \right]^{1/2} \equiv k_2 , & \quad k_s = \left[\frac{\nu}{\mu} \right]^{1/2} \equiv k_1 ,
\end{aligned}
\tag{4.97}
$$

where

$$u_p = \frac{(\Delta - k_s^2) u}{k_p^2 - k_s^2} , \qquad u_s = \frac{(\Delta - k_p^2) u}{k_s^2 - k_p^2} ; \qquad u \in C^\infty(G) . \tag{4.98}$$

Lemma 4.3. The solution to (4.81) satisfies the relation

$$\Delta^n u(x) = k_s^{2n} u_s + k_p^{2n} u_p . \tag{4.99}$$

Indeed, from (4.94) we have

$$\mu^n \Delta^n u + \left(\frac{\nu}{\lambda + 2\mu} \right)^{n-1} \frac{\nu u - \mu \Delta u}{\lambda + \mu} \{ (\lambda + 2\mu)^n - \mu^n \} = \nu^n u .$$

Hence,

$$
\begin{aligned}
\Delta^n u = & \left(\frac{\nu}{\mu} \right)^{n-1} \frac{\lambda + 2\mu}{\lambda + \mu} \left[\Delta u - \frac{\nu}{\lambda + 2\mu} u \right] \\
& - \left(\frac{\nu}{\lambda + 2\mu} \right)^{n-1} \frac{\mu}{\lambda + \mu} \left[\Delta u - \frac{\nu}{\mu} u \right] .
\end{aligned}
\tag{4.100}
$$

Recalling that

$$\left(k_s^2 - k_p^2 \right)^{-1} = \frac{\mu(\lambda + 2\mu)}{\nu(\lambda + 2\mu)}$$

and $u = u_s + u_p$, we obtain from (4.100)

$$\Delta^n u = k_s^{2n} \frac{(\Delta - k_p^2) u}{(k_s^2 - k_p^2)} + k_p^{2n} \frac{(\Delta - k_s^2) u}{(k_p^2 - k_s^2)} ;$$

thus, (4.99) is proved.

$\square$

Theorem 4.15. The general solution to (4.81) lying in $C^2(G)$ satisfies the relation

$$N_x^r(u) = \frac{\sinh(rk_p)}{rk_p} \, u_p(x) + \frac{\sinh(rk_s)}{rk_s} \, u_s(x) \, . \tag{4.101}$$

Proof. Substituting (4.99) into

$$N_x^r(u) = \sum_{n=0}^{\infty} \frac{r^{2n}}{(2n+1)!} \, \Delta^n u(x)$$

yields

$$\begin{aligned}
N_x^r(u) &= \sum_{n=0}^{\infty} \frac{r^{2n}}{(2n+1)!} \left\{ k_s^{2n} u_s(x) + k_p^{2n} u_p(x) \right\} \\
&= \frac{\sinh(rk_s)}{rk_s} \, u_s(x) + \frac{\sinh(rk_p)}{rk_p} \, u_p(x) \, . \qquad \square
\end{aligned}$$

Theorem 4.16. *The mean value relation.* The general solution to (4.81) lying in $C^2(G)$ satisfies the following mean value relation:

$$u(x) = \frac{(W_2 - W_1)N_x^r(u) + \left[\dfrac{\sinh(rk_s)}{rk_s} - \dfrac{\sinh(rk_p)}{rk_p} \right] N^1 u(x)}{\dfrac{\sinh(rk_s)}{rk_s} W_2 - \dfrac{\sinh(rk_p)}{rk_p} W_1} \, . \tag{4.102}$$

Proof. The proof follows from (4.95) which, from (4.101), and the fact that $u = u_p + u_s$, can be rewritten in the form

$$N^1 u(x) = W_1 u_s(x) + W_2 u_p(x) \, . \qquad \square$$

Consider now the following system of equations describing the "thermoelastic vibration":

$$\Delta u_i(x) + \alpha \,\text{grad div}\, u'(x) - \gamma \text{grad}\, \theta = 0 \, , \quad i = 1, 2, 3 \tag{4.103}$$

$$\Delta \theta(x) = 0 \, , \quad x \in G \subset \mathbf{R}^3 \, , \tag{4.104}$$

where $u'(x) = (u_1, u_2, u_3)^{\mathrm{T}}$. Let $u_4(x) = \theta(x)$, and $u(x) = (u'(x), u_4(x))^{\mathrm{T}}$.

We obtain now a mean value relation for (4.103, 104) which generalizes the result (4.102).

Theorem 4.17. The general solution to system (4.103, 104) lying in $C^{\infty}(G)$ satisfies the following mean value relation:

$$u'(x) = N^1 u(x) - \frac{1}{\omega_3} \int_{S(0,1)} \frac{3r\gamma}{3 + \alpha} \, u_4 \nabla s \, d\Omega(s) \tag{4.105}$$

$$u_4(x) = N_x^r(u_4) \, .$$

Proof. From (4.103) it is easy to prove that $\Delta^2 u' = 0$. For the harmonic function $u_4(x)$, we have

$$u_4(x) = N_x^r(u_4) = \frac{3}{4\pi r^3} N^{(V)} u_4(x) , \tag{4.106}$$

where $N^{(V)} u_4(x)$ is the volume integral of u_4 over the ball $V = \{|x - y| < r\}$. Thus, for the harmonic function $\partial u_4 / \partial x_k$ relation (4.106) takes the form

$$\frac{\partial}{\partial x_k} u_4(x) = \frac{3}{4\pi r^3} \int_V \frac{\partial u_4}{\partial x_k} dV = \frac{3 r^2}{4\pi r^3} \int_{S(0,1)} u_4 \frac{x_k}{r} d\Omega(s) , \; k = 1,2,3 ,$$

that is,

$$\frac{\partial}{\partial x_k} u_4(x) = \frac{3}{3\pi r^3} \int_{S(0,1)} u_4 s_k d\Omega(s) , \quad k = 1,2,3 . \tag{4.107}$$

From (4.84) we obtain by using first, $\Delta^2 u'(x) = 0$ and then (4.107),

$$N^1 u'(x) - u'(x) = \frac{r^2(\Delta + \alpha \,\mathrm{grad\,div}) u'(x)}{3 + \alpha} + \sum_{n=2}^{\infty} \frac{r^{2n}}{(2n+1)!(2n+3)}$$

$$\times \left\{ \frac{15 n \gamma \Delta^{n-1} \,\mathrm{grad}\, u_4(x)}{3 + \alpha} - 3(n-1)\Delta^n u'(x) \right\}$$

$$= \frac{r^2 \gamma \,\mathrm{grad}\, u_4(x)}{3 + \alpha} = \frac{\gamma r^2}{3 + \alpha} \frac{3}{4\pi r} \int_{S(0,1)} u_4 \nabla s \, d\Omega(s) ,$$

and (4.105) is proved. □

Note that the proof of the mean value relations is obtained on the basis of expansion of $N_x^r(u)$ and $N^1(u)$. Consequently, they are obtained for sufficiently small r. Generalization to r such that $S(x, r) \subseteq G$ can be obtained as mentioned in the proof of Theorem 3.5.

It is now not difficutl to show that $\|A_\nu\|_1 < 1$ for sufficiently large Re $\{\nu\} \geq \nu_\varepsilon$, ε is a fixed value determining Γ_ε. Thus, for $\nu : \mathrm{Re}\,\{\nu\} \geq \nu_\varepsilon$, (4.81) can be solved by the walk on spheres algorithm. To obtain the solution at $\nu = 0$ we can then use the method of analytical continuation based on the transformation $\nu = \psi(n)$ described in Chap. 2.

4.4 Application to Stochastic Elasticity Problems

4.4.1 The Bending Problem for a Plate Lying on an Elastic Base

Consider a rectangular plate on a stochastic elastic base

$$\Delta^2 u(x) + k u(x) = f , \quad x \in G$$
$$u|_\Gamma = \Delta u|_\Gamma = 0 . \tag{4.108}$$

In testing the reliability of constructions undergoing random vibration it is necessary to solve problems of the type of (4.108) where k is a random field and f is a random load [4.12]. Moreover, it may appear that k and f are dependent random fields. In [4.12] a number of problems of this type are considered such as vibration of structures in gusty winds, flutter, resistance against vibrations, etc. The statistical characteristics of k and f are known [4.12], and the problem is to calculate the various statistical characteristics of the solution of (4.108) which permits one to evaluate the reliability, the probability and intensity of rejection, the mean life, the fatigue curve of an elastic system, etc.

We described above the walk on circles algorithm assuming that k is a constant. If k is a function of $x = (x_1, x_2)$, we can apply the same algorithm if we choose sufficiently small radii $d(\xi_i)$. In this case there will arise an additional bias of the numerical solution. We now estimate this deterministic error.

Consider an equation with variable coefficients

$$\Delta^2 u(x) + c_1(x)\Delta u(x) + c_2(x)u(x) = 0 , \quad x \in G \subset \mathbf{R}^2 .$$

Let $S(x, r) \subset G$ be an arbitrary circle, and let $C(r)$ be the transition matrix of the walk on circles algorithm determined by the corresponding mean value relation given by Theorem 4.4

Lemma 4.4. Let $\beta = \left[1 - \frac{r^2}{4}c_1\right]^{-1}$. Then

$$C(r) = \begin{pmatrix} 1 & \frac{-\beta r^2}{4} \\ \frac{\beta r^2 c_2}{4} & \beta \end{pmatrix} + 0(r^4) \quad \text{as} \quad r \to 0 . \tag{4.109}$$

Proof. Let $k_1^2 \neq k_2^2$ be the roots of the characteristic equation $\lambda^2 + c_1\lambda + c_2 = 0$. Using the expansion

$$I_0(z) = J_0(iz) = \sum_{k=0}^{\infty} \frac{(-1)^k}{(k!)^2} \left(\frac{iz}{2}\right)^{2k}$$

we get

$$I_0(k_i z) = 1 + \frac{(k_i r)^2}{4} + 0(r^4) \quad \text{as} \quad r \to 0 .$$

Substituting this expression into the coefficients of the mean value relations of Theorem 4.4 we obtain (4.109). $\qquad\square$

Consider now the general case of the boundary ∂G with conditions

A) $\quad u|_\Gamma = g_1 , \quad \left.\dfrac{\partial u}{\partial n}\right|_\Gamma = g_2$

B) $\quad u|_\Gamma = g_1 , \quad \left\{\Delta u - \alpha \dfrac{\partial u}{\partial n}\right\}\Big|_\Gamma = g_2 .$

In both cases the transition matrix in the first $N - 1$ steps coincides with the matrix (4.109). In the last circle the transition matrix has the form

$$C(r) = \begin{pmatrix} 1 + 0(r^4) & -\dfrac{r^2}{4}\beta\alpha + 0(r^\gamma) \\ \dfrac{r^2}{4}\beta c_2 + 0(r^4) & \beta\alpha + 0(r^\gamma) \end{pmatrix} \tag{4.110}$$

where $\alpha = 2/r$, $\beta = (1 - r^2 c_1/8)^{-1}$, $\gamma = 3$ in case A) and $\alpha = 1$, $\beta = 1 - r^2 c_1/4 - \frac{\alpha r}{2}(1 - r^2 c_1/8)^{-1}$, $\gamma = 4$ in the case B). Now estimate the deterministic error of the walk on small circles algorithm for the case described by Lemma 4.4.

Theorem 4.18. Let

$$C_k(r) = \tilde{C}_k(r) + R_k = \begin{pmatrix} 1 & -\dfrac{r^2}{4}\beta \\ \dfrac{r^2}{4}\beta c_2 & \beta\alpha \end{pmatrix} + \begin{pmatrix} 0(r^4) & 0(r^4) \\ 0(r^4) & 0(r^4) \end{pmatrix}$$

be the transition matrix on the k-th step. Let $y_0 = (y_1, y_2)$ be the approximate solution of $(u, \Delta u)$ obtained using the exact transition matrix, and $\tilde{y}_0$ be the solution obtained on the basis of the approximate matrix [see (4.109)]:

$$\tilde{C}_k(r) = \begin{pmatrix} 1 & -\dfrac{r^2}{4}\beta \\ \dfrac{r^2}{4}\beta c_2 & \beta\alpha \end{pmatrix}.$$

Then

$$\|y_0 - \tilde{y}_0\| = 0(r^4) \quad \text{as} \quad r \to 0.$$

Proof. Indeed

$$y_0 = \langle C_1(r)C_2(r)\dots C_N(r)g \rangle = \left\langle \prod_{k=1}^{N} \left[\tilde{C}_{N+1-k}(r) + R_{N+1-k} \right] g \right\rangle$$

$$= \left\langle \prod_{k=1}^{N} \tilde{C}_{N+1-k}(r)g + \left(\sum_{k=1}^{N} R_{N+1-k} \prod_{\substack{i=1 \\ i \neq k}}^{N} \tilde{C}_{N+1-i}(r) \right. \right.$$

$$+ \sum_{k=1}^{N} \sum_{j=1}^{N} R_{N+1-k} R_{N+1-j} \prod_{\substack{i=1 \\ i \neq k}}^{N} \tilde{C}_{N+1-i}(r)$$

$$\left. \left. + \prod_{k=1}^{N} R_{N+1-k} \right) g \right\rangle = \left\langle \prod_{k=1}^{N} \tilde{C}_{N+1-k}(r)g \right\rangle + \langle Rg \rangle .$$

Thus

$$y_0 = \tilde{y}_0 + \langle Rg \rangle ,$$

where

$$R = \begin{pmatrix} 0(r^4) & 0(r^4) \\ 0(r^4) & 0(r^4) \end{pmatrix} \quad \text{as} \quad r \to 0 . \qquad \square$$

We now describe the results of numerical solution of stochastic problems of the type (4.108) obtained by the walk on small circles and the randomization principle. To calculate the correlation function, two conditionally independent trajectories of the walk on circles process were constructed per one realization of the random fields k and f.

In the first experiment, f was taken as a constant load acting on a square plate, and k was sampled as an isotropic gaussian random field with the spectral density ($\langle k \rangle = 0$)

$$p(\lambda) = \frac{1}{2\pi\alpha^2} \left[1 + \frac{|\lambda|^2}{\alpha^2} \right]^{-3/2}$$

which corresponds to an exponential correlation function $R(|x|) = \exp\{-\alpha|x|\}$. Simulation of k can be carried out as follows (Sect. 1.2.4):

1) $|\kappa_k| = \alpha[\text{rand}^{-2} + 1]^{1/2}$
2) ω is chosen as an isotropic vector
3) the argument $\theta_k(x) = (|\kappa_k|\omega, x)$ is substituted into the formulas of the type (1.60, 61), for example,

$$k(x) = \xi \cos[\theta_k(x)] + \eta \sin[\theta_k(x)] .$$

Here "rand" is a random number uniformly distributed on $[0,1]$. Using the transition matrix

$$\tilde{C}_k(r) = \begin{pmatrix} 1 & -\dfrac{r^2}{4} \\ \dfrac{r^2}{4} k & 1 \end{pmatrix}$$

of the walk on small circles process we calculated the correlation function $R(|x_1 - x_2|)$ where the points are situated on the line $y = 0$. The calculations for different values of α are shown in Fig. 4.1 where the accuracy varies between 6 and 8%. In the next calculations k and f have the correlation functions $\exp(-\alpha|x|)$ and $[1 + \beta^2|x|^2]^{-3/2}$, respectively. The random fields k and f are dependent: in the argument $\kappa_k = |\kappa_k|\omega$, the isotropic vector ω is the same as in $\kappa_f = |\kappa_f|\omega$ while $|\kappa_k| = \alpha[\text{rand}^{-2} + 1]^{1/2}$ and $|\kappa_f| = -\beta \ln\{\text{rand}_1 \text{rand}_2\}$. The results for different values of β are shown in Fig. 4.2.

In the next experiment the following stochastic problem was solved:

$$\Delta^2 u(x) = \xi(\omega)\delta(x - x_0) , \quad x, x_0 \in G ,$$
$$u|_\Gamma = \Delta u|_\Gamma = 0 ,$$

where G is a unit square, $\xi(\omega)$ is a random variable having a gaussian or exponential distribution. The mean solution $\bar{u}$ for the case when ξ is gaussian with the mean a and the variance b^2 is shown in Fig. 4.3 where $a = 1$, $b = 1$; $a = 2$, $b = 1$; $x_0 = (0.5, 0.7)$, $\alpha = 0.1$ along the line $x_2 = 0.5$; $\varepsilon = 0.01$, the number of trajectories $N = 1000$.

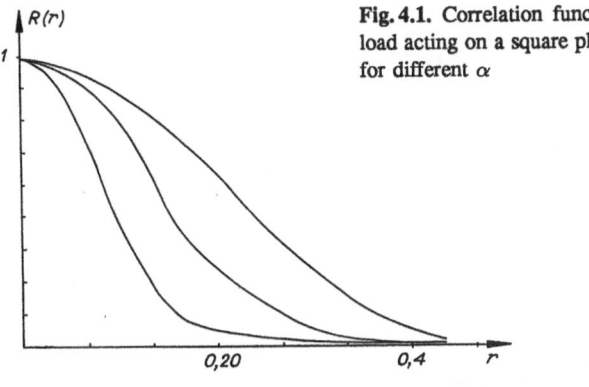

Fig. 4.1. Correlation function of the solution for a constant load acting on a square plate; walk on small circles algorithm for different α

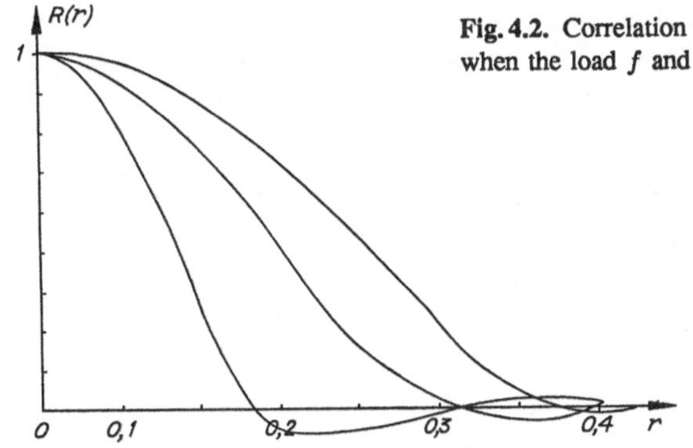

Fig. 4.2. Correlation function as in Fig. 4.1 for different β when the load f and the coefficient k are dependent

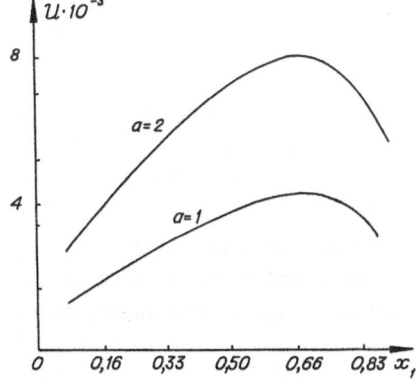

Fig. 4.3. The mean bending of a plate under a gaussian load ($x_2 = 0.5$, $a = 1$, $b = 2$)

These results, obviously, could be obtained by solving the same boundary value problem with the mean load $\bar{\xi} = a$. We now calculate the nonlinear functionals such as the probability $P(u > u_0)$ where u_0 is a cricital level. In Fig. 4.4 we show the function P at two points $P_c(0.5, 0.5)$ and $P_b(0.2, 0.5)$ for gaussian ξ with $a = 1$, $b = 1$ along the line $x_2 = 0.5$. In Fig. 4.5 the same result is shown for $a = 1.5$, $b = 1$. The exponential distribution of ξ with $\alpha = 1$ gives the result shown in Fig. 4.6.

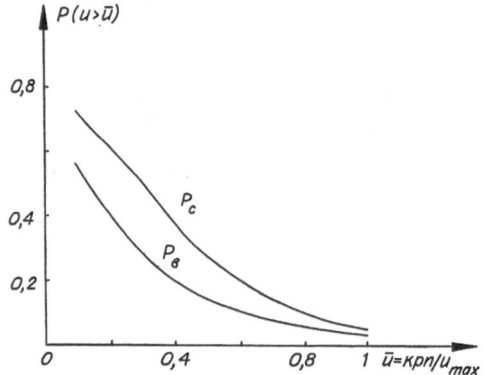

Fig. 4.4. Probability $P(u > u_0/m_{\text{max}})$ of surpassing a critical bending value u at the center (P_c) and near the boundary (P_b) along the line $x_2 = 0.5$, for a gaussian distribution $a = 1$, $b = 1$

4.4.2 Random Loads

Consider a stochastic boundary value problem

$$\Delta^2 u(x) = f(x) , \quad x \in G ,$$
$$u|_\Gamma = \Delta u|_\Gamma = 0 ,$$

where G is a unit square, f is an isotropic gaussian field with the spectral density

$$p(\lambda) = \frac{1}{2\pi\alpha^2} \left[1 + \frac{|\lambda|^2}{\alpha^2} \right]^{-3/2} .$$

The simulation formula was used according to (1.61) with $N = 10$. The correlation function is shown in Fig. 4.7. The points were obtained by the Monte Carlo method (1000 trials), and the crosses were obtained by the method of [4.13]. In this case we also calculated the distribution $P(u > u_0)$ (Fig. 4.8). In Fig. 4.9 we compare probabilities P for random loads whose distributions are close to gaussian ($N = 20$) and not gaussian ($N = 1$). Note that in the case of (1.60) ($N = 1$) one-dimensional distributions are gaussian. The corresponding mean solutions are shown in Fig. 4.10.

188

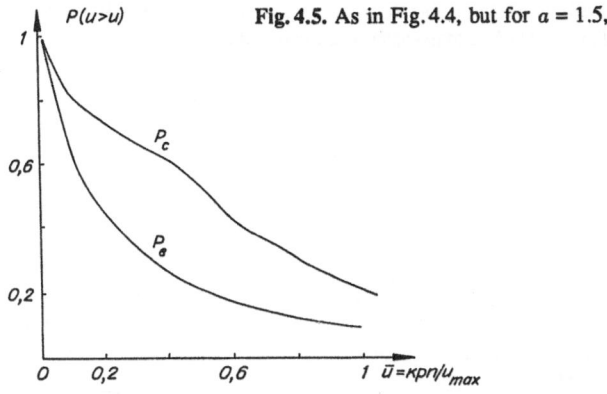

Fig. 4.5. As in Fig. 4.4, but for $a = 1.5$, $b = 1$

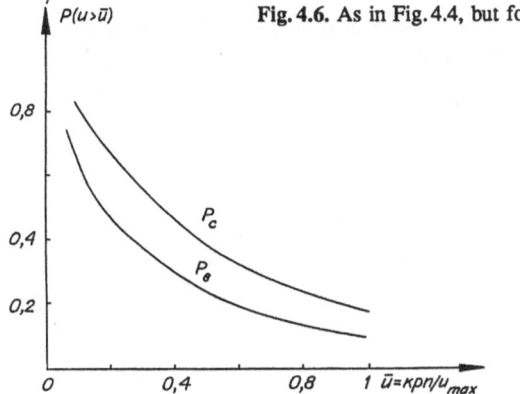

Fig. 4.6. As in Fig. 4.4, but for exponential distribution $\alpha = 1$

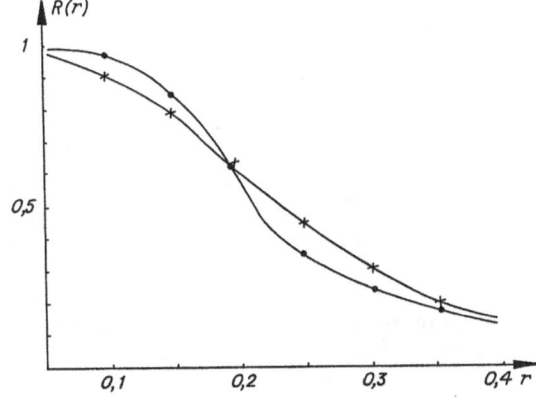

Fig. 4.7. Comparison of the correlation function of Monte Carlo calculations (*circles*) with the method of [4.13] (*crosses*)

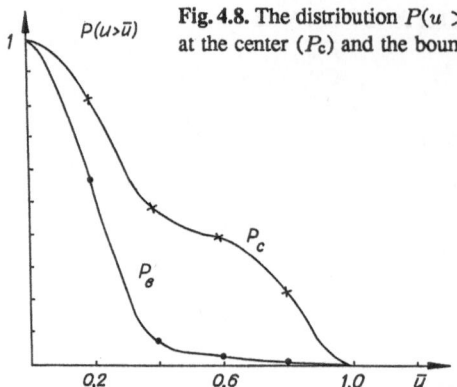

Fig. 4.8. The distribution $P(u > u_0 / u_{max})$ of surpassing a critical bending value at the center (P_c) and the boundary (P_b) of a plate under a random load

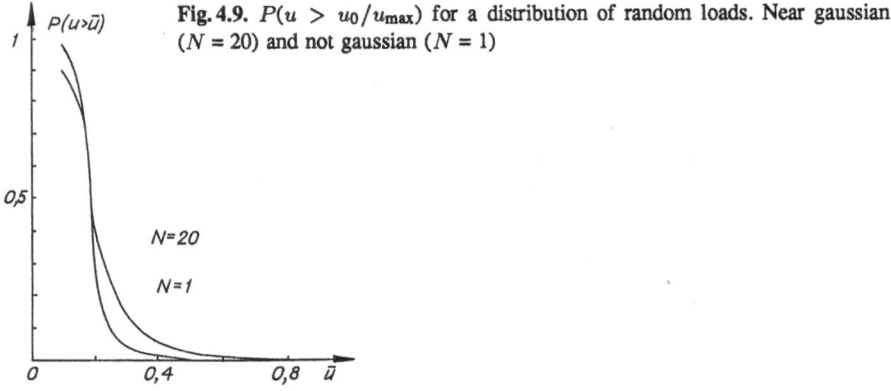

Fig. 4.9. $P(u > u_0 / u_{max})$ for a distribution of random loads. Near gaussian ($N = 20$) and not gaussian ($N = 1$)

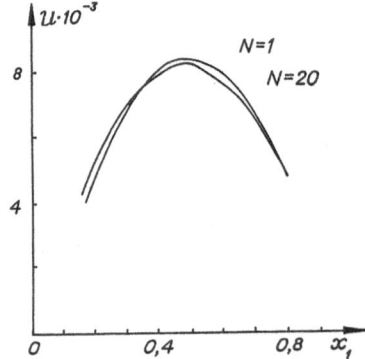

Fig. 4.10. The mean solutions for $N = 20$ and $N = 1$ along the line $x_2 = 0.5$

5. Monte Carlo Algorithms for Solving Diffusion Problems

In this chapter we treat problems of particle diffusion in turbulent velocity fields. Two main situations are considered. In the first one it is assumed that the diffusion process is well described by a deterministic turbulent diffusion equation for the mean concentration. For example, the classical heat equation also describes the Brownian diffusion process. In Sect. 5.1 we describe the walk on boundary algorithm for solving the heat equation and some other parabolic equations. A more general situation arises when probabilistic distributions of concentration of particles moving in a turbulent flow are to be found. Moreover, accurate description of the mean concentrations by a turbulent diffusion equation requires the coefficient of a turbulent diffusion which is defined through Lagrangian characteristics of the particles moving in turbulent flows.

One of the main problems in the second approach is to simulate random velocity fields with the desired probabilistic distributions. For example, we construct an incompressible isotropic random velocity field with Kolmogorov's energy spectrum and simulation formulas for turbulent velocities in the atmospheric boundary layer. This approach is presented in Sect. 5.2. In Sect. 5.3 the Cauchy problem is treated. Some applied diffusion problems are described in Sect. 5.4.

5.1 Walk on Boundary Algorithms for the Heat Equation

In this section random estimates are constructed on the trajectories of the walk on boundary processes $\{(\xi_k, \tau_k)\}$ and $\{(\xi_k^*, \tau_k^*)\}$ defined in Sect. 1.2.2.

5.1.1 Generalization of Isotropic Walk on Boundary Processes to the Nonstationary Case

Let G_i be a bounded simple-connected domain in $\mathbb{R}^n$ with a Ljapunov-continuous boundary ∂G, and let $\bar{G}_i = G_i \cup \partial G$, $G_e = \mathbb{R}^n \backslash \bar{G}_i$. We recall here the notation introduced in Sect. 1.2.2:

$$Q_t = G_i \times (0, t) , \quad \Gamma_t = \partial G \times (0, t) , \quad \Gamma = \partial G \times (-\infty, \infty) ,$$

$d\sigma_\xi$ is the surface element of ∂G, $Z_0(x, t)$ is the fundamental solution of the heat equation; below ξ and ξ' are points of the boundary ∂G.

As in the stationary case, we introduce the simple-layer potential with a density $\mu(\xi, \tau)$:

$$u(x,t) = \int_0^t d\tau \int_{\partial G} Z_0(x - \xi, t - \tau)\mu(\xi, \tau)d\sigma_\xi \ . \tag{5.1}$$

The normal derivative of the potential (5.1) has a discontinuity on the boundary and the jump is described by

$$\lim_{x \to \xi} \frac{\partial u}{\partial n}(x,t) = \pm \frac{1}{2}\mu(\xi, \tau) + \int_0^t d\tau' \int_{\partial G} a^2 \frac{\partial Z_0}{\partial n(\xi)}(\xi - \xi', \tau - \tau')$$
$$\times \mu(\xi', \tau')d\sigma_\xi \ .$$

Here n is the normal vector exterior to G_i, the sign "+" is taken if $x \in G_i$ when $x \to \xi$, and the sign "−" is taken if $x \in G_e$ when $x \to \xi$. The jump of the double-layer potential

$$v(x,t) = \int_0^t d\tau \int_{\partial G} -a^2 \frac{\partial Z_0}{\partial n(\xi)}(x - \xi, t - \tau)\nu(\xi, \tau)d\sigma_\xi \tag{5.2}$$

with a density $\nu(\xi, \tau)$ is given by

$$\lim_{x \to \xi} v(x,t) = \pm \frac{1}{2}\nu(\xi, \tau) + \int_0^t d\tau' \int_{\partial G} -a^2 \frac{\partial Z_0}{\partial n(\xi)}(\xi - \xi', \tau - \tau')$$
$$\times \nu(\xi', \tau')d\sigma_\xi \ ,$$

with the same choice of sign as above in the right-hand side of this equality. We seek the solution of the interior (in G_i) and exterior (in G_e) Dirichlet problem

$$\frac{\partial v}{\partial t} = a^2 \Delta v(x,t) \qquad x \in G_i \quad \text{(interior problem)}$$
$$\text{or} \quad x \in G_e \quad \text{(exterior problem)}$$
$$\qquad\qquad\qquad\qquad\qquad\qquad\qquad t > 0 \tag{5.3}$$
$$v(x,0) = 0$$
$$v(\xi, t) = \psi(\xi, t) \ , \qquad \xi \in \partial G, \quad t > 0$$

in the form of the double-layer potential given by (5.2) with unknown density ν. Then $v(x,t)$ solves (5.3) if ν satisfies the boundary integral equation [5.1]

$$\nu(\xi, \tau) = \lambda \int_0^\tau d\tau' \int_{\partial G} \frac{|\xi - \xi'|}{\tau - \tau'} \cos(\varphi_{\xi'\xi})Z_0(\xi - \xi', \tau - \tau')$$
$$\times \nu(\xi, \tau')d\sigma_{\xi'} + \Psi(\xi, \tau) \tag{5.4}$$

where $\lambda = -1$, $\Psi(\xi, \tau) = 2\psi(\xi, \tau)$ for the interior, and $\lambda = 1$, $\Psi(\xi, \tau) = 2\psi(\xi, \tau)$ for the exterior Dirichlet problem. Here $\varphi_{\xi'\xi} = \angle(n(\xi'), \xi' - \xi)$ is the angle between $n(\xi')$ and $\xi' - \xi$. Analogously, we seek the solution of interior (exterior) Neumann problem

$$\frac{\partial u}{\partial t} = a^2 \Delta u(x,t) \qquad \begin{array}{l} x \in G_i \quad \text{(interior problem)} \\ \text{or} \quad x \in G_e \quad \text{(exterior problem)} \end{array}$$

$$u(x,0) = 0 \qquad\qquad t > 0 \qquad\qquad\qquad\qquad (5.5)$$

$$a^2 \frac{\partial u}{\partial n(\xi)} = \psi(\xi,t), \qquad \xi \in \partial G, \quad t > 0$$

in the form of the simple-layer potential (5.1). The unknown density μ satisfies the equation

$$\mu(\xi,\tau) = \lambda \int_0^\tau d\tau' \int_{\partial G} \frac{|\xi - \xi'|}{\tau - \tau'} \cos(\varphi_{\xi\xi'}) Z_0(\xi - \xi', \tau - \tau')$$
$$\times \mu(\xi', \tau') d\sigma_{\xi'} + \Psi(\xi,\tau) \qquad\qquad\qquad (5.6)$$

where $\lambda = 1$, $\Psi(\xi,\tau) = 2\psi(\xi,\tau)$ for the interior, and $\lambda = -1$, $\Psi(\xi,\tau) = -2\psi(\xi,\tau)$ for the exterior Neumann problem.

Solution of the third boundary value problem

$$\frac{\partial u}{\partial t} = a^2 \Delta u(x,t) \qquad \begin{array}{l} x \in G_i \quad \text{(interior problem)} \\ \text{or} \quad x \in G_e \quad \text{(exterior problem)} \end{array}$$

$$u(x,0) = 0 \qquad\qquad\qquad\qquad t > 0 \qquad\qquad\qquad (5.7)$$

$$a^2 \frac{\partial u}{\partial n(\xi)} + \alpha(\xi,t)u(\xi,t) = \psi(\xi,t), \qquad \xi \in \partial G, \quad t > 0$$

is also sought in the form (5.1). The density $\mu(\xi,\tau)$ then satisfies the integral equation

$$\mu(\xi,\tau) = \lambda \int_0^\tau d\tau' \int_{\partial G} \left\{ \frac{|\xi - \xi'|}{\tau - \tau'} \cos(\varphi_{\xi\xi'}) - 2\alpha(\xi,\tau) \right\}$$
$$\times Z_0(\xi - \xi', \tau - \tau')\mu(\xi', \tau') d\sigma_{\xi'} + \Psi(\xi,\tau) \qquad\qquad (5.8)$$

where $\lambda = 1$, $\Psi(\xi,\tau) = 2\psi(\xi,\tau)$ for the interior, and $\lambda = -1$, $\Psi(\xi,\tau) = -2\psi(\xi,\tau)$ for the exterior problem.

The equations (5.4, 6, 8) are Volterra integral equations of the second kind (with respect to τ) with weak singularity in the kernels; their solutions are as smooth as $\Psi(\xi,\tau)$.

Now we construct the walk on boundary algorithm for (5.3, 5, 9). Assume first that G_i is convex, and let $\{(\xi_k^*, \tau_k^*)\}_{k=0}^\infty$ be an isotropic walk on boundary process with the initial density $\pi(\xi_0^*, \tau_0^*)$ defined in Sect. 1.2.2. We construct first a random estimate on the trajectories $\{(\xi_k^*, \tau_k^*)\}$ for solving (5.3), i.e., a Monte Carlo estimate for calculating the linear functional (5.2) of the solution of the integral equation (5.4). The right-hand side of this equation is defined on $\partial G \times (0, \infty)$; let us prolongate it to $\partial G \times (-\infty, 0)$ by zero. We preserve the same notation for the function defined on $\Gamma = \partial G \times (-\infty, \infty)$. Then (5.4) can be rewritten in the form

$$\nu(\xi, \tau) = \lambda \int_0^T d\tau' \int_{\partial G} \theta(\tau - \tau') \frac{|\xi - \xi'|}{\tau - \tau'} \cos(\varphi_{\xi'\xi})$$
$$\times Z_0(\xi - \xi', \tau - \tau')\nu(\xi, \tau')d\sigma_{\xi'} + \Psi(\xi, \tau) \qquad (5.9)$$

where

$$\theta(\tau - \tau') = \begin{cases} 1, & \text{if } \tau' < \tau \\ 0, & \text{if } \tau' \geq \tau. \end{cases}$$

From the definition of the transition density p [cf. (1.50)] of the isotropic walk on boundary process it follows that

$$p(\xi, \tau \to \xi', \tau') = \theta(\tau - \tau') \frac{|\xi - \xi'|}{\tau - \tau'} \cos(\varphi_{\xi'\xi}) Z_0(\xi - \xi', \tau - \tau') \qquad (5.10)$$

coincides with the kernel of the integral equation (5.9); notice that (5.10) is a density with respect to the variables ξ', τ'. Therefore, it is natural to choose the adjoint Monte Carlo estimates (Chap. 1) for calculating the linear functional (5.2)

$$v(x, t) = \int_{\Gamma} h_{xt}^{(1)}(\xi, \tau)\nu(\xi, \tau)d\tau d\sigma_{\xi},$$

where

$$h_{xt}^{(1)}(\xi, \tau) = \theta(\tau) \left[-a^2 \frac{\partial Z_0}{\partial n(\xi)} (\xi - \xi', \tau - \tau') \right].$$

Thus, choosing the initial density $\pi(\xi_0^*, \tau_0^*)$ of the walk on boundary process $\{(\xi_k^*, \tau_k^*)\}$ such that $\pi(\xi^*, \tau^*) \neq 0$ when $h_{xt}^{(1)}(\xi, \tau) \neq 0$, [in this case $\pi(\xi^*, \tau^*)$ is called consistent with $h_{xt}^{(1)}(\xi, \tau)$] we get a non-biased estimate for $v(x, t)$

$$\eta_{xt}^{(1)} = \sum_{k=0}^{\infty} Q_0 \lambda^k \Psi(\xi_k^*, \tau_k^*),$$

where

$$Q_0 = \frac{h_{xt}^{(1)}(\xi_0^*, \tau_0^*)}{\pi(\xi_0^*, \tau_0^*)}.$$

Let $N_t^* = \max\{k : \tau_k^* > 0\}$. Since $\Psi(\xi, \tau) = 0$ when $\tau < 0$, we obtain

$$\eta_{xt}^{(1)} = \sum_{k=0}^{N_t} Q_0 \lambda^k \Psi(\xi_k^*, \tau_k^*).$$

Consider now the integral equation (5.8). The function

$$p_2(\tau' \to \tau | \xi, \xi') = \frac{|\xi - \xi'| \exp\left\{-\dfrac{|\xi - \xi'|^2}{4a^2(\tau - \tau')}\right\}}{\Gamma\left(\dfrac{n}{2}\right)(\tau - \tau')[4a^2(\tau - \tau')]^{n/2}}, \qquad \tau > \tau' \qquad (5.11)$$

can be considered as a transition density with respect to τ on the interval $[\tau', \infty)$. Simulation according to this density is carried out similarly to (1.52) but with a backward time:

$$\tau = \tau' + \frac{|\xi - \xi'|^2}{4a^2 \gamma_{n/2}} \tag{5.12}$$

Consequently, the function

$$p_3(\xi', \tau' \to \xi, \tau) = p_\xi(\xi' \to \xi) p_2(\tau' \to \tau | \xi, \xi')$$
$$= \frac{|\xi - \xi'|}{\tau - \tau'} \cos(\varphi_{\xi\xi'}) Z_0(\xi - \xi', \tau - \tau') \tag{5.13}$$

is a transition density with respect to $(\xi, \tau) : \xi \in \partial G, \tau \in (\tau', \infty)$.

Let $\{(\xi_k^*, \tau_k^*)\}_{k=0}^\infty$ be a stationary Markov chain with an initial density $\pi(\xi, \tau)$, the transition density $p_3(\xi', \tau' \to \xi, \tau)$ defined on the phase space $\Gamma = \partial G \times (-\infty, \infty)$. The kernel of the integral equation

$$\mu(\xi, \tau) = \lambda \int_\Gamma p_3(\xi', \tau' \to \xi, \tau) \mu(\xi', \tau') d\sigma_{\xi'} d\tau' + \Psi(\xi, \tau) \tag{5.14}$$

can be considered as a transition density with respect to the free variable (ξ, τ) [in (5.14) the function $\Psi(\xi, \tau)$ is prolonged on $\Gamma = \partial G \times (-\infty, \infty)$ as in (5.9)]. Therefore, it is convenient to calculate the linear functional (5.7) of the solution of (5.14) on the basis of the direct estimate. Indeed, let $\{(\xi_k^*, \tau_k^*)\}_{k=0}^\infty$ be the walk on boundary process with the initial density $\pi(\xi, \tau)[\pi(\xi, \tau) \neq 0$ for $\Psi(\xi, \tau) \neq 0]$, and let

$$h_{xt}^{(2)}(\xi, \tau) = \theta(t - \tau) Z_0(x - \xi, t - \tau) .$$

Then the random quantity

$$\eta_{xt}^{(2)} = \sum_{k=0}^\infty Q_0 \lambda^k h_{xt}^{(2)}(\xi_k^*, \tau_k^*) \tag{5.15}$$

is an unbiased estimate for $u(x, t)$ where $Q_0 = \Psi(\xi_0, \tau_0)/\pi(\xi_0, \tau_0)$. As in (5.10), we rewrite (5.15) in the form

$$\eta_{xt}^{(2)} = \sum_{k=0}^{N_t} Q_0 \lambda^k h_{xt}^{(2)}(\xi_k^*, \tau_k^*) , \tag{5.16}$$

where $N_t = \max\{k : \tau_k < t\}$, since $h_{xt}^{(2)}(\xi_k, \tau_k) = 0$ if $\tau_k > t$. Thus, the random walk processes $\{(\xi_k^*, \tau_k^*)\}_{k=0}^\infty$ and $\{(\xi_k, \tau_k)\}_{k=0}^\infty$ for the Dirichlet and the Neumann problems, respectively, differ only by the behavior of the time [5.2].

We turn now to the third boundary value problem (5.7) and the corresponding boundary integral equation (5.8). Direct generalization of the constructed estimates to this case is not satisfactory. Indeed, the random quantity

$$\eta_{xt}^{(3)} = \sum_{k=0}^{N_t} Q^{(k)} \lambda^k h_{xt}^{(2)}(\xi_k, \tau_k) \,, \tag{5.17}$$

where

$$Q^{(0)} = \frac{\Psi(\xi_0, \tau_0)}{\pi(\xi_0, \tau_0)} \,, \quad Q^{(k)} = Q^{(k-1)} \left[1 - \frac{2\alpha(\xi_k, \tau_k)(\tau_k - \tau_{k-1})}{|\xi_k - \xi_{k-1}| \cos\left(\varphi_{\xi_k \xi_{k-1}}\right)} \right]$$

is an unbiased estimate for (5.7) (which follows from the convergence of the Neumann series of the integral equation with the kernel $|p_3|$, Sect. 5.2.5); however, it has two disadvantages. First, it may appear that the variance is infinite, depending on the form of the domain. Second, if the boundary includes "plane parts", then the weights involve singularities of the type $\left[\cos\left(\varphi_{\xi_k \xi_{k-1}}\right)\right]^{-1}$. To overcome the latter difficulty, we change the transition density. Let $x_0 \in G_i \subset \mathbf{R}^3$ be a point such that

$$\inf_{\xi \in \Gamma} \cos \angle(n(\xi), \xi - x_0) = \inf_{\xi \in \Gamma} \cos\left(\varphi_{\xi x_0}\right) > 0 \,.$$

We denote by $p(\omega|\omega_0)$ the density on a unit sphere having a singularity of type $|\omega - \omega_0|^{-\beta}$, $0 < \beta < 1$ near the direction ω_0, where $\omega_0 = (\xi' - x_0)/|\xi' - x_0|$.

We now choose a spherical coordinate system so that ω_0 coincides with the direction x_3. If we sample the latitude angle φ uniformly from $[0, 2\pi]$, and choose the altitude angle ψ according to the density $c_\beta \psi^{-\beta}$, $\psi \in [0, \pi]$, where $c_\beta = \left(\int_0^\pi x^{-\beta} dx\right)^{-1}$, then near ω_0 the density $p(\omega|\omega_0)$ will have the singularity $|\omega - \omega_0|^{-\beta}$. Note that if ω is isotropic, then the density of ψ behaves like $\sin(\psi)/2$, $\psi \in [0, \pi]$, i.e., the singularity is $|\omega - \omega_0|$.

Thus, the density $\cos(\varphi_{\xi x_0}) p(\omega|\omega_0)/|\xi - x_0|^2$ behaves near ξ' like $|\omega - \omega|^{-\beta}$, i.e.,

$$\frac{c_1}{|\xi - \xi'|^\beta} \le \frac{\cos\left(\varphi_{\xi x_0}\right)}{|\xi - x_0|^2} p(\omega|\omega_0) \le \frac{c_2}{|\xi - \xi'|^\beta} \,, \quad c_1, c_2 > 0 \,.$$

The function

$$p_4(\xi', \tau' \to \xi, \tau) = \left[\frac{\cos\left(\varphi_{\xi x_0}\right)}{|\xi - x_0|^2} p(\omega|\omega_0) \right] \left[\frac{4\pi^{1/2} a^2 |\xi - \xi'|}{[4a^2(\tau - \tau')]^{3/2}} \right]$$

$$\times \exp\left\{ -\frac{|\xi - \xi'|^2}{4a^2(\tau - \tau')} \right\}$$

is a density of the transition from (ξ', τ') to (ξ, τ). Simulation according to this density can be carried out in two steps:

1) Choose ω according to $p(\omega|\omega_0)$ and find ξ as a point of intersection of the ray ω with the boundary ∂G.
2) Calculate the new time as $\tau = \tau' + |\xi - \xi'|^2/(4a^2 \gamma_{n/2})$.

The random quantity

$$\eta_{xt}^{(4)} = \sum_{k=0}^{N_t} Q^{(k)} \lambda^k h_{xt}^{(2)} \left(\bar{\xi}_k, \bar{\tau}_k \right) , \qquad (5.18)$$

where $\{(\bar{\xi}_k, \bar{\tau}_k)\}_{k=0}^{\infty}$ is a stationary Markov chain on Γ with initial density $\pi(\xi, \tau)$ and the transition density p_4, is an unbiased estimate for the solution of (5.7). In (5.18)

$$Q_0 = \frac{\Psi(\bar{\xi}_0, \bar{\tau}_0)}{\pi(\bar{\xi}_0, \bar{\tau}_0)} ,$$

$$Q^{(k)} = Q^{(k-1)} \left\{ \frac{|\bar{\xi}_k - x_0|^2 \cos\left(\varphi_{\bar{\xi}_k \bar{\xi}_{k-1}}\right)}{4\pi^{3/2} \left(\bar{\tau}_k - \bar{\tau}_{k-1}\right) \cos\left(\varphi_{\bar{\xi}_k x_0}\right) p(\omega_k | \omega_{k-1})} \right.$$

$$\left. - \frac{2\alpha(\bar{\xi}_k, \bar{\tau}_k)|\bar{\xi}_k - x_0^2| \cos\left(\varphi_{\bar{\xi}_k \bar{\xi}_{k-1}}\right)}{\cos\left(\varphi_{\bar{\xi}_k x_0}\right) p(\xi_k | \omega_{k-1}^*) 4a^2 |\bar{\xi}_k - \bar{\xi}_{k-1}|\pi^{3/2}} \right\} ,$$

where $\omega_{k-1}^* = \left(\bar{\xi}_{k-1} - x_0\right) / |\bar{\xi}_{k-1} - x_0|$. In Sect. 5.2.5 we shall prove that the estimate (5.18) is unbiased and has a finite variance.

Remark. If G is non-convex, the *Remark* of Sect. 3.1.2 is true in the non-stationary case. Indeed, let G_i be a non-convex domain. Denote by $q(\xi', \xi)$ the number of intersections of the line $\xi + \mu(\xi' - \xi)$, $(\mu > 0)$ with the boundary ∂G (not including the point ξ), $\Omega(\xi)$ is the view angle of ∂G from ξ. Then

$$\int_{\partial G} \frac{|\cos\left(\varphi_{\xi'\xi}\right)| d\sigma_{\xi'}}{q(\xi', \xi)\Omega(\xi)|\xi - \xi'|^{n-1}} = 1 .$$

Therefore, all the estimates can obviously be rewritten, as in the stationary case described in Chap. 3, for non-convex domains, by choosing the next boundary point uniformly from $q(\xi', \xi)$ intersections. For example, the estimate for $v(x, t)$, the solution of (5.3), has the form

$$\eta_{xt}^{(1)} = \sum_{k=0}^{N_t^*} Q^{(k)} \lambda^k \Psi(\xi_k^*, \tau_k^*) . \qquad (5.19)$$

Here

$$N_t^* = \max\{k : \tau_k^* > 0\} ; \quad Q^{(0)} = \frac{h_{xt}^{(1)}(\xi_0^*, \tau_0^*)}{\pi(\xi_0^*, \tau_0^*)} ;$$

$$Q^{(k)} = Q^{(k-1)} q(\xi_k^*, \xi_{k-1}^*) \Omega(\xi_{k-1}^*) \frac{\Gamma(n/2) \cos\left(\varphi_{\xi_k^*, \xi_{k-1}^*}\right)}{\pi^{n/2} |\cos\left(\varphi_{\xi_k^*, \xi_{k-1}^*}\right)|}$$

The estimate $\eta_{xt}^{(2)}$ for $u(x, t)$, the solution of (5.5) becomes

$$\eta_{xt}^{(2)} = \sum_{k=0}^{N_t} Q^{(k)} \lambda^k h_{xt}^{(2)}(\xi_k, \tau_k) \,, \tag{5.20}$$

where

$$N_t = \max\{k : \tau_k < t\} \,; \quad Q^{(0)} = \frac{\Psi(\xi_0, \tau_0)}{\pi(\xi_0, \tau_0)} \,,$$

and the weights $Q^{(k)}$ coincide with the corresponding weights of (5.19) where ξ_k^* must be replaced by ξ_k.

5.1.2 The Variance and Cost of the Walk on Boundary Algorithms

All the Monte Carlo estimates constructed in Sect. 5.1.1 are based on Volterra–Fredholm integral equations of the second kind, $\mu = K\mu + \Psi$ or, in more detail

$$\mu(\xi, \tau) = \int_0^\tau d\tau' \int_{\partial G} k(\xi, \tau; \xi', \tau') \mu(\xi', \tau') d\sigma_{\xi'} + \Psi(\xi, \tau) \,. \tag{5.21}$$

The kernels $k(\xi, \tau; \xi', \tau')$ of these equations satisfy the conditions

$$|k(\xi, \tau; \xi', \tau')| \leq c[\tau - \tau]^{-(n/2+\alpha)} |\xi - \xi'|^\beta \exp\left\{ -\frac{\gamma |\xi - \xi'|^2}{\tau - \tau'} \right\}$$

for arbitrary $(\xi, \tau) \in \Gamma_t \equiv \partial G \times (0, t)$, $(\xi', \tau') \in \Gamma_t$. In (5.21), the positive constants c and γ generally depend on t; α, β are some parameters. We denote by $K_{\alpha\beta}$ a class of kernels satisfying this inequality. To investigate the variance and the cost of algorithms presented in Sect. 5.1.1 we first derive some properties of solutions of equations of the type (5.21) whose kernels belong to $K_{\alpha\beta}$.

Property 1. If $k \in K_{\alpha\beta}$, then $k \in K_{\alpha-\varepsilon/2, \beta-\varepsilon}$ for any $\varepsilon > 0$. Indeed

$$\left[|\xi - \xi'| / (\tau - \tau')^{1/2} \right]^\varepsilon \exp\left\{ -\frac{\delta |\xi - \xi'|^2}{\tau - \tau'} \right\} < \infty$$

since the function $x^\varepsilon \exp\{-\delta x\}$ is bounded on $(0, \infty)$ for arbitrary $\delta > 0$. Consequently, from $k \in K_{\alpha\beta}$ we get

$$|k(\xi, \tau; \xi', \tau')| \leq c[\tau - \tau']^{-(n/2+\alpha)} |\xi - \xi'|^\beta \exp\left\{ -\frac{\gamma |\xi - \xi'|^2}{\tau - \tau'} \right\}$$

$$\leq c_1 [\tau - \tau']^{-(n/2+\alpha-\varepsilon/2)} |\xi - \xi'|^{\beta-\varepsilon} \exp\left\{ -\frac{\gamma |\xi - \xi'|^2}{2(\tau - \tau')} \right\}$$

and the property is proved.

This property implies that it is possible to simplify the two-parameter expressions depending on α and β by turning to one parameter.

Property 2. If $k \in K_{\alpha\beta}$, where

$$\delta = \tfrac{1}{2}(\beta - 2\alpha + 1) > 0 \,, \quad \beta + n - 1 > 0 \,, \tag{5.22}$$

and the right-hand side of (5.21) satisfies the condition

$$|\Psi(\xi, \tau)| \leq c_\Psi \tau^a \,, \quad (\xi, \tau) \in \Gamma_t \,, \quad a > -1 \,,$$

where c_Ψ does not depend on ξ and τ, and ∂G is a Ljapunov surface, then the Neumann series for (5.21) converges and

$$|[K^l \Psi](\xi, \tau)| \leq \Gamma(a + 1)c_\Psi \tau^a \frac{(A\tau^\delta)^l}{\Gamma(a + l\delta + 1)} \,, \tag{5.23}$$

where A is a constant not depending on α, β and ∂G.

We now prove this property by using some general results presented in [5.3]. We denote by $\omega_\xi(r)$ the area of a part of ∂G lying inside a ball $B(x, r)$:

$$\omega_\xi(r) = \int_{|\xi - \xi'| < r} d\sigma_{\xi'} \,.$$

Then

$$\int_{\partial G} f(|\xi - \xi'|)d\sigma_{\xi'} = \int_0^\infty f(r)d\omega(r) \,,$$

and

$$\begin{aligned}
|[K\Psi](\xi, \tau)| &\leq c \int_0^\tau d\tau' \int_{\partial G} |\xi - \xi'|^\beta \exp\left\{-\frac{\gamma|\xi - \xi'|^2}{\tau - \tau'}\right\} \frac{\tau'^a d\sigma_{\xi'}}{[\tau - \tau']^{n/2 + \alpha}} \\
&= c \int_0^\tau d\tau' \int_{\partial G} \frac{r^\beta \tau'^a}{[\tau - \tau']^{n/2 + \alpha}} \exp\left\{-\frac{\gamma r^2}{\tau - \tau'}\right\} d\omega_{\xi'} \,.
\end{aligned} \tag{5.24}$$

Since ∂G is a Ljapunov surface, there exist constants $0 < b < B$, such that $br^{n-1} < \omega_\xi(r) < Br^{n-1}$ and, therefore,

$$\left[\omega_\xi(r)B^{-1}\right]^{1/n-1} \leq r \leq \left[\omega_\xi(r)b^{-1}\right]^{1/n-1} \,.$$

From this we get

$$\begin{aligned}
|[K\Psi](\xi, \tau)| &\leq c_\Psi c \int_0^\tau d\tau' \int_0^\infty (\tau - \tau')^{-n/2 - \alpha} \left[\omega_\xi(r)/b\right]^{\beta/n-1} \\
&\quad \times \tau'^a \exp\left\{-\gamma \frac{\omega_\xi(r)}{\tau - \tau'}\right\} d\omega_\xi(r) \,.
\end{aligned}$$

Using a new variable $\varrho = \gamma[\omega_\xi(r)B^{-1})^{2/(n-1)}]/(\tau - \tau')$ we transform the right-hand side to

$$c(b, B, \beta, \gamma) \int_0^\tau (\tau - \tau')^{\delta - 1} \tau'^a \int_0^\infty \varrho^{\beta + n - 3/2} \exp\{-\varrho\}d\varrho \,,$$

hence,

$$|[K\Psi](\xi,\tau)| \leq c_\Psi c B(a+1,\delta)\tau^{a+\delta} ,$$

where $B(.,.)$ is the Euler Beta-function. Recurrent application of (5.24) gives

$$|[K^j\Psi](\xi,\tau)| \leq c_\Psi c^j B(a+1,\delta)\ldots B(a+(j-1)\delta,\delta)\tau^{a+j\delta} ,$$

$j = 2,3,\ldots$. Substituting $B(x,y) = \Gamma(x)\Gamma(y)/\Gamma(x+y)$ yields

$$|[K^j\Psi](\xi,\tau)| \leq c_\Psi c^j \frac{\Gamma(a+1)\Gamma(\delta)}{\Gamma(a+1+\delta)} \frac{\Gamma(a+1+\delta)}{\Gamma(a+1+2\delta)}$$
$$\ldots \frac{\Gamma(a+1+(l-1)\delta)}{\Gamma(a+1+l\delta)}\tau^{a+l\delta} = c_\Psi \Gamma(a+1)\tau^a \frac{(A\tau^\delta)^l}{\Gamma(a+1+l\delta)} ,$$

where $A = c\Gamma(\delta)$. Thus, the Neumann series converges, since

$$\sum_{k=0}^\infty \frac{a^k}{\Gamma(\delta k)} < \infty$$

for arbitrary positive a and δ. The property 2 is proved.

Property 3. Let

$$k(\xi,\tau;\xi',\tau') = \sum_{i=1}^m k_i(\xi,\tau;\xi',\tau') .$$

Suppose that $k_i \in K_{\alpha,\beta}$, $i = 1,\ldots,m$, and all k_i are weak polar kernels, i.e., the condition (5.22) holds. Assume also that $|\Psi(\xi,\tau)| \leq c_\Psi$, $(\xi,\tau) \in \Gamma_t$. Then the Neumann series for (5.21) converges.

Indeed, from (5.24) we have

$$|[K\Psi](\xi,\tau)| \leq \sum_{i=1}^m |[K_i\Psi](\xi,\tau)| \leq \sum_{i=1}^m c_\Psi B(1,\delta_i)c_i\tau^{\delta_i} ,$$

where K_i is the integral operator generated by the kernel k_i and $\delta_i = (\beta_i - 2\alpha_i + 1)/2 > 0$, $i = 1,2,\ldots,m$. From this we get

$$|[K\Psi](\xi,\tau)| \leq \sum_{i_1=1}^m \ldots \sum_{i_l=1}^m c_\Psi c_{i1} \ldots c_{il} B(1,\delta_{i1}) B(1+\delta_{i1},\delta_{i2})$$
$$\ldots B(1+\delta_{i1}+\ldots+\delta_{il-1},\delta_{il})\tau^{\delta_{i1}+\delta_{i2}+\ldots+\delta_{il}}$$
$$= \sum_{i_1=1}^m \ldots \sum_{i_l=1}^m c_\Psi c_{i1} \ldots c_{il} \frac{\Gamma(1)\Gamma(\delta_{i1})\ldots\Gamma(\delta_{il})}{\Gamma(1+\delta_{i1}+\ldots\delta_{il})}$$
$$\times \tau^{\delta_{i1}+\delta_{i2}+\ldots+\delta_{il}} ,$$

Hence

$$|[K^l\Psi](\xi,\tau)| \leq c_\Psi (mc_{\max}\gamma)^l \frac{\tau^{l\delta^0}}{\Gamma(1+l\delta_{\min})} ,$$

where $\delta_{\min} = \min_i \delta_i$, $c_{\max} = \max_i c_i$, $\delta^0 = \delta_{\min}$ if $0 < \tau \le 1$ and $\delta^0 = \delta_{\max} = \max_i \delta_i$ if $\tau > 1$; $\gamma = \max_i \Gamma(\delta_i)$. This completes the proof.

We now turn to prove that the estimates $\eta_{xt}^{(i)}$, $i = 1, 2, 3, 4$ are unbiased and have finite variances.

Non-biasedness of these estimates follows from the convergence of Neumann series for (5.4, 6, 8). Indeed, $|\cos(\varphi)| < c|\xi - \xi'|^\lambda$, where φ equals to $\varphi_{\xi\xi'}$ or to $\varphi_{\xi'\xi}$, $0 < \lambda < 1$, since ∂G is a Ljapunov surface. Hence, the kernels of (5.4, 6) belong to $K_{1,1+\lambda}$ (weak polar kernels), and from the property 2 we conclude that the Neumann series converge if the right-hand side is bounded. Next, the kernel of (5.8) is represented as a sum of two kernels: the first kernel belongs to $K_{1,1+\lambda}$, and the second one lies in $K_{0,0}$. Thus by the property 3 the Neumann series for (5.8) converges.

To prove that the variance of $\eta_{xt}^{(1)}$ is finite, it is sufficient to show that the Neumann series for the equation

$$\varphi^*(\xi, \tau) = \int_0^\tau d\tau' \int_{\partial G} \frac{k^2(\xi, \tau; \xi', \tau')}{p_3(\xi', \tau' \to \xi, \tau)} \varphi^*(\xi', \tau') d\sigma_{\xi'} + \frac{\left(h_{xt}^{(1)}(\xi, \tau)\right)^2}{\pi(\xi, \tau)}$$

converges. Here $k(\xi, \tau; \xi', \tau')$ is the kernel of (5.4), p is the transition density defined by (5.10). Convergence of the Neumann series follows now from the fact that $k^2/p = k \in K_{1,1+\lambda}$ (of course, for π consistent with $h_{xt}^{(1)}$). The proof of variance finiteness of the estimate $\eta_{xt}^{(2)}$ is fully analogous.

To prove the variance finiteness of $\eta_{xt}^{(3)}$, it is sufficient to prove that the Neumann series for the equation

$$\varphi(\xi, \tau) = \int_0^\tau d\tau' \int_{\partial G} \frac{k^2(\xi', \tau'; \xi, \tau)}{p_3(\xi', \tau' \to \xi, \tau)} \varphi^*(\xi', \tau') d\sigma_{\xi'} + \frac{\Psi^2(\xi, \tau)}{\pi(\xi, \tau)} \, ,$$

converges, where $k(\xi', \tau'; \xi, \tau)$ is the kernel of (5.8). Now,

$$k_1 = \frac{|\xi - \xi'|}{\tau - \tau'} \cos\left(\varphi_{\xi\xi'}\right) Z_0(\xi - \xi', \tau - \tau') \, ,$$

$$k_2 = -4\alpha(\xi, \tau) Z_0(\xi - \xi', \tau - \tau') \, ,$$

$$k_3 = \frac{4\alpha^2(\xi, \tau)(\tau - \tau')}{|\xi - \xi'| \cos\left(\varphi_{\xi\xi'}\right)} Z_0(\xi - \xi', \tau - \tau') \, .$$

Note that k_1 and k_2 are weak polar kernels, while k_3 becomes weak polar if additional assumptions are made. Indeed, if $\cos(\varphi_{\xi\xi'}) \sim |\xi - \xi'|$, then $k_3 \in K_{-1,-2}$, and hence the condition (5.22) is not satisfied. Therefore, we assume that the boundary ∂G satisfies the condition

$$\cos(\varphi_{\xi\xi'}) > c|\xi - \xi'|^\lambda \, , \quad \xi, \xi' \in \partial G \, , \quad 1 > \lambda_1 > 0 \, .$$

Then $k_3 \in K_{-1,-1-\lambda_1}$, therefore the kernel k_3 is weak polar if $1 > \lambda_1 > 0$. Thus, for those boundaries the variance of $\eta_{xt}^{(3)}$ is finite.

Now we show that the variance of $\eta_{xt}^{(4)}$ is finite if the Ljapunov parameter λ is larger than 1/2. It suffices to prove that k^2/p_4 is a weak polar kernel where k is the kernel of (5.8). Now,

$$c_1[\tau - \tau']^{-3/2}|\xi - \xi'|^{1-\beta} \exp\left\{-\frac{|\xi - \xi'|^2}{4a^2(\tau - \tau)}\right\} \leq p_4(\xi', \tau' \to \xi, \tau)$$

$$\leq c_2[\tau - \tau']^{-3/2}|\xi - \xi'|^{1-\beta} \exp\left\{-\frac{|\xi - \xi'|^2}{4a^2(\tau - \tau')}\right\},$$

therefore

$$k^2/p_4 = k_1(\xi'\tau'; \xi, \tau) + k_2(\xi', \tau'; \xi, \tau) + k_3(\xi', \tau'; \xi, \tau),$$

where $k_1 \in K_{2,1+2\lambda+\beta}$, $k_2 \in K_{1,1+\beta}$, $k_3 \in K_{0,\beta-1}$. Consequently, the kernels k_1, k_2 and k_3 are all weak polar if

$$2\lambda + \beta > 2, \quad \lambda + \beta > 1. \tag{5.25}$$

The second inequality in (5.25) can be satisfied for arbitrary $\lambda > 0$ by appropriate choice of $\beta \in (0, 1)$, while the first inequality is satisfied for $\lambda > 1/2$.

We now turn to the estimation of the cost of the walk on boundary algorithms. To this end we use Theorem 1.4 where the average number of steps of the process $\{(\xi_k, \tau_k)\}$ was estimated as $MN_t \equiv n_t \leq c_1 t$. Note also that the variance of N_t is estimated as $DN_t \leq c_2 t$ since, in the context of Theorem 1.4, $DN'_t = \sigma^2 t/\mu^3 + 0(t)$ as $t \to \infty$ and $N'_t \geq N_t$.

Consider the estimate $\eta_{xt}^{(1)}$. Let $d(x)$ be the distance from the point x to the boundary ∂G. Then

$$|Q_0| = \left|\frac{\eta_{xt}^{(1)}(\xi_0^*, \tau_0^*)}{\pi(\xi_0^*, \tau_0^*)}\right| \leq \frac{c_4}{[d(x)]^{n+1}}$$

if $\pi(\xi_0^*, \tau_0^*) > c_3 > 0$ for all $(\xi^*, \tau^*) \in \Gamma_t$. From this we get, assuming that $\Psi(\xi, \tau)$ is bounded,

$$M\left[\eta_{xt}^{(1)}\right]^2 \leq \frac{c_5 t^2}{[d(x)]^{2(n+1)}}. \tag{5.26}$$

Choosing $\pi > c > 0$ and bounded $\Psi(\xi, \tau)$, an analogous estimation can be obtained for $\eta_{xt}^{(2)}$. This implies that the cost of the walk on boundary algorithms has the asymptotics $0(t^3)$ as $t \to \infty$.

Let us now consider the stabilization method for solving stationary boundary value problems and compare its efficiency with that of stationary walk on boundary algorithms described in Chap. 3.

Let us consider a pair of problems in a bounded domain G:

$$\frac{\partial u}{\partial t} = \Delta u(x, t) + f(x), \quad x \in G, \quad t > 0$$

$$u(x, 0) = 0 \qquad\qquad x \in G,$$

$$u(\xi, t) = \psi(\xi), \qquad\qquad \xi \in \partial G, \quad t > 0,$$

and

$$\Delta v(x) + f(x) = 0, \quad x \in G; \quad v(\xi) = \psi(\xi), \quad \xi \in \partial G.$$

Then the function $\bar{u}(x,t) = u(x,t) - v(x)$ satisfies the conditions

$$\frac{\partial \bar{u}}{\partial t} = \Delta \bar{u}(x,t) \quad x \in G, \quad t > 0$$

$$\bar{u}(x,0) = -v(x) \quad x \in G,$$

$$\bar{u}(\xi,t) = 0, \qquad \xi \in \partial G, \quad t > 0.$$

Assume that

$$\bar{u}(x,t) = \sum_{k=1}^{\infty} c_k \exp\{-\lambda_k t\}\, \varphi_k(x), \quad \Delta \bar{u}(x,t) = \sum_{k=1}^{\infty} d_k \exp\{-\lambda_k t\}\, \varphi_k(x),$$

where $\{\lambda_k\}_{k=1}^{\infty}$ and $\{\varphi_k\}_{k=1}^{\infty}$ are the eigen-values and eigen-functions of the Dirichlet problem for the Laplace equation, respectively; c_k and d_k are the Fourier coefficients of expansions of v and f, respectively. Then

$$\|\bar{u}(\cdot,t)\|_{L_2(G)} \le \exp\{-\lambda_1 t\}\, \|v\|_{L_2(G)},$$

$$\|\Delta \bar{u}(\cdot,t)\|_{L_2(G)} \le \exp\{-\lambda_1 t\}\, \|f\|_{L_2(G)},$$

where it is assumed that $|\lambda_1| \le |\lambda_2| \le \dots$. By the imbedding theorem [5.1] and from $|\bar{u}|_{\partial G} = 0$ we get the following uniform estimation

$$\|\bar{u}(\cdot,t)\|_{C(G)} \le c \exp\{-\lambda_1 t\}.$$

This implies that to achieve $\|\bar{u}(\cdot,t)\|_{C(G)} < \varepsilon$, it is necessary to take $t \cong |\ln \varepsilon|$. Thus, from $MN_t \cong t$, $D\eta_{xt}^{(i)} \simeq t^2$ we conclude that the cost of our algorithms behaves as

$$T_\varepsilon = 0(|\ln \varepsilon|^3/\varepsilon^2) \quad \text{as} \quad \varepsilon \to 0.$$

5.1.3 Diffusion Equation in a Half-space. Direct Monte Carlo Scheme

The walk on boundary algorithms presented in Sect. 5.1.1 are based on boundary integral equations equivalent to the original differential problem. Such integral equations can be obtained for some differential equations with variable coefficients [5.3]. However, this approach is applicable only to a simple geometry, in particular, to a half-space.

Let us consider in a half-space $G = \{(x_1, x_2, x_3) : x_3 > 0\}$ the following boundary value problem:

$$\frac{\partial c_1}{\partial t} - \sum_{i=1}^{3} \mu_i \frac{\partial^2 c_1}{\partial x_i^2} + \sum_{i=1}^{3} u_i(x,t) \frac{\partial c_1}{\partial x_1} + \sigma(x,t)c_1 = f(x,t),$$

$$c_1(x,0) = 0; \quad -\frac{\partial c_1}{\partial x_3} + \alpha c_1 = 0; \quad x \in \Gamma = \partial G.$$

Using the change of variables

$$c_1(x,t) = \exp\{\alpha x_3\} c\,(\lambda_1 x_1, \lambda_2 x_2, \lambda_3 x_3, t)\,,$$

where $\lambda_i = [\mu/\mu_i]^{1/2}$, ($\mu$ is a constant), we write

$$\frac{\partial c}{\partial t} - \mu \Delta c + \sum_{i=1}^{3} a_i(x,t) \frac{\partial c}{\partial x_i} + b(x,t)c = g(x,t)\,,$$

$$c\,(x,t) = 0\,, \quad \frac{\partial c}{\partial x_3} = 0\,; \quad x \in \Gamma = \partial G\,,$$

(5.27)

where

$$a_i(xt) = \lambda_i u_i \left(\frac{x_1}{\lambda_1}, \frac{x_2}{\lambda_2}, \frac{x_3}{\lambda_3}, t \right)\,, \quad i = 1,2\,;$$

$$a_3(x,t) = \lambda_3 u_3 \left(\frac{x_1}{\lambda_1}, \frac{x_2}{\lambda_2}, \frac{x_3}{\lambda_3}, t \right) - 2\mu_3 \alpha \lambda_3\,;$$

$$b(x,t) = \sigma \left(\frac{x_1}{\lambda_1}, \frac{x_2}{\lambda_2}, \frac{x_3}{\lambda_3}, t \right) - \mu_3 \alpha^2 + u_3 \left(\frac{x_1}{\lambda_1}, \frac{x_2}{\lambda_2}, \frac{x_3}{\lambda_3}, t \right)\,.$$

Thus, we shall construct the walk on boundary algorithm for solving (5.27). Let

$$Z_0^{(\alpha)}(x,t,x',t') = \frac{\exp\left\{ -\sum_{i=1}^{3} \frac{R_i^2(x,t,x',t')}{[4\mu(t-t')]} - \beta(t-t') \right\}}{[4\pi\mu(t-t')]^{-3/2}}\,,$$

where $R_i(x,t,x',t') = x_i - x_i' - \alpha_i(t-t')$. This function satisfies the equation

$$\frac{\partial Z_0^{(\alpha)}}{\partial t} - \mu \Delta_x Z_0^{(\alpha)} + \sum_{i=1}^{3} \alpha_i \frac{\partial Z_0^{(\alpha)}}{\partial x_i} + \beta Z_0^{(\alpha)} = \delta(x - x', t - t')$$

(5.28)

for arbitrary $x,t,x',t' \in \mathbf{R}^3$. We introduce the function

$$c_0 = Z_0^{(\alpha)}(x,t,x',t') + \hat{Z}_0^{(\alpha)}(x,t,x',t')\,,$$

where in the first term the following parameters are taken

$$\alpha_i = a_i' = a_i(x',t')\,, \quad i = 1,2,3\,, \quad \beta = b' = b(x',t')\,,$$

and

$$\hat{Z}_0^{(\alpha)}(x,t,x',t') = [4\pi\mu(t-t')]^{-3/2} \exp\left\{ -\sum_{i=1}^{2} \frac{R_i^2(x,t,x',t')}{[4\mu(t-t')]} \right.$$
$$\left. + \frac{(x_3 + x_3' + a_3(x_1',t')(t-t'))^2}{4\mu(t-t')} - b(t-t') \right\}\,,$$

204

where in $R_i(x, t, x', t')\alpha_i = a_i$, $i = 1, 2$. This function satisfies (5.28) with $\alpha_i = a_i$, $i = 1, 2$, $\alpha_3 = -a_3$, $\beta = b'$ and the right-hand side $\delta(x_1 - x_1', x_2 - x_2', x_3 + x_3', t - t')$.

It is clear that $\partial c_0/\partial x_3 = 0$ on the boundary $x_3 = 0$. Thus, c_0 solves the problem

$$\frac{\partial c_0}{\partial t} - \mu \Delta c_0 + \sum_{i=1}^{3} a_i' \frac{\partial c_0}{\partial x_i} + b' c_0 = \delta(x - x', t - t') + \delta(x - \hat{x}', t - t')$$

$$+ 2a_3' \frac{\partial Z_0}{\partial x_3}(x, t, x', t') , \quad \hat{x} = (x_1, x_2, -x_3) ; \tag{5.29}$$

$$\frac{\partial c_0}{\partial x_3} = 0 , \quad x \in \Gamma = \partial G ; \quad c_0(x, t, x', t') = 0 \quad \text{for} \quad t < t' .$$

This equation is obtained by adding equations of type (5.28) for Z_0 and $\hat{Z}_0$.

We seek the solution of (5.27) in the form of a volume potential with unknown density $q(x', t')$ and the kernel $c_0(x, t, x', t')$:

$$c(x, t) = \int_0^t dt' \int_G c_0(x, t, x', t')q(x', t')dx' .$$

To satisfy (5.27), q must be chosen from

$$q(x, t) + \int_0^t dt' \int_G k(x', t'x, t)q(x', t')dx' = g(x, t) \tag{5.30}$$

where

$$k(x', t'x, t) = \sum_{i=1}^{3} [a_i(x, t) - a_i(x', t')] \frac{\partial c_0}{\partial x_i}(x, t, x', t')$$

$$+ [b(x, t) - b(x', t')]c_0(x, t, x', t') + 2a_3(x', t')$$

$$\times \frac{\partial \hat{Z}_0}{\partial x_3}(x, t, x', t') . \tag{5.31}$$

Substituting c_0 in (5.31) yields

$$k(x', t'x, t) = \sum_{i=1}^{3} \frac{(a_i - a_i')[x_i - x_i' - a_i'(t - t')]}{2\mu(t - t')} Z_0$$

$$- \sum_{i=1}^{3} \frac{(a_i - a_i')[x_i - \hat{x}_i' - \hat{a}_i'(t - t')]}{2\mu(t - t')} \hat{Z}_0$$

$$- \frac{a_3'[x_3 + x_3' + a_3'(t - t')]}{\mu(t - t')} \hat{Z}_0 + (b - b')(Z_0 + \hat{Z}_0)$$

$$- \varrho Z_0 \left[\sum_{i=1}^{3} (a_i - a_i') \frac{\alpha_i}{2\mu(t - t')} \right] - \hat{\varrho} \hat{Z}_0$$

$$\times \left[\sum_{i=1}^{3}(a_i - a_i')\beta_i/2\mu(t - t') - 2a_3'\beta_3/[2\mu(t - t')] \right]$$
$$+ (b - b')(Z_0 + \hat{Z}_0) .$$

Here

$$\varrho = |x - x' - a'(t - t')| ; \qquad \hat{\varrho} = |x - \hat{x}' - \hat{a}'(t - t')| ;$$
$$\alpha_i = \frac{x_i - x_i' - a_i'(t - t')}{\varrho} ; \qquad \beta_i = \frac{x_i - \hat{x}_i' - \hat{a}_i'(t - t')}{\varrho} .$$

We now construct the Markov chain $\{(x_n, t_n)\}_{n=0}^{\infty}$ for solving (5.31). Let $\pi_0(x, t)$ be the initial density of the chain, and define its transition density as

$$r(x', t', x, t) = p_t(t|x', t')p_x(x|x', t', t)$$

where

$$p_t(t|x', t') = 2[(\tau - t')(t - t')]^{-1/2} , \quad t \in [t', \tau] ,$$
$$p_x(x|x', t', t) = \frac{\pi^{1/2}}{2}[4\mu(t - t')]^{-1/2}\exp\{b'(t - t')\}(\varrho Z_0 + \hat{\varrho}\hat{Z}_0) .$$

Let $N \equiv N_\varepsilon = \min\{n : t_n + \varepsilon > t\}$. Then the random quantity

$$\xi = \sum_{n=0}^{N_\varepsilon} Q_n c_0(x, t, x_n, t_n)$$

is an ε-biased estimate for $c(x, t)$. Simulation of the transition according to $r(x', t', x, t)$ can be carried out by two steps: First, simulate a new time $t = (T - t')\alpha^2$ and calculate an auxiliary vector $y = x' + 2[\mu(t - t')]^{1/2}\gamma_2\omega$, $\gamma_2 = -\ln(\alpha_2\alpha_3)$ is a random variable with a Gamma distribution, whose parameter is 2; $\alpha_1, \alpha_2, \alpha_3$ are independent random variables uniformly distributed on $(0, 1)$, ω is a unit isotropic vector. Next, calculate the new point

$$x = y = (y_1, y_2, y_3) , \quad \text{if} \quad y_3 > 0 ,$$
$$x = \hat{y} = (y_1, y_2, -y_3) , \quad \text{if} \quad y_3 < 0 .$$

5.1.4 Adjoint Scheme

Let $x' = (x_1', x_2', \ldots, x_n')$. In the half-space $x_n' > 0$ consider the problem

$$\frac{\partial c}{\partial t'} - \mu\Delta_{x'}c(x', t') + \sum_{i=1}^{3} u_i(x', t')\frac{\partial c}{\partial x} + \sigma(x', t')c = f(x', t') ,$$

$$c(x', 0) = \varphi(x') , \quad -\frac{\partial c}{\partial x_n'} + \alpha c(x', t') = 0 , \quad x_n' = 0 .$$

(5.32)

Let

$$Z_0(x, t, x', t') = \frac{\exp\left\{-\sum_{i=1}^{3}\dfrac{|x' - x' + v(x, t)(t - t')|^2}{4\mu(t - t')} - \sigma(t - t')\right\}}{[4\pi\mu(t - t')]^{-n/2}}$$

if $t' < t$, and $Z_0(x, t, x', t') = 0$ if $t' \geq t$. Here $v(x, t)$ is an arbitrary vector function $v(x, t) = (v_1(x, t), \ldots, v_n(x, t))$ such that $v_n(x, t) = 0$. In particular, one may take

$$v_i(x, t) = u_i(x, t) , \quad i = 1, \ldots, n - 1 ; \quad v_n \equiv 0 .$$

For brevity, in this section we shall use the notation $v'_i = v_i(x', t')$. Direct evaluation yields

$$\frac{\partial Z_0}{\partial t'} - \mu \Delta_{x'} Z_0 - \sum_{i=1}^{3} v_i(x, t) \frac{\partial Z_0}{\partial x'_i} + \sigma(x, t) Z_0 = \delta(x - x', t - t') .$$

From this equation and by (5.32) we get

$$\mu \Delta_{x'} c' = \frac{\partial c'}{\partial t'} + \sum_{i=1}^{n} u'_i \frac{\partial c'}{\partial x'_i} + \sigma' c' - f' ,$$

$$\mu \Delta_{x'} Z_0 = -\frac{\partial Z_0}{\partial t'} - \sum_{i=1}^{n} v_i(x, t) \frac{\partial Z_0}{\partial x'_i} + \sigma(x, t) Z_0 - \delta(x - x', t - t') .$$

Consequently,

$$\mu(Z_0 \Delta_{x'} c' - c' \Delta_{x'} Z_0) = Z_0 \left(\frac{\partial c'}{\partial t'} + \sum_{i=1}^{n} u'_i \frac{\partial c'}{\partial x'_i} + \sigma' c' - f' \right)$$

$$- c' \left(-\frac{\partial Z_0}{\partial t'} - \sum_{i=1}^{n} v_i(x, t) \frac{\partial Z_0}{\partial x'_i} + \sigma(x, t) Z_0 - \delta(x - x', t - t') \right)$$

$$= \frac{\partial}{\partial t}(Z_0 c') + Z_0 \sum_{i=1}^{n} u'_i \frac{\partial c'}{\partial x'_i} + c' \sum_{i=1}^{n} v_i(x, t) \frac{\partial Z_0}{\partial x'_i}$$

$$+ (\sigma' - \sigma) Z_0 c' - Z_0 f' + c' \delta(x - x', t - t') .$$

From this, using

$$Z_0 u'_i \frac{\partial c'}{\partial x'_i} = \frac{\partial}{\partial x'_i}(Z_0 u'_i c') - c' \frac{\partial}{\partial x'_i}(Z_0 u'_i) = \frac{\partial}{\partial x'_i}(Z_0 u'_i c') - c' Z_0 \frac{\partial u'}{\partial x'_i}$$

$$- c' u'_i \frac{\partial Z_0}{\partial x'_i} ,$$

we obtain

$$\mu(Z_0 \Delta_{x'} c' - c' \Delta_{x'} Z_0) = \frac{\partial}{\partial t'}(Z_0 c') + \sum_{i=1}^{n} \frac{\partial}{\partial x'_i}(Z_0 u'_i c') - (\text{div } u') Z_0 c'$$

$$+ c' \delta(x - x', t - t') + c' \sum_{i=1}^{n}(v_i - u'_i) \frac{\partial Z_0}{\partial x'_i} + (\sigma - \sigma') Z_0 c' - Z_0 f' .$$

Integrating this equality with respect to x' and t' over $G \times (t - \tau)$ we obtain by Green's formula

$$\int_{t-\tau}^{t} dt' \int_{\Gamma} \left(Z_0 \mu \frac{\partial c'}{\partial n_{x'}} - \mu c' \frac{\partial Z_0}{\partial n_{x'}} \right) d\Gamma_{x'} = \int_{t-\tau}^{t} dt' \int_{G} \frac{\partial}{\partial t'} (Z_0 c') dx'$$

$$+ \int_{t-\tau}^{t} dt' \int_{G} \sum_{i=1}^{n} \frac{\partial}{\partial x_i'} (Z_0 u_i' c') dx' + \int_{t-\tau}^{t} dt' \int_{G} c' \sum_{i=1}^{n} (v_i - u_i') \frac{\partial Z_0}{\partial x_i'} dx'$$

$$+ \int_{t-\tau}^{t} dt' \int_{G} (\sigma - \sigma' - \operatorname{div} u') Z_0 c' dx' - \int_{t-\tau}^{t} dt' \int_{G} f' Z_0 dx' + c(x,t) .$$

$$(5.33)$$

By the Ostrogradsky formula

$$\int_{t-\tau}^{t} dt' \int_{G} \sum_{i=1}^{n} \frac{\partial}{\partial x_i} (Z_0 u_i' c') dx' = \int_{t-\tau}^{t} dt' \int_{\Gamma} Z_0 c' (u', n_{x'}) d\Gamma_{x'}$$

$$= \int_{t-\tau}^{t} dt' \int_{\Gamma} Z_0 c' u_n' d\Gamma_{x'} .$$

But

$$\int_{t-\tau}^{t} dt' \int_{G} \frac{\partial}{\partial t'} (Z_0 c') dx' = \int_{G} Z_0(x,t,x',t') c(x',t') dx'$$

$$- \int_{G} Z_0(x,t,x',t-\tau) c(x',t-\tau) dx' = - \int_{G} Z_0(x,t,x',t-\tau)$$

$$\times c(x',t-\tau) dx' .$$

Therefore, using

$$\frac{\partial c'}{\partial n_{x'}} = -\frac{\partial c'}{\partial x_n'} = -\alpha c' ,$$

we obtain from (5.33)

$$c(x,t) = \int_{t-\tau}^{t} dt' \int_{\Gamma} \left(-\mu \frac{\partial Z_0}{\partial n_{x'}} - \mu \alpha Z_0 \right) c' d\Gamma_{x'}$$

$$+ \int_{G} Z_0(x,t,x',t-\tau) c(x',t-\tau) dx' - \int_{t-\tau}^{t} dt' \int_{\Gamma} Z_0 c' u_n' d\Gamma_{x'}$$

$$+ \int_{t-\tau}^{t} dt' \int_{G} c' \left[- \sum_{i=1}^{n} (v_i - u_i') \frac{\partial Z_0}{\partial x_i'} + (\sigma - \sigma' + \operatorname{div} u') Z_0 \right] c' dx'$$

$$+ \int_{t-\tau}^{t} dt' \int_{G} f' Z_0 dx' .$$

Integrating this equation with respect to τ over $[0, \tau_0]$ we obtain

$$\tau_0 c(x,t) = \int_{t-\tau}^{t} dt' [\tau_0 - (t-t')] \int_{\Gamma} \left(-\mu \frac{\partial Z_0}{\partial n_{x'}} - \mu \alpha Z_0 - u'_n Z_0 \right) c' d\Gamma_{x'}$$

$$+ \int_{t-\tau}^{t} dt' \int_{G} \left\{ Z_0 + [\tau_0(t-t')] \left[-\sum_{i=1}^{n} (v_i - u'_i) \frac{\partial Z_0}{\partial x'_i} \right. \right.$$

$$+ (\sigma - \sigma' + \operatorname{div} u') Z_0 \Big] \Big\} c' dx'$$

$$+ \int_{t-\tau}^{t} dt' [\tau_0 - (t-t')] f(x',t') Z_0(x,t,x',t') dx' .$$

This equation can be rewritten as follows

$$c(x,t) = \frac{1}{\tau_0} \int_{t-\tau_0}^{t} dt' [\tau_0 - (t-t')] \int_{\Gamma} \left(-\mu \frac{\partial Z_0}{\partial n_{x'}} - \mu \alpha Z_0 - u'_n Z_0 \right) c' d\Gamma_{x'}$$

$$+ \frac{1}{\tau_0} \int_{t-\tau_0}^{t} dt' [1 + (\tau_0 - (t-t')) k(x,t,x',t')] Z_0 c' dx'$$

$$+ F_{\tau_0}(x,t) \tag{5.34}$$

using the notation

$$k(x,t,x',t') = \sum_{i=1}^{n} \frac{(v_i - u'_i)}{2\mu(t-t')} \left[x'_i - x_i + v_i(t-t') \right] + \sigma - \sigma' + \operatorname{div} u' ;$$

$$F_{\tau_0}(x,t) = \frac{1}{\tau_0} \int_{t-\tau_0}^{t} dt' \int_{G} \left[\tau_0 - (t-t') \right] f(x',t') Z_0(x,t,x',t') dx' .$$

Now, to construct the random estimate for solving (5.34), we use a density on Γ proportional to $-\mu \partial Z_0/\partial n_x$. Since

$$-\mu \frac{\partial Z_0}{\partial n_{x'}} = \frac{|x' - x + v(x,t)(t-t')| \cos \psi}{2(t-t')[4\pi\mu(t-t')]^{n/2}} \exp \left\{ \frac{|x' - x + v(x,t)(t-t')|^2}{4\mu(t-t')} \right\}$$

the density can be represented as a product of a density of time and conditional density of a spatial point (t is fixed). We introduce a change of variables

$$x' + v(x,t)(t-t') = x'' , \quad t' = t''$$

which generates a one-to-one mapping $G \leftrightarrow G$, since $v_n(x,t) = 0$. Introduce the notation

$$\tilde{f}(x'',t'') = f(x'' - v(x,t)(t-t''), t'') .$$

Then (5.34) takes the form

$$c(x,t) = \frac{1}{\tau_0} \int_{t-\tau_0}^{t} dt'' \int_{\Gamma} [\tau_0 - (t-t'')] \left[\frac{x_n}{2(t-t'')} - \alpha \mu - \tilde{u}'_n \right]$$

$$\times \exp \left\{ -\frac{|x'' - x|^2}{4\mu(t-t'')} - \sigma(x,t)(t-t'') \right\} \tilde{c}(x'',t'') d\Gamma_{x''}$$

209

$$+ \frac{1}{\tau_0} \int_{t-\tau_0}^{t} dt'' \int_G [1 + (\tau_0 - (t - t''))\tilde{k}(x, t, x'', t'')]$$

$$\times \frac{\exp\left\{ -\dfrac{|x'' - x|^2}{4\mu(t - t'')} - \sigma(x,t)(t - t'') \right\} \tilde{c}(x'', t'')}{[4\pi\mu(t - t'')]^{n/2}}$$

$$\times dx'' + F_{\tau_0}(x, t) . \tag{5.35}$$

It is now convenient to rewrite (5.35) formally in a more general form

$$\varphi(x_0) = \int_X k(x_0, x_1)\varphi(T(x_0, x_1))dx_1 + f(x_0) ,$$

where $T(x_0, \cdot)$ is a one-to-one mapping $X \Leftrightarrow X$. Then

$$\varphi(x_0) = \int_X\!\!\int_X k(x_0, x_1)\delta(x_2 - T(x_0, x_1))\varphi(x_2)dx_2 dx_1 + f(x_0) .$$

Assume that it is desired to calculate a linear functional $J = (\varphi, h)$. We construct a Markov chain $\{x_n\}$ on the phase space X with an initial density π_0 and a transition density $r(x \to x')$. Then x_0 is sampled according to $\pi_0(x)$, $x_0 \to x_1$ is constructed according to $r(x \to x')$, then $x_2 = T(x_0, x_1)$. Next x_3 is sampled according to $r(x \to x')$ and $x_2 = T(x_0, x_1)$, etc.: $x_{2n+2} = T(x_{2n}, x_{2n+1})$. Let

$$Q^{(0)} = \frac{h(x_0)}{\pi(x_0)} , \quad Q^{(n)} = Q^{(n-1)}\frac{k(x_{2n-2}, x_{2n-1})}{r(x_{2n-2}, x_{2n-1})} ,$$

$$\xi_n = \sum_{k=0}^{n-1} Q^{(k)} f(x_{2k}) + Q^{(n)}\varphi(x_{2n}) .$$

Then $M\xi_n = J = (\varphi, h)$.

Now let us introduce a density

$$r(x, t, x'', t'') = p\, r_\Gamma(x, t, x'', t'')\chi_\Gamma(x'') + q r_G(x, t, x'', t'')\chi_G(x'') ,$$

where

$$p r_\Gamma = \frac{\tau_0 - \tau}{\tau_0} \frac{x_n}{2\tau} \frac{\exp\{-\varrho^2/4\mu\tau\}}{[4\pi\mu\tau]^{n/2}} ;$$

$$q r_G = \frac{1}{\tau_0} \frac{\exp\{-\varrho^2/4\mu\tau\}}{[4\pi\mu\tau]^{n/2}} ; \quad \tau = t - t'' , \quad \varrho = |x - x''| ,$$

and χ_Γ, χ_G are indicator functions of Γ and G, respectively. The function $r(x, t, x'', t'')$ is a density with respect to x'', t'', $x'' \in G$, $t'' \in [t - \tau_0]$, since the function $c \equiv 1$ satisfies the original differential equation if $u \equiv 0$, $\sigma \equiv 0$, $v \equiv 0$, $f \equiv 0$, $\alpha \equiv 0$. Therefore,

$$p = \int_0^{\tau_0} d\tau \int_\Gamma \frac{\tau_0 - \tau}{\tau_0} \frac{x_n}{2\tau} \frac{\exp\{-\varrho^2/4\mu\tau\}}{[4\pi\mu\tau]^{n/2}} d\Gamma_{x''} ,$$

$$q = \int_0^{\tau_0} d\tau \int_\Gamma \frac{1}{\tau_0} \frac{\exp\{-\varrho^2/4\mu\tau\}}{[4\pi\mu\tau]^{n/2}} , \qquad p + q = 1 .$$

To simulate the transition $(x, t) \to (x', t')$ according to the density $r(x, t, x'', t'')$, we use the rejection method and the composition technique.

First, we simulate a random point x'', t'' according to the right-hand side of the inequality

$$q r_G \chi_G(x'') \le \frac{1}{\tau_0} \frac{\exp\{-\varrho^2/4\mu\tau\}}{[4\pi\mu\tau]^{n/2}}$$

and calculate $\tau = \tau_0 \alpha$, $x'' = x + (2\mu\tau)^{1/2}\eta$, where α is a random number, uniformly distributed on $(0, 1)$, and η is a standard Gaussian random number. If $x''_n > 0$, then x'' and $t'' = t - \tau$ are the desired samples, otherwise (probability of this event is equal to $q = 1 - p$) we start to simulate the point according to the density

$$r_\Gamma = \frac{\tau_0 - \tau}{\tau_0 p} \frac{x_n}{2\tau} \frac{\exp\{-\varrho^2/4\mu\tau\}}{[4\pi\mu\tau]^{n/2}} .$$

We introduce an angle ψ by the equality $x_n = \varrho \cos\psi$. Then

$$r_\Gamma = \frac{\Gamma(n/2)\cos\psi}{\pi^{n/2}\varrho^{n-1}} \frac{\tau_0 - \tau}{\tau_0 p} \frac{\varrho^n}{2\tau} \frac{\exp\{-\varrho^2/4\mu\tau\}}{[4\pi\mu\tau]^{n/2}\Gamma(n/2)} .$$

This representation shows that the simulation can be carried out as follows: first the point $x'' \in \Gamma$ is chosen isotropically, next τ is sampled according to the density

$$\frac{\tau_0 - \tau}{\tau_0 p} \frac{\varrho^n}{2\tau} \frac{\exp\{-\varrho^2/4\mu\tau\}}{[4\pi\mu\tau]^{n/2}\Gamma(n/2)} .$$

Here it is convenient to introduce a new random variable $\theta = \varrho^2/4\mu\tau$. Let $\theta_0 = \varrho^2/4\mu\tau_0$. Then the density of θ has the form

$$\left(1 - \frac{\theta_0}{\theta}\right) \frac{\exp\{-\theta\}}{\Gamma(n/2)} \theta^{(n/2-1)} \frac{d\theta}{2p} , \qquad \theta \in [\theta_0, \infty) .$$

Having constructed θ, we then calculate $\tau = \varrho^2/4\mu\theta = t - t''$; hence, $t'' = t - \varrho^2/4\mu\theta$.

It remains to calculate the weights. Let

$$q(x, t, x'', t'') = \begin{cases} 1 - 2\tau\dfrac{\alpha\mu + \tilde{u}''_n}{x_n} , & x'' \in \Gamma , \\ 1 + (\tau_0 - \tau)\tilde{k}(x, t, x'', t'') , & x'' \in G . \end{cases}$$

Then $Q^{(0)} = 1$, $Q^{(n)} = Q^{(n-1)} q(x_{n-1}, t_{n-1}, x_n, t_n)$.

Thus, we have described how to construct the transition inside the domain G. If $x \in \Gamma$, we use the condition

$$\left\{\frac{\partial c}{\partial n} + \alpha c\right\}\bigg|_{\Gamma} = 0 \, .$$

Approximating near Γ the derivative in the direction of the exterior normal we get

$$\frac{\partial c}{\partial n} + \alpha c \approx \frac{c_0 - c_\varepsilon}{\varepsilon} + \alpha c_0 = 0 \, , \quad c_0 = \frac{c_\varepsilon}{1 + \alpha\varepsilon} \, .$$

Hence, $x'' = x + (0, \ldots, 0, \varepsilon)$, and the weight is multiplied by $q = 1/(1 + \alpha\varepsilon)$.

Let us now describe how $F_{\tau_0}(x, t)$ can be taken into account. For simplicity we consider the case when $f(x, t) = \delta(x - x_0)q(t)$. Then

$$F_{\tau_0}(x, t) = \frac{1}{\tau_0} \int_{t-\tau_0}^{t} dt' [\tau_0 - (t - t')]q(t')$$

$$\times \exp\left\{-\frac{\varrho^2}{4\mu(t - t')} - \sigma(t - t')\right\} [4\pi\mu(t - t')]^{-1} \, .$$

Now,

$$\varrho^2 = |x_0 - x + v(t - t')|^2 = |x_0 - x|^2$$
$$+ 2(x_0 - x, v(x, t))(t - t') + |v|^2(t - t')^2 \, .$$

Putting $\tau = t - t'$, $\xi = \tau/\tau_0$, we get

$$F_{\tau_0}(x, t) = \frac{1}{\tau_0} \int_{t-\tau_0}^{t} [4\pi\mu(t - t')]^{-1}[\tau_0 - (t - t')]q(t')$$

$$\times \exp\left\{-\frac{|x - x_0|^2}{4\mu(t - t')} - \frac{(x_0 - x, v)}{2\mu} - \frac{|v|^2(t - t')^2}{4\mu} - \sigma(t - t')\right\} dt'$$

$$= \exp\left\{-\frac{(x_0 - x, v)}{2\mu}\right\} \int_0^{\tau_0} \frac{(\tau_0 - \tau)q(t - \tau)}{4\pi\mu\tau}$$

$$\times \exp\left\{-\frac{|x - x_0|^2}{4\mu} - \left(\sigma + \frac{v^2}{4\mu}\right)\tau\right\} \frac{d\tau}{\tau_0}$$

$$= \exp\left\{-\frac{(x_0 - x, v)}{2\mu}\right\} \int_0^1 \frac{(1 - \xi)}{4\pi\mu\xi} q(t - \tau_0\xi) \exp\left\{-\frac{a}{\xi} - b\xi\right\} d\xi \, ,$$

where $a = |x - x_0|^2/4\pi\mu\tau_0$; $b = \tau_0(\sigma + v^2/4\mu)$. The last integral can be easily calculated using, e.g., Simpson's formula. Here it is necessary to estimate beforehand the quantities a and b. Indeed, if $q(t) \equiv 1$, then

$$\int_0^1 \frac{1 - \xi}{\xi} \exp\left\{-\frac{a}{\xi}\right\} d\xi \leq \int_a^\infty \frac{\exp\{-\xi\}}{\xi} d\xi \, .$$

Consequently, for $a = |x - x_0|/4\mu\tau_0 > 10$ the value of $F_{\tau_0}(x, t)$ is close to zero. This implies that $F_{\tau_0}(x, t)$ must be calculated only in the neighborhood of the point x_0, namely, inside the ball $B(x_0, 2[10\mu\tau]^{1/2})$.

5.1.5 Nonhomogeneous Case

As in the stationary problems (Chap. 3) we treat the nonhomogeneous problems by reducing to the homogeneous case and applying the double randomization principle. We present here two methods which permit us to carry out this reduction. The first approach is based on the application of the volume potentials. The second approach exploits the Green's function for the original problem

$$
\frac{\partial u}{\partial t} = a^2 \Delta u(x,t) + f(x,t) , \quad x \in G , \quad t > 0 ,
$$

$$
u(x,0) = \varphi(x) , \qquad\qquad x \in G ,
$$

$$
Bu(\xi,t) = \psi(\xi,t) , \qquad\qquad \xi \in \partial G , \quad t > 0 ,
$$

(5.36)

where $G = G_i$ for the interior problem, and $G = G_e$ for the exterior one, $B = I$ for the Dirichlet problem, and $B = a^2 \partial/\partial n$ for the Neumann problem, and $B = a^2 \partial/\partial n - \alpha I$ for the third boundary value problem.

Describe the first approach. Let $\bar{f}(x,t)$ and $\bar{\varphi}(x)$ denote prolongation of f and φ on $\mathbf{R}^n \times (0,\infty)$ and $\mathbf{R}^n$, respectively. If f satisfies the Hölder condition, and φ is continuous, then the function

$$
\begin{aligned}
u_1(x,t) = & \int_0^t dt' \int_{\mathbf{R}^n} \bar{f}(x',t') Z_0(x - x', t - t') dx' \\
& + \int_{\mathbf{R}^n} \bar{\varphi}(x') Z_0(x - x', t) dx'
\end{aligned}
$$

(5.37)

solves the problem

$$
\frac{\partial u_1}{\partial t} = a^2 \Delta u_1(x,t) + f(x,t) , \quad x \in G , \quad t > 0 ,
$$

$$
u(x,0) = \varphi(x) , \qquad\qquad x \in G .
$$

Let $u_2(x,t) = u(x,t) - u_1(x,t)$. From (5.37) we get

$$
\frac{\partial u_2}{\partial t} = a^2 \Delta u_2(x,t) + f(x,t) , \quad x \in G , \quad t > 0 ,
$$

$$
u_2(x,0) = \varphi(x) , \qquad\qquad x \in G .
$$

$$
Bu_2(\xi,t) = \psi(\xi,t) - Bu_1(\xi,t) = \psi_1(\xi,t) , \quad \xi \in \partial G , \quad t > 0 .
$$

Consequently, to calculate $u(x,t) = u_1 + u_2$, it is necessary to calculate also $Bu_1(\xi,t)$. To this end, we use the double randomization technique. Let

$$
J_1(x,t) = \int_0^t dt' \int_{\mathbf{R}^n} \bar{f}(x',t') Z_0(x - x', t - t') dx'
$$

$$
J_2(x,t) = \int_{\mathbf{R}^n} \bar{\varphi}(x') Z_0(x - x', t) dx' .
$$

Then the random quantity

213

$$\chi_1^{(0)}(x,t) = t\bar{f}(x + 2a[(t-\theta)\gamma_{n/2}]^{1/2}\omega, t-\theta)$$

is an unbiased estimate for J_1, i.e., $M\chi_1^{(0)}(x,t) = J_1(x,t)$. Here θ is a random number uniformly distributed in $[0,t]$, ω is a unit isotropic vector in $\mathbf{R}^n$. Analogously,

$$J_2(x,t) = M\chi_2^{(0)}(x,t),$$

where

$$\chi_2^{(0)}(x,t) = \bar{\varphi}\left(x + 2a[+\gamma_{n/2}]^{1/2}\omega\right).$$

To calculate $\psi_1(\xi,t)$, the estimate for the derivatives are used:

$$\frac{\partial J_1}{\partial x_i}(x,t) = M\chi_1^{(i)}, \quad \frac{\partial J_1}{\partial x_i}(x,t) = M\chi_1^{(i)}, \quad i = 1,\ldots,n,$$

where

$$\chi_1^{(i)} = -\frac{2\Gamma\left(\dfrac{n+1}{2}\right)}{\Gamma(n/2)} t^{1/2}\bar{f}\left(x + 2a\left(t - \frac{\theta^2}{t}\right)^{1/2}\left[\frac{\gamma_{n+1}}{2}\right]^{1/2}, \frac{\theta^2}{t}\right)\omega_i,$$

$$\chi_2^{(i)} = -\frac{\Gamma\left(\dfrac{n+1}{2}\right)}{\Gamma\left(\dfrac{n}{2}\right)} t^{-1/2}\bar{\varphi}\left(x + 2a[\gamma_{n+1}/2]^{1/2}\omega, \frac{\theta^2}{t}\right)\omega_i,$$

and ω_i are the coordinates of the isotropic vector ω. Note that $\chi_2^{(i)}$ has a singularity at $t = 0$ which must be taken into account when choosing the initial density $\pi(\xi,\tau)$.

Consider the second approach. Solution of the problem (5.36) can be written as $u(x,t) = u_1(x,t) + u_2(x,t)$, where the functions u_1 and u_2 satisfy the conditions

$$\begin{aligned}
\frac{\partial u_1}{\partial t} - a^2\Delta u_1(x,t) &= f(x,t), & x &\in G, & t &> 0, \\
u_1(x,0) &= \varphi(x), & x &\in G, & & \\
Bu_1(\xi,t) &= 0, & \xi &\in \partial G, & t &> 0; \\
\frac{\partial u_2}{\partial t} - a^2\Delta u_2(x,t) &= 0, & x &\in G, & t &> 0, \\
u_2(x,0) &= 0, & x &\in G, & & \\
Bu_2(\xi,t) &= \psi(\xi,t), & \xi &\in \partial G, & t &> 0.
\end{aligned}$$

(5.38)

To reduce the problem to the homogeneous case, we represent u_1 in the form

$$u_1(x,t) = \int_0^t dt' \int_G f(x',t')G(x,t-t';x')dx'$$
$$+ \int_G \varphi(x')G(x,t;x')dx' \tag{5.39}$$

where $G(x,t;x')$ is the Green's function for (5.36), i.e.,

$$\frac{\partial G}{\partial t} - a^2 \Delta_x G(x,t;x') = \delta(x-x',t), \quad x \in G, \quad t > 0,$$
$$G(x,0;x') = 0, \qquad\qquad\qquad x \in G, \quad x \neq x'$$
$$B\,G(\xi,t;x') = 0, \qquad\qquad\qquad \xi \in \partial G, \quad t > 0.$$

We seek G in the form

$$G(\xi,t;x') = Z_0(x-x',t) - g(\xi,t;x')$$

where g solves the problem

$$\frac{\partial g}{\partial t} = a^2 \Delta_g(x,t;x'), \qquad\qquad x \in G, \quad t > 0,$$
$$g(x,0;x') = 0, \qquad\qquad\qquad\quad x \in G, \tag{5.40}$$
$$B\,g(\xi,t;x') = BZ_0(\xi-x',t), \quad \xi \in \partial G, \quad t > 0.$$

Thus, u_2 and g are solutions of homogeneous problems (5.38) and (5.40), respectively. Therefore, it is possible to use the corresponding estimates presented in Sect. 5.1.1. Using then the double randomization technique we can calculate (5.37) and, finally, obtain an unbiased estimate for $u = u_1 + u_2$. Note that if G is unbounded, it is necessary to make some additional assumptions about the behavior of f and φ as $|x| \to \infty$, to provide finiteness of J_1, J_2, and the integrals in (5.39). For example, it is sufficient to assume that the functions f and φ are estimated by $\exp\{bx^2\}$, where b is a positive constant; then the solution of the first boundary value problem has an asymptotics $0(|x|^{-2})$ as $|x| \to \infty$, while the asymptotics of the second and third boundary value problem is $0(|x|^{-1})$.

5.1.6 Calculation of Derivatives

Consider the problem of calculation of derivatives (with respect to the normal vector at a boundary point $\xi \in \partial G$) of the first boundary value problem

$$\frac{\partial u}{\partial t} = a^2 \Delta u(x,t) + f(x,t), \quad x \in G \subset \mathbf{R}^n, \quad t > 0,$$
$$u(x,0) = \varphi(x), \qquad\qquad\qquad x \in G, \tag{5.41}$$
$$u(\xi,t) = 0, \qquad\qquad\qquad\qquad \xi \in \partial G, \quad t > 0.$$

This problem cannot be solved by direct differentiation of the double-layer potential since the corresponding integral diverges (exactly as in the stationary case described in Chap. 3). We present here another approach based on a representation discussed in [5.1]. Let $G(x, t; x')$ be the Green's function for (5.41). Then

$$u(x, t) = \int_0^t dt' \int_G f(x', t')G(x, t - t'; x')dx' + \int_G \varphi(x')G(x, t; x')dx' \ .$$

Let $g(\xi, t; x') = G(x, t; x') - Z_0(x - x', .t)$. Then

$$u(x, t) = \int_0^t dt' \int_G f(x', t')g(x, t - t'; x')dx' + \int_G \varphi(x')g(x, t; x')dx'$$
$$+ \int_0^t dt' \int_G f(x', t')Z_0(x - x', t - t')dx'$$
$$+ \int_G \varphi(x')Z_0(x - x', t)dx' \ . \tag{5.42}$$

The function $g(x, t; x')$ satisfies the conditions

$$\frac{\partial g}{\partial t} = a^2 \Delta g(x, t; x') \ , \qquad x \in G \ , \quad t > 0 \ ,$$
$$g(x, 0; x') = 0 \ , \qquad x \in G \ , \tag{5.43}$$
$$g(\xi, t; x') = -Z_0(\xi - x', t) \ , \quad \xi \in \partial G \ , \quad t > 0 \ .$$

Let $G' = \mathbf{R}^n \backslash (G \cup \partial G)$, $Q = G \times (0, \infty)$, $Q' = G' \times (0, \infty)$. We use a prolongation of $g(\xi, t; x')$ inside of Q': $g(\xi, t; x') = -Z_0(\xi - x', t)$, $(x, t) \in Q'$. Since $x' \notin G'$, the function $g(\xi, t; x')$ satisfies in Q' the conditions

$$\frac{\partial g}{\partial g} = a^2 \Delta g(x, t; x') \ , \quad x \in G \ , \quad t > 0 \ ,$$
$$g(x, 0; x') = 0 \ , \qquad x \in G \ ,$$
$$a^2 \frac{\partial g}{\partial n(\xi)}(\xi, t; x') = -a^2 \frac{\partial Z_0}{\partial n(\xi)}(\xi - x', t) = \psi(\xi, t; x') \ , \quad \xi \in \partial G \ , \quad t > 0 \ .$$

Therefore, the function $g(\xi, t; x')$ can be represented in Q' in the form of a simple-layer potential

$$g(\xi, t; x') = \int_0^t d\tau \int_G Z_0(x - \xi, t - \tau)\mu(\xi, \tau; x')d\sigma_\xi \ , \tag{5.44}$$

where the unknown density $\mu(\xi, \tau; x')$ on $\Gamma = \partial G \times (0, \infty)$ solves the equation

$$\mu(\xi, \tau; x') = \lambda \int_0^t d\tau' \int_G \frac{|\xi - \xi'|}{\tau - \tau'} \cos(\varphi_{\xi\xi'}) Z_0(\xi - \xi', \tau - \tau')\mu(\xi', \tau'; x')d\sigma_{\xi'}$$
$$+ \Psi(\xi, \tau; x') \ , \tag{5.45}$$

where $\lambda = 1$, $\Psi(\xi, \tau; x') = 2\psi(\xi, \tau; x')$ if G' is an interior domain (with respect to ∂G), and $\lambda = -1$, $\Psi = -2\psi$, if G' is an exterior domain. Note that the

216

representation (5.44) written for $g(\xi, t; x')$ in Q' is also true for $(x, t) \in Q$, since the potential (5.44) satisfies the conditions (5.43). Consequently, we obtain from (5.42)

$$
u(x, t) = \int_0^t dt' \int_G f(x', t') \left[\int_0^{t-t'} d\tau \int_{\partial G} Z_0(x - \xi', t - t' - \tau) \right.
$$
$$
\left. \times \mu(\xi', \tau; x') d\sigma_{\xi'} \right] dx' + \int_G \varphi(x') \left[\int_0^t d\tau \int_{\partial G} Z_0(x - \xi', t - \tau) \right.
$$
$$
\left. \times \mu(\xi', \tau; x') d\sigma_{\xi'} \right] dx' + J_1(x, t) + J_2(x, t) , \qquad (5.46)
$$

where

$$
J_1(x, t) = \int_0^t dt' \int_G f(x', t') Z_0(x - x', t - t') dx' ,
$$
$$
J_2(x, t) = \int_G \varphi(x') Z_0(x - x', t) dx' .
$$

Differentiating (5.46) along $n(\xi)$ yields

$$
a^2 \frac{\partial u}{\partial n(\xi)} = \lim_{x \to \xi} a^2 \sum_{i=1}^n \frac{\partial u}{\partial x_i}(x, t) n_i(\xi) = \int_0^t dt' \int_G f(x', t') \left[-\frac{\lambda}{2} \mu(\xi', \tau; x') \right.
$$
$$
\left. + \int_0^{t-t'} d\tau \int_{\partial G} a^2 \frac{\partial Z_0}{\partial n(\xi)}(\xi - \xi', t - t' - \tau) \mu(\xi', \tau; x') d\sigma_{\xi'} \right] dx'
$$
$$
+ \int_G \varphi(x') \left[-\frac{\lambda}{2} \mu(\xi', \tau; x') + \int_0^t d\tau \int_{\partial G} a^2 \frac{\partial Z_0}{\partial n(\xi)}(\xi - \xi', t - \tau) \right.
$$
$$
\left. \times \mu(\xi', \tau; x') d\sigma_{\xi'} \right] dx' + a^2 \left(\frac{\partial J_1}{\partial n(\xi)} + \frac{\partial J_2}{\partial n(\xi)} \right) .
$$

Thus, $a^2[\partial u / \partial n(\xi)](\xi, t)$ can be calculated as a linear functional of the integral equation (5.45) by using the double randomization technique.

For simplicity, let us consider the case when $\varphi(x) = 0$. Then it is sufficient to calculate the integral

$$
J_3(\xi, t) = \int_0^t dt' \int_G f(x', t') \left[\int_0^{t-t'} dt' \int_{\partial G} a^2 \frac{\partial Z_0}{\partial n(\xi)} \right.
$$
$$
\left. \times (\xi - \xi', t - t' - \tau') \mu(\xi', \tau'; x') d\sigma_{\xi'} \right] dx'
$$

since the construction of estimates for $\int_Q f \mu dx \, dt$ and $\partial J_1 / \partial n$, $\partial J_2 / \partial n$, was considered in Sect. 5.1.5. Let $p_0(x', t')$ be a density on $G \times (0, t)$ such that $p_0(x', t') \ne 0$ if $f(x', t') \ne 0$. Consider a random estimate

$$\eta_{\xi t} = \sum_{k=0}^{N^*_{t-t'}} \lambda^k Q^{(k)} \Psi(\xi_k^*, \tau_k^*; x') \,,$$

where $\{(\xi_k^*, \tau_k^*)\}$ is the walk on boundary process with initial density

$$\pi(\xi^*, \tau^*) = -2a^2 \frac{\partial Z_0}{\partial n(\xi)}(\xi^* - \xi, t - t' - \tau^*)$$

$$= \frac{|\xi - \xi^*|}{(t - t' - \tau^*)} \cos(\varphi_{\xi^* \xi}) Z_0(\xi^* - \xi, t - t' - \tau^*)$$

and with the same transition density $p(\xi, \tau \to \xi', \tau')$ used in Sect. 5.1.1, and

$$Q^{(0)} = -\frac{\cos(\varphi_{\xi \xi_0^*})}{2\cos(\varphi_{\xi_0^* \xi})} \,, \qquad Q^{(k)} = Q^{(k-1)} \frac{\cos(\varphi_{\xi_{k-1}^* \xi_k^*})}{\cos(\varphi_{\xi_k^* \xi_{k-1}^*})} \,.$$

Thus, the random quantity $\eta_{\xi t}$ depends on random parameters $\{(x', t', \omega^*)\}$, where ω^* is the trajectory $\{(\xi_k^*, \tau_k^*)\}$. By definition

$$M_{\omega^*}\{\eta_{\xi t}|(x', t')\} = \int_0^{t-t'} dt' \int_{\partial G} a^2 \frac{\partial Z_0}{\partial n(\xi)}(\xi - \xi', t - t' - \tau')$$
$$\times \mu(\xi', \tau'; x') d\sigma_{\xi'} \,,$$

therefore

$$J_3(x, t) = M\left\{ f(x', t') p_0^{-1}(x', t') \eta_{\xi t} \right\} \,.$$

Statement. The variance of the estimate $f(x', t') p_0^{-1}(x', t') \eta_{\xi t}$ is finite, if the following conditions hold.

1) ∂G is a Ljapunov surface.
2) There exists $\delta > 0$ such that $f(x) = 0$ for

$$x \in \partial G_\delta = \left\{ x' \in G : \inf_{x'' \in \partial G} |x' - x''| < \delta \right\} \,. \tag{5.47}$$

3) $\displaystyle \int_0^t dt' \int_G \frac{f^2(x', t')}{p_0(x', t')} dx' < \infty.$

Proof. From condition 2 we get

$$\sup_{x' \in G \backslash \partial G} |\psi(\xi, \tau; x')| < c$$

where c depends on δ and the domain G, hence

$$M\left\{\eta_{\xi t}^2 | (x', t')\right\} < c_1 \,, \quad (x', t') \in (G \backslash \partial G_\delta) \times (0, t) \,. \tag{5.48}$$

Now,

$$M \left(\frac{f(x', t')}{p_0(x', t')} \eta_{\xi t} \right)^2 = M_{(x', t')} M_{\omega^*} \left\{ \frac{f^2(x', t') \eta_{\xi t}^2}{p_0^2(x', t')} \Big| (x', t') \right\}$$

$$= M_{(x', t')} \left[\frac{f^2(x', t')}{p_0^2(x', t')} M_{\omega^*} \left\{ \eta_{\xi t}^2 | (x', t') \right\} \right] .$$

From this we get, using (5.47, 48),

$$M \left(\frac{f(x', t')}{p_0(x', t')} \eta_{\xi t} \right)^2 = \int_0^t dt' \int_{G \backslash \partial G} \frac{f^2(x', t')}{p_0(x', t')} M \left\{ \eta_{\xi t}^2 | (x', t') \right\} dx'$$

$$\leq c_1 \int_0^t dt' \int_G \frac{f^2(x', t')}{p_0(x', t')} dx' < \infty$$

which proves the statement. $\qquad\qquad\qquad\qquad\qquad\qquad\qquad\qquad\qquad\square$

Note that (5.46) can be used not only to calculate the solution near the boundary but also in the general case, because the asymptotics is $u \approx 0(|x|^{-1})$ as $|x| \to \infty$ while in the case of standard double layer potential representation $u \approx 0(|x|^{-2})$ as $|x| \to \infty$.

5.2 The Walk Inside the Domain Algorithms

In this section we continue to handle deterministic non-stationary problems and turn to the walk inside the domain algorithms: in Sect. 5.2.1 we construct the walk on spheres method for solving the Cauchy problem, and in Sect. 5.2.2 we apply the Laplace transform to utilize the walk on spheres algorithms constructed in Chap. 3.

5.2.1 Cauchy Problem

Consider the classical Cauchy problem in a layer $(0, T) \times \mathbf{R}^n$:

$$\frac{\partial u}{\partial t} - \mu \Delta u(x, t) = \sum_{i=1}^n b_i(x, t) \frac{\partial u}{\partial x_i} + c(x, t) u = f(x, t) , \qquad (5.49)$$

$$u(x, 0) = \varphi(x) ,$$

and make the assumption that there exists a unique solution to this problem. For example, this assumption is true if the following conditions are satisfied [5.4]:

1) The functions $b_i(x, t)$, $i = 1, \dots, n$ are bounded and the Hölder condition is satisfied:

$$|b_i(x, t) - b_i(x', t')| \leq M_1 \left(|x - x'|^\lambda + |t - t'|^{\lambda/2} \right) , \qquad \lambda \in (0, 1] .$$

2) The derivatives $\partial b_i(x, t)/\partial x_i$ are continuous and bounded in H.

3) The functions $c(x, t)$ and $f(x, t)$ are bounded in H and in $\bar{H} = [0, t] \times \mathbf{R}^n$, respectively, and $f(x, t)$, as a function of x, satisfies the Hölder inequality.
4) $\varphi(x)$ is continuous and bounded in $\mathbf{R}^n$.

Let

$$Z(x, t; x', t') = \theta(t - t')[4\pi\mu(t - t')]^{-n/2} \exp\{-R_b^2(x, x')/4\mu(t - t')\} ,$$
$$Z^{(\alpha)}(x, t; x', t') = Z(x, t; x', t') - [4\pi\mu\alpha]^{-n/2} ,$$

where $R_b(x, t) = |x - x' - (t - t')b(x, t)|$ (α is a parameter).

We now define the following sets:

$$B_\alpha(x, t) = \{(x', t') \in \mathbf{R}^{n+1} ; \quad t' < t, \quad Z^{(\alpha)}(x, t; x', t') > 0\} ,$$
$$B_\alpha^{(\beta)}(x, t) = \{x', t' \in \mathbf{R}^{n+1} ; \quad t' - \beta < t' < t, \quad Z^{(\alpha)}(x, t; x', t') > 0\}$$

β being a parameter.

Obviously, a section of $B_\alpha(x, t)$ with the plane $t' = t - \tau$ ($0 < \tau < \alpha$) is a ball in $\mathbf{R}^n$ with the center at the point $x(\tau) = x - \tau b(x, t)$, of radius $R(\tau) = [2\tau\mu n \ln(\alpha/\tau)]^{1/2}$. We say that a pair of numbers $0 < \beta \le \alpha$ is admissible if $B_\alpha^{(\beta)}(x, t) \subset H$.

The following mean value relation can be obtained by simple generalization of a result presented in [5.2]:

$$u(x, t) = \int_{B(x(\beta), R(\beta))} Z^{(\alpha)}(x, t; x', t' - \beta)dx'$$

$$+ \int_{\partial B_\alpha^{(\beta)}(x, t)} \left(-\mu \frac{\partial Z^{(\alpha)}}{\partial n_{x'}}\right) u(x', t')d\sigma_{x'} dt'$$

$$+ \int_{B_\alpha^{(\beta)}(x, t)} k(x, t; x', t')Z^{(\alpha)}(x, t; x', t')u(x', t')dx' dt'$$

$$+ \int_{B_\alpha^{(\beta)}(x, t)} Z^{(\alpha)}(x, t; x', t')f(x', t')dx' dt' , \tag{5.50}$$

where $x(\beta) = x - \beta b(x, t)$, $B(x, t)$ is a ball of radius r centered at x, and

$$k(x, t; x', t') = \frac{b_i(x, t) - b_i(x', t')}{2\mu(t - t')} [x_i' - x_i - (t - t')b(x, t)]$$

$$+ \frac{\partial b_i}{\partial x_i}(x', t') - c(x', t') .$$

Taking in (5.50) the limit as $\alpha \to \infty$ ($\beta = t$ fixed) we get [keeping in mind that the maximum-minimum principle for $u(x, t)$ holds]

$$u(x, t) = \int_0^t dt' \int_{\mathbf{R}^n} k(x, t; x', t')Z(x, t; x', t')u(x', t')dx' + F(x, t) , \tag{5.51}$$

where

$$F(x, t) = \int_{\mathbf{R}^n} Z(x, t; x', 0)\varphi(x')dx' + \int_0^t dt' \int_{\mathbf{R}^n} Z(x, t, x', t')f(x', t')dx' .$$

We now prove that the Neumann series for the integral equation (5.51) converges in $C(\bar{H})$. Note that by the definition of F

$$\|F\|_{C(\bar{H})} \le T\|f\|_{C(\bar{H})} + \sup_{\mathbf{R}^n} |\varphi(x)| ,$$

i.e., $F \in C(\bar{H})$. We recall (Sect. 5.1.2) that kernels satisfying the condition

$$|k(x,t,x',t')| \le c\,(t-t')^{-(1-\nu+n/2)} \exp\left\{ -\gamma \frac{R_b^2(x,x')}{t-t'} \right\} \tag{5.52}$$

(for some positive constants c, ν, γ) are weak polar, i.e., the Neumann series for an integral operator with such kernels converges. Indeed, from

$$\int_0^t dt' \int_{\mathbf{R}^n} t'^{a}(t-t')^{-(1-\nu+n/2)} \exp\left\{ -\gamma \frac{R_b^2(x,x')}{t-t'} \right\} dx'$$

$$= \left(\frac{\pi}{\gamma}\right)^{n/2} B(a+1,\nu)t^{a+\nu}$$

we conclude that for kernels k satisfying (5.52) the following inequality holds

$$|[K^l f](x,t)| \le c_0^l \sup_H |f(x',t')| B(l-1)\nu+1,\nu)\dots B(1,\nu)t^{l\nu} ,$$

$l = 1, 2, \dots$. Consequently,

$$\left| \sum_{l=1}^\infty [K^l f](x,t) \right| \le \sup_H |f(x',t')| \sum_{l=0}^\infty \frac{[c_0 t^\nu \Gamma(\nu)]^l}{\Gamma(l\nu+1)} < \infty .$$

It remains to note that the kernel kZ of the integral equation is represented as a sum of two weak polar kernels, therefore it is also weak polar. The functions $b_i(x,t)$, $c_i(x,t)$ and $\partial b_i/\partial x_i$ are bounded, therefore

$$|k(x,t,x';t')|Z(x,t;x',t') \le \left[c_1 + c_2 \frac{R_b(x,x')}{t-t'} \right] (t-t')^{-n/2}$$

$$\times \exp\left\{ -\frac{R_b^2(x,x')}{t-t'} \right\} ,$$

and hence

$$\frac{R_b(x,x')}{t-t'} \exp\left\{ -\frac{R_b^2(x,x')}{4\mu(t-t')} \right\} \le c_3(t-t')^{-1/2} \exp\left\{ -\frac{R_b^2(x,x')}{8\mu(t-t')} \right\}$$

since the function $x^\alpha \exp\{-\varepsilon x^2\}$ is bounded for an arbitrary $\varepsilon > 0$. Thus, convergence of the Neumann series for (5.51) is proved.

Now let us present the corresponding random estimates. Let $p(t \to t')$ be a transition density such that $p(t \to t') \ge c_4 > 0$, $t \in (0,t')$, and $p(t \to t') = 0$ if $t' \ge t$ or $t' < 0$; $q(t) = 1 - \int_0^t p(t \to t')dt' > 0$, $q(t) < 1$. Then the transition $(x,t) \to (x',t')$ according to the density

$$r(x, t, x', t') = p(t \to t') Z(x, t, x'; t')$$

is simulated as follows:

1) Sample a random number γ uniformly distributed in $[0, 1]$.
2) If $\gamma < q(t)$, then a termination occurs.
3) If $\gamma \geq q(t)$, then sample t' from $[0, t]$ according to the density $p(t \to t')/[1 - q(t)]$, and calculate

$$x' = x - b(x, t)(t - t') + [4\mu(t - t')\gamma_{n/2}\omega]^{1/2} .$$

Let $\{(x_k, t_k)\}_{k=0}^{N}$ be a Markov chain [starting from (x_0, t_0)] with the transition density $r(x, t, x'; t')$; N is the random number of the termination state. Then

$$\xi(x_0, t_0) = \sum_{k=0}^{N} Q^{(k)} F(x_k, t_k) \tag{5.53}$$

is an unbiased estimate for $u(x_0, t_0)$, the solution of (5.49). In (5.53)

$$Q^{(0)} = 1 , \quad Q^{(k)} = Q^{(k-1)} \frac{k(x_{k-1}, t_{k-1}; x_k, t_k)}{p(t_{k-1} \to t_k)} .$$

If we replace in (5.53) $F(x_k, t_k)$ by its unbiased estimate

$$\xi_k^F = \varphi \left(x_k - b(x_k, t_k) t_k + [4\mu t_k \gamma_{n/2}]^{1/2} \omega_k \right)$$
$$+ t_k f \left(x_k - \tau_k b_k + [4\mu t_k \gamma_{n/2}]^{1/2} \omega_k, \tau_k \right)$$

where τ_k is uniformly distributed in $[0, t_k]$, then

$$\xi'(x_0, t_0) = \sum_{k=0}^{N} Q^{(k)} \xi_k^F$$

is an unbiased estimate for $u(x_0, t_0)$.

We now prove that the variance of ξ' is finite. It is sufficient to prove that the Neumann series for an integral equation with the kernel

$$\frac{k^2 Z^2}{r} = k^2(x, t; x', t') \frac{Z(x, t; x', t')}{p(t \to t')} \tag{5.54}$$

converges. The kernel (5.54) is weak polar. Indeed,

$$\frac{k^2 Z}{p} \leq c_5 \left(\frac{|b(x, t) - b(x', t')| R_b(x, x')}{2\mu(t - t')} + c_6 \right)^2 Z(x, t; x', t')$$
$$= k_1(x, t; x', t') + k_2(x, t; x', t') + k_3(x, t; x', t') ,$$

where

$$k_1 = c_5 \frac{|b(x, t) - b(x', t')| R_b^2(x, x')}{4\mu^2(t - t')^2} Z(x, t; x', t') ;$$

$$k_2 = c_5 c_6 \frac{|b(x,t) - b(x',t')| R_b(x,x')}{\mu(t-t')} Z(x,t;x',t') ;$$

$$k_3 = c_5 c_6^2 Z(x,t;x',t') .$$

The kernels k_2 and k_3 are obviously weak polar. The proof of the fact that k_1 is weak polar too, follows from calculations

$$k_1(x,t;x',t') \leq c_7 \frac{(|x-x'|^\lambda + (|t-t'|^{\lambda/2})^2}{(t-t')^2} R_b^2(x,x') Z$$

$$\leq c_7 \left[R_b^\lambda + |b(x,t)|(t-t')^\lambda + |t-t'|^{\lambda/2} \right]^2 \frac{R_b^2(x,x') Z(x,t;x',t')}{\mu(t-t')} .$$

Here we used an inequality of the type $(a+b)^\lambda \leq a^\lambda + b^\lambda$ for $a > 0$, $b > 0$, $\lambda \in [0,1]$, the Hölder condition, and the inequality

$$\sup_{(x',t')} \frac{[R_b(x,x')]^{\mu_1}}{(t-t')^{\mu_1/2}} \exp \left\{ -\varepsilon \frac{R_b^2(x,x')}{(t-t')} \right\} \leq c_8 ,$$

where the constant c_8 depends only on μ_1 and ε. Thus the kernel (5.54) is weak polar, and the variance of $\xi'(x_0,t_0)$ is finite.

5.2.2 Use of the Laplace Transform

In this section we apply the Laplace transform to construct Monte Carlo algorithms for solving nonstationary problems on the basis of Monte Carlo estimates for stationary boundary value problems described in Chap. 3 and by using a special technique for constructing the inverse Laplace transform.

We now present the general scheme and start with the equation

$$R_k[u(x,t)] - L_m[u(x,t)] = f(x,t) , \qquad (x,t) \in \mathbf{R}^n \times [0,T] , \qquad (5.55)$$

where $R_k[u]$ is a linear differential operator with constant coefficients of the type:

$$R_k[u] = p_0 \frac{\partial^k u}{\partial t^k} + p_1 \frac{\partial^{k-1} u}{\partial t^{k-1}} + \ldots + p_{k-1} \frac{\partial u}{\partial t} ,$$

and $L_m[u]$ is a linear differential operator in $\mathbf{R}^n$ of order $2m$ with real coefficients $a_\alpha(x)$. Suppose that m differential operators $B_j(x,D)$ are given, and the order of B_j is equal to $m_j(\leq 2m - 1)$. Let us consider a boundary value problem in $G \subset \mathbf{R}^n$:

$$R_k[u(x,t)] - L_m[u(x,t)] = f(x,t) ,$$

$$B_j(x,D)u|_{\partial G} = g_j , \quad j = 1,2,\ldots,m , \qquad (5.56)$$

$$u(x,0) = \varphi_0(x) , \quad \frac{\partial u}{\partial t}(x,0) = \varphi_1(x),\ldots, \frac{\partial^{n-1} u}{\partial t^{n-1}}(x,0) = \varphi_{n-1}(x) .$$

We shall assume that $g_j \in C^\infty(\partial G)$, and $\{B_j\}_{j=1}^m$, ∂G satisfy the conditions which provide unique solvability of (5.56) [5.5]. Moreover, we assume that there exist Laplace transforms (with respect to t) of the functions $u(x,t)$ and of all derivatives in (5.56). We introduce the notation

$$U(x,t) = \int_0^\infty u(x,t)\exp\{-st\}dt \equiv \mathcal{L}(u) . \tag{5.57}$$

Under quite general assumptions, when $\mathcal{L}(\partial^l u / \partial x^l)$ exist we obtain

$$\frac{\partial U}{\partial x}(x,s) = \int_0^\infty \frac{\partial u}{\partial x}(x,t)e^{-st}dt, \dots, \frac{\partial^l u}{\partial x^l}(x,s) = \int_0^\infty \frac{\partial^l u}{\partial x^l}(x,t)e^{-st}dt ,$$

$$\mathcal{L}\left(\frac{\partial^r u}{\partial t^r}(x,t)\right) = s^r U(x,s) - s^{r-1}\varphi_0(x) - s^{r-2}\varphi_1(x) - \dots - \varphi_{r-1}(x) .$$

Suppose also that there exist Laplace transforms of the functions $f(x,t)$ and $B_j u|_{\partial G}$:

$$\mathcal{L}(f(x,t)) = F(x,t) , \quad \mathcal{L}(g_j) = G_j(x,s) .$$

Now the Laplace transform of (5.56) gives

$$-R_k[s]U(x,s) + L_m[U(x,s)] = -F(x,s) - F_0(x,s) , \tag{5.58}$$

where

$$F_0(x,s) = \sum_{i=0}^{k-1} R_i(s)\varphi_{k-i-1}(x) ,$$

$$R_i[s] = p_0 s^i + p_1 s^{i-1} + \dots + p_i , \quad i = 0, \dots, k ,$$

and

$$B_j U|_{\partial G} = G_j , \quad j = 1, 2, \dots, m . \tag{5.59}$$

Now we assume that it is possible to derive a local integral equation equivalent to (5.58, 59) (Chap. 1). Then the walk inside the domain algorithms permit us to construct the Laplace images for all desired values of the parameter s simultaneously.

It is known that the construction of the inverse Laplace transforms is a numerically unstable procedure [5.6]. We apply a regularization method which is well adjusted to the dependent sampling technique where the images $U(x,s_1), \dots, U(x,s_l)$ are calculated simultaneously. Let us describe this method in detail. Changing the variables $z = \exp(-t)$, we obtain from (5.57) that

$$U(x,s) = \int_0^\infty u(x,t)\exp(-st)dt = \int_0^1 z^{s-1}u(x,-\ln(z))dz . \tag{5.60}$$

Using a discrete approximation to (5.60) we get [5.6]

$$U(x,s) = \sum_{i=1}^{n} w_i z_i^{s-1} u(x - \ln(z_i)) + R_n(u,s) , \qquad (5.61)$$

where w_i, z_i and $R_n(u,s)$ are the weights, the nodes and the remainder of the quadrature formula, respectively. Taking $s = 1,2,\ldots,n$ in (5.61), we derive a system of linear equations for the unknowns $u(x, -\ln(z_i))$, where $a_{sj} = w_j z_j^{s-1}$, $(s,j = 1,2,\ldots,n)$ are the entries of the matrix of this system. Notice that the matrix $A = (a_{sj})_{s,j=1}^{n}$ can be represented as a product of a diagonal matrix with entries w_i and the Wandermond matrix $(z_j^{s-1})_{s,j=1}^{n}$. The inverse matrix $A^{-1} = (a_{ij}^*)_{i,j}^{n}$ can be obtained explicitly:

$$a_{ij}^* = (\Delta_n w_i)^{-1}(-1)^{i+j}\sigma_{n-j} \prod_{\substack{n \geq p > k \geq 1 \\ k,p \neq i}} (z_p - z_k) .$$

Here Δ_n is the determinant of the Wandermond matrix

$$\Delta_n = \prod_{n \geq i > k \geq 1} (z_i - z_k) , \qquad \sigma_{n-j}^{(i)} = \Sigma_1(z_{k_1} z_{k_2} \cdots z_{k_{n-j}}) ,$$

Σ_1 denotes a sum over $\{1 \leq k_1 < \ldots k_{n-j} \leq n; k_1,\ldots,k_{n-j} \neq i\}$. Thus,

$$u(x,t_i) = \sum_{s=1}^{n} a_{is}^* U(x,s) - \sum_{s=1}^{n} a_{is}^* R_n(u,s) , \qquad t_i = -\ln(z_i) .$$

We now turn to the question of error analysis. Denote by $\tilde{u}(x,t_i)$ the approximate solution. Then

$$\tilde{u}(x,t_i) - u(x,t_i) = R_i + D_i , \qquad (5.62)$$

$$R_i = \sum_{s=1}^{n} a_{is}^* R_n(u,s) , \qquad D_i = \sum_{s=1}^{n} a_{is}^* \Delta(x,s) , \qquad i = 1,\ldots,n , \qquad (5.63)$$

and $\Delta(x,s)$ denotes the error of calculation of the Laplace images. As is seen from (5.62, 63), it is desired to construct a quadrature formula as accurate as possible [to minimize the first term in (5.62)]. On the other hand, it is also necessary to minimize the second term in (5.62), namely, the error D_i due to Monte Carlo calculations. To this end we use the dependence sampling technique. In this case the errors $\Delta(x,s)$ are correlated: $\Delta(x,s) \approx \delta(x)$, therefore

$$D_i \approx \delta(x)\varrho_i , \qquad \varrho_i = \sum_{s=1}^{n} a_{is}^* .$$

The important feature of the Wandermond matrix is that the values ϱ_i are small while the absolute values of a_{is}^* are very large (due to sign alternation). Below we show the values of ϱ_i for the Gauss approximation formula, $n = 12$:

i	1	2	3	4	5	6	7	8	9	10	11
ϱ_i	−0.63	0.97	−1.27	1.6	−1.99	2.5	−3.21	4.31	−6.15	9.8	−19.3

and ($\varrho_{12} = 6.1$).

We now give some practical recommendations. It is convenient to introduce a weight function $\psi(t)$ such that the function $f_1(x,t) = \psi(t)u(x,t)$ is slowly varying on an interval $[t_1, t_n]$, where $t_1 = -\ln(z_1)$, $t_n = -\ln(z_n)$. In this case, $R_i \to 0$, and the relative error $\tilde{u}(x, t_i)$ is approximately equal to $c x \delta(x)\varrho_i$. In particular, if $u(x,t) \approx \exp(-\alpha t)$, then $\psi(t) \approx \exp(\alpha t)$, and

$$\int_0^\infty f_1(x,t)\exp(-st)dt = \int_0^\infty u(x,t)\exp[-(s-\alpha)t]dt = U(x, s-\alpha) .$$

Therefore,

$$u(x, t_i) = \exp(\alpha t_i) \sum_{s=1}^n a_{is}^* U(x, s-\alpha) , \quad i = 1, \ldots, n .$$

Now let us consider simple examples.

1) The heat equation.

Consider the following boundary value problem for the heat equation ($x \in R^3$):

$$\frac{\partial u}{\partial t} = \Delta u - qu + f , \quad q \geq 0 , \quad (x,t) \in D = G \times (0,T) ,$$

$$u(x,0) = \varphi_0(x) , \quad x \in \bar{G} ,$$

$$u|_{\partial G} = v(x,t) , \quad (x,t) \in \partial G \times (0,T) , \quad v(x,0) = \varphi_0(x) .$$

Let $U(x,s) = \mathcal{L}(u(x,t))$, $F(x,s) = \mathcal{L}(f(x,t))$, $V(x,s) = \mathcal{L}(v(x,t))$. Laplace transformation of the boundary value problem yields

$$\Delta U(x,s) - c(s)U(x,s) = g(x,s) , \quad U(x,s)|_{\partial G} = v(x,s) , \qquad (5.64)$$

where $c(s) = q + s$, $g(x,s) = -\varphi_0(x) + F(x,s)$. The problem (5.64) can be solved by the walk on spheres algorithm (Sect. 3.2) or by the walk on boundary algorithm (Sect. 3.1) simultaneously for a given set of values of the parameters $s : s_1, s_2, \ldots, s_l$.

2) The wave equation.

Consider the boundary value problem

$$\frac{\partial^2 u}{\partial t^2} = \Delta u - qu + f , \quad q \geq 0 , \quad (x,t) \in D = G \times (0,T) ,$$

$$u(x,0) = \varphi_0(x) , \quad \frac{\partial u}{\partial t}(x,0) = \varphi_1(x) , \quad x \in \bar{G} ,$$

$$u(x,t) = v(x,t) , \quad v(x,0) = \varphi_0(x) , \quad (x,t) \in \partial G \times (0,T) .$$

Laplace transformation of this problem also yields (5.64) where $c(s) = q^2 + s^2$, $g(x, s) = -s\varphi_0(x) - \varphi_1(x) + F(x, s)$, therefore the same algorithm could be applied.

3) Vibration of a thin plate.
The boundary value problem

$$\frac{\partial^2 u}{\partial t^2} + \Delta\Delta u = f , \quad (x, t) \in D = G \times (0, T) ,$$

$$u(x, 0) = \varphi_0(x) , \quad \frac{\partial u}{\partial t}(x, 0) = \varphi_1(x) ,$$

$$u|_{\partial G} = v_1(x, t) , \quad \Delta u|_{\partial G} = v_2(x, t)$$

is reduced obviously to

$$\Delta\Delta u(x, s) = s^2 U(x, s) = -s\varphi_0(x) - \varphi_1(x) + F_1(x, s) ,$$

$$U(x, s)|_{\partial G} = V_1(x, s) , \quad \Delta V(x, s)|_{\partial G} = V_2(x, s) ,$$

so the algorithms of Chap. 4 can be applied. Note that the walk on spheres method is applicable if Re $\{\lambda_i\} \geq 0$, $i = 1, 2$, where λ_1, λ_2 are the roots of the characteristic equation $\lambda^2 + s^2 = 0$ (Sect. 4.2), so it works for all values of s. If the solution can be represented as $U(x, t) = tu_1(x, t)$, then it is convenient to use $U_1(x, s) = \mathcal{L}(u_1)$. In this case the random estimate for

$$U_1(x, s) = -\int_0^s U(x, p)dp$$

has the form

$$\xi_1(x, s) = \sum_{k=0}^N Q^{(k)} F_p(s - p)g[\xi_k + \varrho_k(p)\omega_k] - Q^{(N)}(p)\theta(s - p)V(Q^*, p) ,$$

where p is uniformly distributed on $[0, n]$, and $\theta(t)$ is the Heaviside step function.

Notice that this kind of averaging as well as averaging over s_0 leads to smoothing out of the error behavior (as a function of s); in turn, this results in improved accuracy when the inverse Laplace transforms are calculated. A detailed discussion can be found in [5.7]. We also compared this method with the method based on probabilistic representation for the telegraph equation described at the end of Sect. 1.1. Good results were obtained for dimensionless time of order 100 when the parameter α of the telegraph equation was taken as 9/160.

5.3 Particle Diffusion in Random Velocity Fields

In this section we are not directly concerned with a boundary value problem. However we study the same diffusion process which is described by a parabolic turbulent diffusion equation. This equation describes in fact only the mean concentration field which can be derived from the statistics of the motion of a single particle. The second moment, which is of particular interest, requires consideration of the joint statistics of the motions of two particles. As mentioned in [5.8] [cf. 5.9, 10] the Lagrangian Monte Carlo model, in which many particle trajectories are generated in order to evaluate ensemble average properties of the concentration field directly, in principle overcomes many of the limitations of the other approaches.

5.3.1 Particle Diffusion in Local-Isotropic Velocity Fields

The case of a local-isotropic velocity field is the most investigated [5.11] and therefore provides a good test for the algorithms constructed below.

Now let us consider the general formulation of the problem. There exist two main approaches. The first approach is based on the numerical solution of the motion equation for an ensemble of particles in a three-dimensional random velocity field [5.12, 13] $u(x,t) = U(x,t) + u'(x,t)$:

$$\frac{dX_\alpha(t)}{dt} = u(X_\alpha(t), t) , \quad X_\alpha(0) = x_{0\alpha} . \tag{5.65}$$

Here X_α represents the Lagrangian coordinates of the particle with index α, $\alpha = 1, 2, \ldots$, starting at $t = 0$ from the point $x_{0\alpha}$; $U(x,t) = \langle u(x,t) \rangle$ is the mean velocity field, $u'(x,t)$ is the fluctuation velocity field. In this approach, the input data are the mean field $U(x,t)$ and the fluctuations u'. In practice, it is necessary to calculate various functionals of the random trajectories $X_\alpha(t)$. For example, the mean concentration $c(x,t)$ from a unit instant point source placed at a point $x_{0\alpha}$ can be represented as a distribution density (of x) of a random point $X_\alpha(t)$:

$$c(x,t) = \langle \delta(X_\alpha(t) - x) \rangle \tag{5.66}$$

where δ is the Dirac function.

Note that (5.66) presents one-point distribution, i.e., a simplest statistical characteristic. More complicated are the multipoint distributions

$$p(X_1, X_2, \ldots, X_n, V_1, V_2, \ldots, V_n; x_1, x_2, \ldots, x_n, t_1, t_2, \ldots, t_n)$$

of random points $X_1(t_1), \ldots, X_n(t_n)$ with velocities $V_1(t_1) = u(X_1(t_1), t_1), \ldots,$ $V_n(t_n) = u(X_n(t_n), t_n)$. The one-point Lagrangian correlation matrix presents an example of important Lagrangian statistical characteristics:

$$R_L(t_1, t_2 | x_{01}) = \langle [V_1(t_1) - \bar{V}_1(t_1)][V_1(t_2) - \bar{V}_1(t_2)]^T \rangle ,$$

where $V_1(t) = u(X_1(t), t)$ is the Lagrangian velocity of the particle at a random point $X_1(t)$, $\bar{V}_1(t) = \langle \bar{V}_1(t) \rangle$, the superscript "T" means conjugation of a row vector. More generally, a two-point Lagrangian correlation tensor

$$R_{2L}(t_1, t_2 | x_1, x_2) = \langle [V_1(t_1) - \bar{V}_1(t_1)][V_2(t_2) - \bar{V}_2(t_2)]^{\mathsf{T}} \rangle$$

can be considered. Below, we shall consider also other statistical characteristics, e.g., the dispersion tensor, the velocity structural functions, the mean squared separation between two particles, etc. Calculation of these characteristics in this approach is based on the numerical solution of (5.65) for a fixed sample of the field $u(x, t)$, and then averaging over all realizations of $u(x, t)$.

In the second approach, it is assumed that mutual distributions of the set of random Lagrangian velocities $V_1(t), \ldots, V_n(t)$ of fixed particles $X_1, \ldots, X_n$, and the Lagrangian trajectories are found by solving the problem

$$\frac{dX_\alpha(t)}{dt} = V_\alpha(t), \quad X_\alpha(0) = x_{0\alpha}. \tag{5.67}$$

The approaches described have both advantages and disadvantages. It is clear that construction of $X_\alpha(t)$ according to (5.67) is significantly simpler because, in this case, we have to simulate a random vector process, while in the first approach it is necessary to simulate a vectorial random field. Note that only information about the statistical characteristics of the Euler velocity field $u(x, t)$ is available; measurement of mutual distributions for $V_\alpha(t)$ is practically not possible.

Formally, there exists the obvious connection between statistical characteristics of $u(x, t)$ and $V_\alpha(t)$:

$$V_\alpha(t) = u\left(\int_0^t V_\alpha(t') dt', t \right).$$

However, this relation is too complicated to use it in practice. Therefore, in the second approach it is necessary to use non-strict hypotheses about the structure of the statistical characteristics of $V_\alpha(t)$. To test these hypotheses one can utilize the first approach which is completely correct but its computational cost is higher.

In this chapter we use the first approach in the case of incompressible isotropic gaussian random fluctuations with a Kolmogorov energy spectrum. Lagrangian correlation and the structural functions of the velocity are calculated.

We present a detailed analysis of the mean squared separation of two particles and calculate the constant a in the Richardson law:

$$K(l_*) = a\varepsilon^{-1/3} l_*^{4/3}. \tag{5.68}$$

The Model of an Incompressible Isotropic Random Field. Suppose that the Taylor hypothesis is satisfied: $u'(x, t) = v(x - Ut)$; $v(x)$ is a gaussian isotropic incompressible random field with the Kolmogorov energy spectrum:

$$E(\lambda) = \begin{cases} c\bar{\varepsilon}^{2/3}\lambda^{-5/3} \,, & \lambda \geq \lambda_{\min} \,, \\ 0 & , \quad \lambda < \lambda_{\min} \,, \end{cases} \tag{5.69}$$

where the mean flow U is assumed to be constant, $\bar{\varepsilon}$ is the mean turbulent energy dissipation rate, $\lambda_{\min} = 2\pi/L$, L is the external scale of turbulence, c is a universal constant (≈ 1.4 [5.11], or ≈ 1.6).

Assume that $U = 0$, then $u(x,t) = u'(x,t) = v(x)$. Simulation of such a random field can be carried out by the formula (Chap. 1) [5.14]

$$v(x) = \left[\frac{\langle v^2 \rangle}{2}\right]^{1/2} [(\xi \times \omega)\cos\lambda(\omega, x) + (\eta \times \omega)\sin\lambda(\omega, x)] \tag{5.70}$$

where

$$\left[\frac{\langle v^2 \rangle}{2}\right]^{1/2} = \int_0^\infty E(\lambda)d\lambda = \frac{3}{2}c\bar{\varepsilon}^{2/3}\lambda_{\min}^{-2/3} = c_1\bar{\varepsilon}^{2/3}L^{2/3} \tag{5.71}$$

is the mean energy, $c_1 = 3c(2\pi)^{-2/3}/2$, ξ, η are two standard independent gaussian vectors, ω is a unit isotropic vector in $\mathbf{R}^3$, λ is random variable distributed on $(0, \infty)$ according to the density

$$p(\lambda) = 2\frac{E(\lambda)}{\langle v^2 \rangle} \,, \quad \lambda \in (0, \infty) \,. \tag{5.72}$$

It should be noted that the field (5.70) has the desired spectrum; its one-point distributions are gaussian but not its multipoint distributions. Therefore, we shall use the following modification. Let

$$\text{supp}\, E(\lambda) = \bigcup_{i=1}^n \Delta_i \,, \quad \Delta_i \cap \Delta_j = \varnothing \quad (i \neq j) \tag{5.73}$$

be a partition of the support of $E(\lambda)$ and let

$$u_{i*}^2 = \int_{\Delta_i} E(\lambda)d\lambda \,, \quad p_i = \begin{cases} \dfrac{E(\lambda)}{u_{i*}^2} \,, & \lambda \in \Delta_i \\ 0 & , \quad \lambda \notin \Delta_i \,. \end{cases}$$

Then the random field

$$v(x) = \sum_{i=1}^n u_{i*}[\xi_i \times \omega_i)\cos\lambda_i(\omega_i x) + (\eta_i \times \omega_i)\sin\lambda_i(\omega_i, x)]$$

also has the desired spectrum $E(\lambda)$ and approaches a gaussian field as n increases. Here ξ_i and η_i are standard independent gaussian vectors, ω_i are independent unit isotropic vectors, λ_i are random variables distributed in Δ_i according to the density $p_i(\lambda)$. We choose the partition (5.73) such that

$$u_{i*}^2 = n^{-1} \int_0^\infty E(\lambda) d\lambda = \frac{\langle v^2 \rangle}{2n} , \quad i = 1, \ldots, n ,$$

where

$$\Delta_i = \left[\lambda_{\min} \left(1 - \frac{i-1}{n} \right)^{-3/2} , \lambda_{\min} \left(1 - \frac{i}{n} \right)^{-3/2} \right] ,$$

$$i = 1, \ldots, n-1 , \quad \Delta_n = [\lambda_{\min} n^{3/2}, \infty) .$$

The simulation formulas have the form

$$\lambda_i = \lambda_{\min} \left(1 - \frac{i - \alpha_i}{n} \right)^{-3/2} ,$$

where α_i, $i = 1, \ldots, n$ are independent random variables uniformly distributed on [0,1]. Thus, the desired random field is simulated by

$$v_n(x) = \sum_{i=1}^n u_{i*} [(\xi_i \times \omega_i) \cos \theta_i(x) + (\eta_i \times \omega_i) \sin \theta_i(x)] , \tag{5.74}$$

where $\theta_i(x) = \lambda_i (\omega_i, x)$, $u_{i*} = [\langle v^2 \rangle / 2n]^{1/2}$.

One-Point Statistical Characteristics. Consider the motion of a particle in a stationary, homogeneous random field $u(x, t) = U + u'(x, t)$ where U is the mean velocity (U does not depend on x, t since u' is stationary and homogeneous)

$$\frac{dX(t)}{dt} = u(X(t), t) , \quad X(0) = x . \tag{5.75}$$

Let $Y(t) = X(t) - x$ be the displacement vector, and let $Y'(t) = Y(t) - \langle Y(t) \rangle = Y(t) - Ut$ be the fluctuation of the displacement. The dispersion tensor is described by the well-known Taylor formula

$$D_{ij}(t) \equiv \langle Y_i'(t) Y_j'(t) \rangle = \int_0^t (t - \tau) \left[B_{ij}^{(L)}(\tau) + B_{ji}^{(L)}(\tau) \right] d\tau \tag{5.76}$$

where

$$B_{ij}^{(L)}(\tau) = \langle V_i'(t + \tau) V_j'(t) \rangle , \quad i, j = 1, 2, 3$$

is the Lagrangian velocity correlation tensor, and

$$V(t) = u(X(t), t) , \quad V'(t) = V(t) - \langle V(t) \rangle .$$

We define the Lagrangian time scales

$$T_{ij} = \frac{1}{2} \int_0^\infty \left[R_{ij}^{(L)}(t) + R_{ji}^{(L)}(t) \right] dt , \tag{5.77}$$

where

$$R_{ij}^{(L)}(\tau) = B_{ij}^{(L)}(\tau)[\langle(u_i')^2\rangle\langle(u_j')^2\rangle]^{-1/2}$$

is the standardized Lagrangian correlation function $R_{ii}^{(L)}(0) = 1$. It is known that, under some general assumptions, e.g., finiteness of the integrals (5.77) and

$$\int_0^\infty t R_{ij}^{(L)}(r)dt < \infty ,$$

the asymptotic formula

$$D_{ij}(t) = 2[\langle(u_i')^2\rangle\langle(u_j')^2\rangle]^{1/2}T_{ij}t \tag{5.78}$$

is true, where $t \gg \max_{i,j} |T_{ij}| \equiv T_L$. Thus, in the homogeneous case there exists the very important relation (5.78) for large values of t; but for small times there is no simple relation of this type. Indeed, the general formula (5.76) includes very complicated and indeterminate functions $B_{ij}^{(L)}$. Note that it is often assumed that $B_{ij}^{(L)}$ has an exponential form, however, this is not always the case.

In the case of local homogeneous turbulence it is possible to obtain more detailed results about $D_{ij}(t)$. To do this, we use a moving coordinate system to eliminate the large scale velocity fluctuations [5.11]. Let

$$V^{(s)}(t) = X(t) - x - u(x,0)t = Y'(t) - u'(x,0)t ,$$
$$V^{(s)}(t) = V(t) - u(x,0)$$

and denote by $D_{ij}^{(L)}$ the Lagrangian structure tensor

$$D_{ij}^{(L)}(t) = \langle V_i^{(s)}(t)V_j^{(s)}(t)\rangle .$$

Then the following relation holds [5.11]:

$$B_{ij}^{(L)}(0)t^2 - D_{ij}(t) = \langle Y_i^{(s)}(t)Y_j^{(s)}(t)\rangle = \int_0^t (t-\tau)D_{ij}^{(L)}(\tau)d\tau . \tag{5.79}$$

According to Kolmogorov's hypothesis, in the inertial range $\tau_\eta \ll t \ll T_0$ the dependence (5.79) has a universal form, and [5.12]

$$\langle Y_i^{(s)}(t)Y_j^{(s)}(t)\rangle = \tfrac{1}{3}c_0\bar{\varepsilon}t^3\delta_{ij} \tag{5.80}$$

where c_0 is a universal constant, δ_{ij} is the Kronecker symbol. Formulas (5.79, 80) provide a representation of $D_{ij}(t)$:

$$D_{ij}(t) = \langle u_i'(x,0)u_j'(x,0)\rangle t^2 - \tfrac{1}{3}c_0\bar{\varepsilon}t^3\delta_{ij} , \qquad \tau_\eta \ll t \ll T_0 .$$

Analogous arguments show that

$$D_{ij}^{(L)}(t) = \langle V_i^{(s)}(t)V_j^{(s)}(t)\rangle = c_0\bar{\varepsilon}t^3\delta_{ij} , \qquad \tau_\eta \ll t \ll T_0 .$$

232

Theory of Relative Diffusion. Let $X_1 = X_1(t)$, $X_2 = X_2(t)$ be coordinates of a pair of particles started at $t = 0$ at a fixed point $x_{01} = X_1(0)$, $x_{02} = X_2(0)$ and moving in the isotropic random field (5.74). Consider the vector of relative displacement $l(t) = X_2(t) - X_1(t)$ and denote by $\varrho^2(t)$ the mean squared distance between $X_1(t)$ and $X_2(t)$: $\varrho(t) = [\langle l^2(t) \rangle]^{1/2}$. It should be noted that the realistic random velocity fields are in fact local–isotropic, i.e., the difference of the velocity vectors at two points in the inertial range is well described by the corresponding velocity difference of type (5.74). The quantities $\langle v^2/2 \rangle$ and L in (5.71) are then the energy and the external scale of the turbulent fluctuations in the inertial range where the following assumptions hold: fluctuations are isotropic, $\bar{\varepsilon}$ is constant, and the spectrum has the form (5.69).

One of the important characteristics of the inner range is the inner Kolmogorov scale $\eta = (\nu^3/\bar{\varepsilon})^{1/4}$, where ν is the kinematic viscosity coefficient. The inner scale of the model (5.74) is equal to zero because of (5.69). However, in the numerical implementation of (5.74) there exists a minimal scale of order $\eta_n = L n^{-3/2}$. If the value of the initial distance $l_0 = l(0) = x_{02} - x_{01}$ is close to the quantity η_n then the influence of this scale must be taken into account. Below, we investigate $\varrho(t)$ for large times or for relatively large initial distances l_0, so it is possible to neglect the influence of the scales η and η_n.

Let $\tau_3 = \tau_3(l_0)$ be a time such that the probability that $|l(t)| \gg l_0$ for $t > \tau_3$ holds, is large, i.e., it is assumed that $\varrho(t)$ does not depend on l_0 if $t > \tau_3$. If $l_0 \ll L$ then it is possible to define τ_1 such that the probability that $|l(v)| \ll L$ for $t < \tau_1$ holds, is large. Finally, define τ_0 such that $|l(t)| \gg L$ for $t > \tau_0$ (it is clear that $\tau_0 > \tau_1$). The interval (τ_3, τ_1) is called intermediate, (τ_3, τ_0) quasiasymptotic, and (τ_0, ∞) asymptotic [5.11]. It should be noted that the interval (τ_3, τ_1) is not empty if $l_0 \ll L$. We define $\tau(l_0)$ so that the probability that $|l(t)| \gg \eta$ is large if $t > \tau(l_0)$. The general theory [5.11] predicts that

$$\varrho^2(t) = g\bar{\varepsilon}t^3 \tag{5.81}$$

in the interval $t \in (\tau_3, \tau_1)$, if $l_0 \gg \eta$. In (5.81) g is a universal constant. This formula is also true for $l_0 \lesssim \eta$ in the interval $\max\{\tau_3(l_0), \tau(l_0)\} < t < \tau_1$. For $t > \tau_0$, the following asymptotics hold [5.11]:

$$\varrho^2(t) = 2[\langle (u_1')^2 \rangle T_1 + \langle (u_2')^2 \rangle T_2 + \langle (u_3')^2 \rangle T_3]t ,$$

where T_i, $i = 1, 2, 3$ are the Lagrangian time scales of the corresponding velocity components. Note that the relation (5.81) is closely related to Richardson's law. In this law, one considers diffusion of N particles $X_\alpha(t)$, $\alpha = 1, \ldots, N$. The effective diameter of the cloud of particles is defined by

$$l_*(t) = \left[\frac{1}{N(N-1)} \sum_{\alpha \neq \beta} \langle l_{\alpha\beta}^2(t) \rangle \right]^{1/2} ,$$

where $l_{\alpha\beta}(t) = |X_\alpha(t) - X_\beta(t)|$ is the distance between the particles with coordinates X_α and X_β. The quantity

$$K = \frac{1}{6} \frac{d}{dt} l_*^2(t) \tag{5.82}$$

is called the virtual coefficient of turbulent diffusion of the cloud. If for every pair X_α and X_β the conditions mentioned above hold [i.e., when (5.81) is true], then, obviously,

$$l_*^2(t) = g\bar{\varepsilon}t^3 , \quad t \in (\tau_3, \tau_1) . \tag{5.83}$$

From this we get by using (5.82) that $K = g\bar{\varepsilon}t^2/2$, so, taking (5.83) into account, we obtain Richardson's law

$$K(l_*) = a\bar{\varepsilon}^{1/3} l_*^{4/3} ,$$

where $a = g^{1/3}/2$.

Many authors have tried to obtain the constants g and a: in [5.11, 15] it was suggested that $a \approx 1$, in [5.16], $a \approx 0.1$; *Tatarsky* [5.17] estimated $g \approx 0.06$, $a \approx 0.2$, *Kraichnan* [5.11] $g \approx 2.4$; in [5.18] the value of g was taken as 0.4, in [5.19] $g \approx 0.6$. Note that the dispersion of these values is connected with the fact that in the experimental measurements of the constants it is difficult to obtain, simultaneously with $K(l_*)$, the value of $\bar{\varepsilon}$. In our numerical model we can define the turbulence characteristics $\bar{\varepsilon}$ and L and calculate the constants g and a to a higher accuracy (than the experimental values): $g = 0.15c^{3/2}(1 \pm 0.1)$, $a = 0.23c^{1/2}(1 \pm 0.03)$ (see the numerical results below).

Numerical Results. It is convenient to test the described technique in the case of isotropic turbulence on the basis of a comparison with the theoretical results mentioned above. Consider equation (5.65) with the right-hand side of (5.74)

$$\frac{dX(t)}{dt} = v_n(X(t), t) , \quad X(0) = x_0 . \tag{5.84}$$

Suppose that L, $\langle v^2 \rangle$, and $\bar{\varepsilon}$ are the input characteristics fully defining the isotropic turbulence. Note that in view of (5.71), only two of these parameters are independent. Let us choose them, for example, as $\langle v^2 \rangle$ and L. We introduce the dimensionless variables

$$\tilde{X}_i = \frac{X_i}{L} , \quad i = 1, 2, 3 ; \quad \tilde{t} = \frac{t}{T} ,$$

where $T = L[\langle v^2 \rangle]^{-1/2}$, and rewrite (5.84) in dimensionless form

$$\frac{d\tilde{X}}{d\tilde{t}} = \frac{1}{(2n)^{1/2}} \sum_{i=1}^{n} [a_i \cos(\theta_i(\tilde{X})) + b_i \sin(\theta_i(\tilde{X}))] \equiv \tilde{v}_n(\tilde{X}) \tag{5.85}$$

where

$$a_i = (\xi_i \times \omega_i) , \quad b_i = (\eta_i \times \omega_i) , \quad \theta_i(\tilde{X}) = \tilde{\lambda}_i(\omega_i, \tilde{X}) ,$$

$$\tilde{\lambda}_i = 2\pi \left[1 - \frac{i - \alpha_i}{n} \right]^{-3/2} , \quad i = 1, \dots, n ,$$

and the random parameters $\alpha_i, \xi_i, \eta_i, \omega_i$ are the same as in (5.74).

We have calculated the following dimensionless one-point statistical characteristics:

1) $\tilde{R}_{ij}^{(L)}(\tilde{t})$ – the Lagrangian correlation tensor
2) $\tilde{D}_{ij}(\tilde{t})$ – the dispersion tensor
3) $\tilde{D}_{ij}^{(L)}(\tilde{t})$ – the Lagrangian structure tensor of the velocity.

The dimensional quantities were then calculated from the relations

$$R_{ij}^{(L)}(t) = \langle v^2 \rangle \tilde{R}_{ij}^{(L)} \left(\frac{t}{T} \right) , \quad D_{ij}(t) = L^2 \tilde{D}_{ij} \left(\frac{t}{T} \right) ,$$

$$D_{ij}^{(L)}(t) = \langle v^2 \rangle \tilde{D}_{ij}^{(L)} \left(\frac{t}{T} \right) .$$

The tensors 1–3 were obtained by numerical solution of (5.84) by the Euler method with a uniform step Δt. The right-hand side of (5.85) is a sum of harmonics with various periods, the minimal period is $2\pi / \tilde{\lambda}_n \le n^{-3/2}$. From this, we chose Δt so that

$$\sigma = [\langle |X(t + \Delta t) - X(t)| \rangle]^{1/2} < n^{-3/2} .$$

Since the quantity σ is of the order Δt, it is reasonable to take $\Delta t = n^{-3/2}/m$, where $m \ge 1$ is a parameter (positive integer). Thus we have two parameters of the model: n and m. To obtain stable numerical values of the tensors 1–3, the calculations were carried out for $n = 2, 4, 8, 16$ and $m = 1, 2, 3$. In Fig. 5.1, the function $\tilde{R}_{11}^{(L)}(\tilde{t})$ is shown for $n = 16$, $m = 1$ and $m = 3$, the absolute statistical error varies between 0.022 and 0.016. $N = 4000$ samples of the random field were used. The results do not change significantly when one varies the step from Δt to $\Delta t / 3$. However, the calculations carried out for $n = 2, 4, 8, 16$ (Fig. 5.2) show that stable results can be obtained only for $n = 8$ and $n = 16$. Perhaps this is related to the fact that in order to evaluate $\tilde{R}_{11}^{(L)}(\tilde{t})$ accurately enough, it is important to simulate a gaussian (multidimensional) random field, i.e., it is not sufficient to preserve the form of the spectrum [note that $\tilde{v}_n$ has the desired spectrum (5.69) for arbitrary $n = 1, 2, \dots$]. In Fig. 5.3 the function $\tilde{D}(\tilde{t}) = \sum_1^3 D_{ii}(\tilde{t})$ is shown for $n = 2, 4, 8, 16$. These results agree with the linear law (5.72). Note that here multidimensional normality is also important. The coefficient $K = 0.5 \, dD/dt$ depends on n; the corresponding values for $n = 9$ and $n = 16$ are approximately equal to 0.075.

We now describe the results for the Lagrangian structure tensor $\tilde{D}_{11}^{(L)}(\tilde{t})$. In Fig. 5.4 this function is shown for $n = 16$, $m = 1$ and $m = 3$, and in Fig. 5.5 for $n = 8$, $m = 2$ and $n = 16$, $m = 1$.

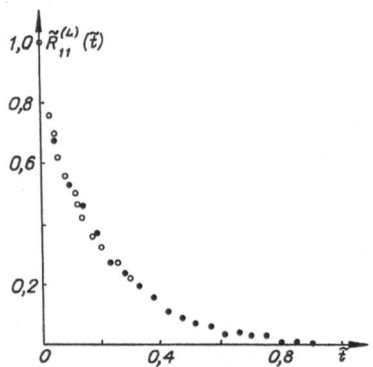

Fig. 5.1. Lagrangian correlation tensor $\tilde{R}_{ij}^{(L)}(\tilde{t})$ ($n = 16$); *circles, m = 1, points, m = 3*

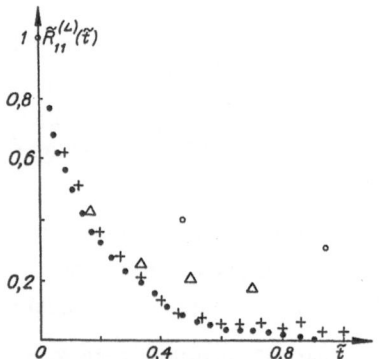

Fig. 5.2. Lagrangian correlation tensor $\tilde{R}_{ij}^{(L)}(\tilde{t})$ for different $n = 2, 4, 8, 16$; *circles, triangles, plusses, points*, respectively

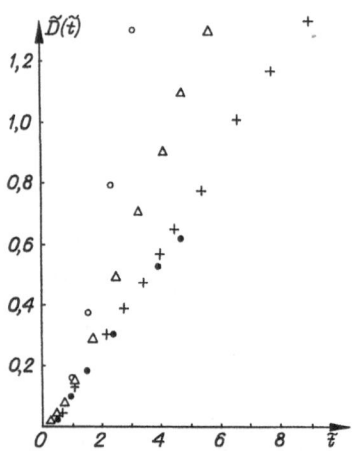

Fig. 5.3. Dispersion tensor $\tilde{D}(\tilde{t})$ calculated with different values of $n = 2, 4, 9, 16$; *circles, triangles, plusses, points,* respectively

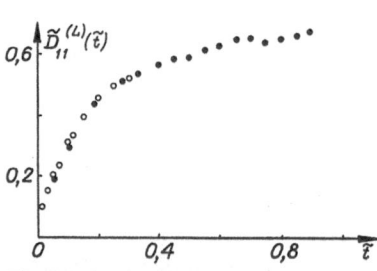

Fig. 5.4. Lagrangian structure tensor of the velocity $\tilde{D}_{ij}^{(L)}(\tilde{t})$ for $n = 16$, $m = 1$ (*points*) and $m = 3$ (*circles*)

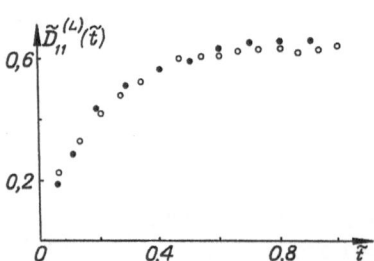

Fig. 5.5. *Points* as in Fig. 5.4, *circles* $n = 8$, $m = 2$

The relative statistical error is less than 10%, the number of samples is $N = 4000$. Comparison of Figs. 5.4 and 5.5 shows that it is sufficient to take $n = 8$. Note that $\tilde{D}_{11}^{(L)}(\tilde{t})$ does not change significantly above $\tilde{t} \approx 0.7$ and is close to 0.66. In Fig. 5.6 the function $[\ln(\tilde{R}_{11}^{(L)}]/3$ is shown. Here we have used a logarithmic scale to investigate the possibility of approximating the Lagrangian correlation function by $\exp(-t/T_L)$, where T_L is a Lagrangian time scale; Fig. 5.6 shows that the function can be approximated by $\exp(-5.2\tilde{t})$, i.e., $T_L = 0.19$. It should be noted that this approximation can be used in the interval $0.15 \leq \tilde{t} \leq 0.7$ (the relative error is approximately 10%).

We now describe the results of computation of the function

$$\varrho(t) = [\langle X_1(t) - X_2(t)| \rangle]^{1/2} \, ,$$

where X_1, X_2 are two simultaneous solutions of (5.75) with the initial values $x_{01} = X_1(0)$, $x_{20} = X_2(0)$. The numerical results are presented in the dimensionless form $\tilde{\varrho}(\tilde{t})$; then $\varrho(t) = L^2 \tilde{\varrho}(t/T)$. In Fig. 5.7, the function $[\tilde{\varrho}(\tilde{t})]^{2/3}$ is shown for $n = 2, 4, 8, 16$, the initial distance between the particles is $l_0 = 0.05$, the integration step Δt is chosen as described above, the number of random realizations is $N = 1000$. The relative error varies between 3 and 8%. Stable results are obtained already for $n = 2$, i.e., the behavior of $\varrho(t)$ weakly depends on whether the multidimensional velocity distribution is gaussian, provided that the initial distance $\varrho(0)$ is small. However, the situation changes if $\tilde{\varrho}(0)$ is not so small, as shown in Fig. 5.8, where $[\tilde{\varrho}(\tilde{t})]^{2/3}$ is presented for $n = 2$ and $n = 16$, for various initial distances $\tilde{\varrho}(0) = 0.05, 0.1, 0.25, 0.5$.

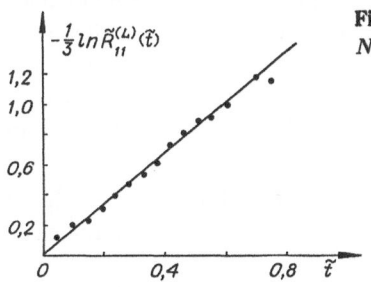

Fig. 5.6. The function $-\ln[\tilde{R}_{11}^{(L)}(\tilde{t})]/3$ for $n = 16$, $m = 1$, $N = 4000$ as a function of $\tilde{t}$

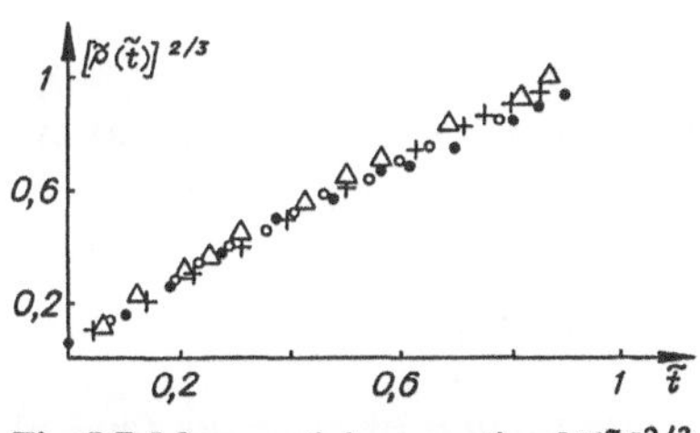

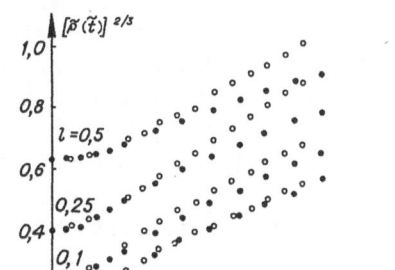

Fig. 5.7. Mean particle separation $[\tilde{\varrho}(\tilde{t})]^{2/3}$ for $n = 2, 4, 8, 16$; *circles, triangles, plusses*, and *points*, respectively

Fig. 5.8. The function $[\tilde{\varrho}(\tilde{t})]^{2/3}$ for $n = 2$ (*circles*) and $n = 16$ (*points*) for several initial distances l_0

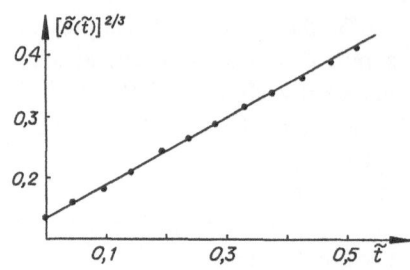

Fig. 5.9. Richardson's constant a from the slope of $[\tilde{\varrho}(\tilde{t})]^{2/3}$ vs. $\mathscr{F}$ for $n = 16$, $m = 1$, $N = 4000$

Let us compare the behavior of $\tilde{\varrho}(\tilde{t})$ with the theoretical results described above. Consider the variant $n = 16$ in Fig. 5.8. In the interval $0 \le \tilde{t} \le 0.5$ the numerical results agree well with (5.81). If the initial distance $\tilde{\varrho}(0)$ increases, this interval decreases, so that it nearly disappears at $\tilde{\varrho}(0) = 0.5$. Note that the value of $\tilde{t}$ where the cubic law is no longer valid [i.e., $\tilde{\varrho}^{2/3}(\tilde{t})$ becomes linear] does not significantly depend on $\tilde{\varrho}(0)$ and is approximately equal to 0.5. The numerical results obtained permit us to evaluate the constant g in (5.83) and Richardson's constant a. In Fig. 5.9, the function $[\tilde{\varrho}(\tilde{t})]^{2/3}$ is shown for $n = 16$, $m = 1$, $N = 4000$, the relative error is not larger than 2%. From this we can obtain $g = 0.246(1 \pm 0.1)$, $a = 0.313(1 \pm 0.03)$, (if $c = 1.4$). In general, $g = 0.15c^{3/2}(1 \pm 0.1)$, $a = 0.23c^{1/2}(1 \pm 0.03)$.

Thus, the numerical statistical model of a gaussian incompressible isotropic velocity field with the Kolmogorov spectrum of (5.69) shows good agreement with the classical theory of local-isotropic turbulence. This model can be used to test theories describing Lagrangian statistical characteristics. It should be noted again that the model suggested involves an integer parameter n, the number of harmonics which must be chosen subject to the characteristics to be calculated. For example, the mean-squared distance between two particles can be calculated stably already for $n = 2, 4$; however, when the Lagrangian structure function is calculated, it is necessary to take $n = 8, 16$. In general, n must be increased to obtain a numerically stable results (v_n converges to a gaussian field as $n \to \infty$) [5.20]. Theoretically, $v_n(x)$ has no "viscous interval" for arbitrary $n = 1, 2, \ldots$; however, numerical implementation of v_n generates non-zero "viscous interval" since v_n in (5.74) involves a finite number of harmonics. The ratio of the external scale to the inner scale of the turbulence can be estimated by $n^{3/2}$, so the Reynolds number can be estimated as $Re_n \ge (L/\eta_n)^{4/3} \approx n^2$.

Note finally that the form of the spectrum, incompressibility, and the isotropy are not limitations of the method, and the algorithm is easily extended to any spectra, compressible fields, and random fields with desired anisotropy.

5.3.2 Calculation of Statistical Characteristics of a Cloud

Consider now the problem of calculating the concentration of particles (and its variance) moving in a random velocity field $u(x) = (u_1, u_2, u_3)$. Assume that the particles are moving in a half-space $\mathbf{R}^3_{(+)} = \{(x, y, z) : z \ge 0\}$. Let $f(x, y, z, t)$ be the space-time distribution density of the source of particles, i.e., the sources emit $f(x,y,z,t)\Delta t \Delta x \Delta y \Delta z$ particles in the volume $\Delta x \Delta y \Delta z$ per interval of time Δt. In particular, f may represent a finite number of instant sources. A particle is moving according to (5.65). We assume that the particle may be absorbed, i.e., the medium is specified by an absorption coefficient such that Q, the "weight" of the particle is decreased according to $Q' = Q \exp[-\sigma(x,y,z)\Delta t]$ during a small time interval Δt. The particle weight can be interpreted as the admixture concentration in the neighborhood of the particle. More exactly, the weight of the particle satisfies the equation (considered along the particle trajectory) $X(t)$ [5.14]:

$$\frac{dQ}{dt} + \sigma(X(t),t)Q(t) = 0 ,$$

$$Q|_{t=t_0} = 1 .$$

(5.86)

We shall denote the solution of the system (5.65, 86) by

$$X_\alpha(t|x_{0\alpha}, t_0)$$

$$Q(t|x_{0\alpha}, t_0)$$

$(\alpha = 1, 2, \ldots)$, where $x_{0\alpha}$ indicates the initial point of α-th particle: $x_{0\alpha} = X_\alpha(0)$. This sytem describes the particle motion inside the domain $\mathbf{R}^3_{(+)}$. We describe the interaction with the boundary $\Gamma = \{(x,y,z) : z = 0\}$. If after the transition

$$X_\alpha(t + \Delta t) = X_\alpha(t) + u_\alpha(X(t),t)\Delta t ,$$

$$Q(t + \Delta t) = Q(t) \exp\{-\sigma(X(t)\Delta t\}$$

it happens that $X_3(t + \Delta t) < 0$, then we take

$$X_3'(t + \Delta t) = -X_3(t + \Delta t) ,$$

$$Q'(t + \Delta t) = (1 - p_0)Q(t + \Delta t) .$$

Here p_0 denotes the probability that the particle is absorbed by Γ. Consequently, the weight of the absorbed particle equals $p_0 Q(t + \Delta t)$.

Now we define the concentration of particles in the neighborhood of a phase point (x,y,z,t) as

$$c(r,t) = \int_0^\infty dt_0 \int_{\mathbf{R}^3_{(+)}} dr_0 f(r_0, t_0)Q(t|r_0, t_0)\delta(r - X(t|r_0, t_0)) ,$$

(5.87a)

where $r = (x,y,z)$, $r_0 = (x_0, y_0, z_0)$.

Notice that $c(r,t)$ is a random function since u is a random field. Consequently, the following functionals are also random:

$$J_g = \int_0^\infty dt \int_{\mathbf{R}_+} dr g(r,t)c(r,t) ,$$

(5.87b)

where $g(r,t)$ is an arbitrary function. From (5.87) we obtain, using (5.86)

$$J_g = \int_0^\infty dt_0 \int_{\mathbf{R}_{(+)}^3} dr_0 f(r_0, t_0 \left\{ \int_0^\infty dt g(X(t|r_0, t_0), t) Q(t|r_0, t_0) \right\} . \quad (5.88)$$

Consider the following special cases:

1) $g(r, t) = \chi_D(r)\delta(t - t^*)$, where

$$\chi_D(r) = \begin{cases} 1, & \text{if } r \in D \\ 0, & \text{if } r \notin D . \end{cases}$$

In this case (5.87) defines the total concentration in the domain D at the time instant t^*.

2) If $g(r, t) = \chi_D(r)\chi_\Delta(t)$, where $\Delta = [t_1^*, t_2^*]$, then (5.87) gives the total concentration in the domain D integrated over the time interval Δ.

3) If $g(r, t) = \sigma(r)\chi_D(r)\chi_\Delta(t)$, then (5.87) determines the concentration of particles absorbed in D during the time interval Δ.

To calculate the statistical characteristics of the functionals J_g we introduce a network $t_k = k\Delta t$, $k = 0, 1, \dots$. Then

$$J_g = \int_0^\infty dt_0 \int_{\mathbf{R}_{(+)}^3} dr_0 f(r_0, t_0) \sum_{k=0}^\infty g(X(t_k|r_0, t_0), t_k) Q(t_k|r_0, t_0) \Delta t . \quad (5.89)$$

The trajectory of the particle $X(t_k|r_0, t_0)$ and the weight $Q(t_k|r_0, t_0)$ at a transition are determined as described above, where the initial positions and weights are taken as $X(t_{k_0}|r_0, t_0) = r_0$, where k_0 is the integer part of $t_0/\Delta t$, and $Q(t_{k_0}|r_0, t_0) = 1$, while $Q = 0$ if $k < k_0$. We define the concentration $c_s(\gamma, \Delta)$ of particles absorbed by a region $\gamma \in S$ during a time interval $\Delta = [t_1^*, t_2^*]$ as

$$c_s(\gamma, \Delta) = \int_0^\infty dt_0 \int_{\mathbf{R}_{(+)}^3} dr_0 f(r_0, t_0) \sum_{k \in K(r_0, t_0)} p_0 Q(t_k|r_0, t_0) , \quad (5.90)$$

where $K(r_0, t_0)$ denotes a set of indices of time instances $t_k \in \Delta$ such that the trajectory $X(t|r_0, t_0)$ intersects the region γ:

$$K(r_0, t_0) = \{k : t_k \in \Delta, (X_1(t_k|r_0, t_0), X_2(t_k|r_0, t_0)) \in \gamma\} ,$$

$$X_3(t_k|r_0, t_0) + u_3(X(t_k|r_0, t_0), \Delta t\, k)\Delta t \leq 0 .$$

We are interested in the following statistical characteristics: $\langle J_g \rangle$, the mean concentration, $D_g = \langle J_g^2 \rangle - \langle J_g \rangle^2$, the variance, $\Pr\{J_g > crc\}$ – the probability that J_g exceeds the level of the critical concentration crc. Analogous statistical characteristics could be defined for the functional $c_s(\gamma, \Delta)$.

Now, applying the double randomization technique to the integral (5.89) we construct an algorithm of calculation of $\langle J_g \rangle$ and D_g. This means that we shall calculate the integral of the type (5.90) using one random node.

240

Eulerian Statistical Characteristics. In what follows, we shall assume that $g(x,t) \equiv 0$ if $t > T$, where $T = t_2^*$ is the last time instant of "measurements" of the functional J_g. We denote by $p(r_0, t_0)$ a density in $\mathbf{R}_{(+)}^3 \times (0, \infty)$ which is consistent with $f(r_0, t_0)$.

Algorithm A. (for calculating $\langle J_g \rangle$)
1) Construct a realization of the given random field $u(x, t)$.
2) Choose a random point (r_0, t_0) from the density $p(r_0, t_0)$ and calculate the initial weight $Q^{(0)} = f(r_0, t_0)/p(r_0, t_0)$.
3) Simulate the trajectory $X^{(0)}(t_k) = X(t_k|r_0, t_0)$ and calculate the weights $Q^{(0)}(t_k) = Q(t_k|r_0, t_0)$, $k = k_0, k_0+1, \ldots$, where k_0 is the integer part $t_0/\Delta t$.
4) Calculate the estimate for the sum in (5.89)

$$\xi_g^{(0)} = \sum_{k=k_0}^{\infty} g(X^{(0)}(t_k), t_k) Q^{(0)}(t_k) \Delta t \ .$$

5) Set $S = S + Q^{(0)} + \xi_g^{(0)}$.
6) Set $L = L + 1$; if $L \leq N$ then go to step 1, otherwise to 7.
7) Set $S = S/N$.

Here N is the number of trials.

The algorithm for calculating $\langle J_g^2 \rangle$ is analogous except that in step 2 we choose two independent points $(r_0^{(1)}, t_0^{(1)})$, $(r_0^{(2)}, t_0^{(2)})$, and calculate

$$Q_0^{(1)} = \frac{f(r_0^{(1)}, t_0^{(1)})}{p(r_0^{(1)}, t_0^{(1)})} \ , \quad Q_0^{(2)} = \frac{f(r_0^{(2)}, t_0^{(2)})}{p(r_0^{(2)}, t_0^{(2)})} \ ;$$

in step 3: we construct two conditionally independent trajectories starting at $(r_0^{(1)}, t_0^{(1)})$ and $(r_0^{(2)}, t_0^{(2)})$; in step 4: we calculate, on these two trajectories, the quantities: $\xi_g^{(1)}$ and $\xi_g^{(2)}$; in step 5: we calculate $S = S + Q_0^{(1)} Q_0^{(2)} \xi_g^{(1)} \xi_g^{(2)}$.

Algorithm B. (for calculating the probability $\Pr\{J_g > \mathrm{crc}\}$)
0) $S = 0$, $l = 1$.
1) As in algorithm A.
2) Choose n independent points $(r_0^{(i)}, t_0^{(i)})$, $i = 1, \ldots, n$ from the density $p(r_0, t_0)$ and calculate

$$Q_0^{(i)} = \frac{f(r_0^{(i)}, t_0^{(i)})}{p(r_0^{(i)}, t_0^{(i)})} \ , \quad i = 1, \ldots, n \ .$$

3) Simulate n trajectories $X^i(t_k)$ starting at the point $(r_0^{(i)}, t_0^{(i)})$, $i = 1, \ldots, n$ (the velocity u is fixed).
4) Calculate

$$\xi_g^{(i)} = \sum_{k=k_0^{(i)}}^{[T/\Delta t]} g(X^{(i)}(t_k), t_k) Q^{(i)}(t_k) \Delta t \ ,$$

where $k_0^{(i)} = [t_0^{(i)}/\Delta t]$ ([a] here means the integer part of a).

5) Calculate

$$\eta_n = \frac{1}{n} \sum_{i=1}^{n} Q_0^{(i)} \xi_g^{(i)} .$$

6) If $\eta_n > $ crc, then set $S = S + 1$, $l = l + 1$.
7) If $l \leq N$, then go to step 1.
8) Set $S = S/N$.

The obtained value S is the desired approximation to $\Pr \{ J_g > \text{crc} \}$.
Let us consider now the following important case, when

$$f(r_0, t_0) = \sum_{i=1}^{n} Q_i \delta(r_0 - r_0^{(i)}) \delta(t_0 - t_0^{(i)}) . \tag{5.91}$$

In this case we obtain from (5.89)

$$J_g = \sum_{j=1}^{m} Q_j \sum_{k=k_0}^{\infty} g(X^{(j)}(t_k), t_k) Q^{(j)}(t_k) \Delta t , \tag{5.92}$$

where $X^{(j)}(t_k) = X(t_k | r_0^{(j)}, t_0^{(j)})$, $k_0^{(j)} = [t_0^{(j)}/\Delta t]$, $Q^{(j)}(t_k) = Q(t_k | r_0^{(j)}, t_0^{(j)})$.
We present here the algorithm for calculating $\langle J_g \rangle$, D_g and $\Pr \{ J_g > \text{crc} \}$.

Algorithm C. (for $\langle J_g^m \rangle$)
  0) Set $S = 0$, $l = 1$.
  1) Construct a sample of u.
  2) Simulate m trajectories $X^j(t_k)$ and calculate the weights $Q^{(j)}(t_k)$, $j = 1, \ldots, m$.
  3) Calculate J_g from (5.92).
  4) Set $S = S + \xi_g$, $l = l + 1$.
  5) If $l \leq N$ then go to step 1.
  6) Set $S = S/N$.

Remark. The scheme of calculating $\langle J_g^2 \rangle$ is the same, the only difference is that in step 4 J_g^2 must be stored. Analogously, to calculate $\Pr \{ J_g > \text{crc} \}$, step 4 must be changed as follows: if $J_g > $ crc, then $S = S + 1$, $l = l + 1$.

Let us describe the scheme of calculating $c_s(\gamma, \Delta)$ from (5.90) for the source (5.91). Substituting of (5.91) into (5.90) yields

$$c_s(\gamma, \Delta) = \sum_{j=1}^{m} Q_j \sum_{k \in K_j} p_0 Q^{(j)}(t_k) , \tag{5.93}$$

where $K_j = K(r_0^{(j)}, t_0^{(j)})$, $Q^{(j)}(t_k) = Q(t_k | r_0^{(j)}, t_0^{(j)})$, $j = 1, \ldots, m$.

Algorithm D. (for $\langle c_s(\gamma, \Delta) \rangle$)
   0) Set $S = 0$, $l = 1$.
   1) Construct a sample of $u(x, t)$.
   2) Simulate m trajectories $X^j(t_k)$ and calculate the weights $Q^{(j)}(t_k)$, $j = 1, \ldots, m$; $k = 0, 1, \ldots, [T/\Delta t]$.
   3) Calculate $c_s(\gamma, \Delta)$ from (5.93).
   4) Set $S = S + c_s(\gamma, \Delta)$; $l = l + 1$.
   5) If $l \leq N$, then go to step 1.
   6) Set $S = S/N$.

   To calculate the variance of c_s and $\Pr\{c_s(\gamma, \Delta) > \mathrm{crc}\}$ several obvious changes must be introduced (see the previous *Remark*).
   In practical implementation of these algorithms some modifications are made to decrease the computing time. We illustrate it below for $\langle J_g \rangle$ and $\langle J_g^2 \rangle$.

Modified Algorithm for $\langle J_g \rangle$
   0) Set $S = 0$, $l = 1$.
   1) Construct a sample of $u(x, t)$.
   2) Choose a random point from a density $p(r_0, t_0)$; $s = 0$; $Q_0 = f(r_0, t_0)/p(r_0, t_0)$.
   3) Set $X = r_0$; $t = t_0$, $Q = 1$.
   4) If $t > T$, go to step 8.
   5) Set $s = s + g(x, t)Q\Delta t$.
   6) Set $Q = Q \times \exp[-\sigma(x)\Delta t]$, $X = X + u(X, t)\Delta t$.
   7) If $X_3 < 0$, then $Q = Q(1 - p_0)$, $X_3 = -X_3$; go to step 4.
   8) Set $S = S + sQ_0$; $l = l + 1$.
   9) If $l \leq N$, then go to step 1.
   10) Set $S = S/N$.

Modified Algorithm for $\langle J_g^2 \rangle$
   0-7) The points 0-7 are not changed.
   8) Choose (r_1, t_1) from a density $p(r_1, t_1)$, $s_1 = 0$, $Q_1 = f(r_1, t_1)/\Phi(r_1, t_1)$.
   9) Set $x = r_1$, $t = t_1$, $Q = 1$.
   10) If $t > T$ then go to step 14.
   11) Set $s_1 = s_1 + g(x, t)Q\Delta t$.
   12) Set $Q = Q\exp(-\sigma(x)\Delta t)$, $x = x + u(x, t)\Delta t$, $t = t + \Delta t$.
   13) If $x_3 < 0$, then $Q = Q(1 - p_0)$, $x_3 = -x_3$; go to step 10.
   14) Set $S = S + Q_0 Q_1 s s_1$; $l = l + 1$.
   15) If $l \leq N$ then go to step 1.
   16) Set $S = S/N$.

Lagrangian Statistical Characteristics. So far we described algorithms for calculating Euler statistical characteristics of the particle concentration. There is currently great interest in Lagrangian statistical characteristics, for instance, the mean and the variance of the geometrical center of a cloud, concentration fluctuations relative to the center, etc.

First, let us define the center of a cloud for a fixed velocity field by

$$0_\alpha^{(t)} = \int_{\mathbf{R}^3_{(+)}} c_1(r,t) x_\alpha dr \;, \quad \alpha = 1,2,3 \;, \tag{5.94}$$

where

$$c_1(r,t) = c(r,t) \left[\int_{\mathbf{R}^3_{(+)}} c(r,t) dr \right]^{-1} \;.$$

We define the variance relative to the center as

$$\sigma_\alpha^2 = \int_{\mathbf{R}^3_{(+)}} c_1(r,t)(x_\alpha - 0_\alpha)^2 dr \;, \quad \alpha = 1,2,3 \tag{5.95}$$

and the variance of the upper part of the cloud (relative to the center) as

$$\sigma(t) = \int_D c_1(r,t)(x_3 - 0_3)^2 dr \;, \tag{5.96}$$

where $D = \{(x,y,z) : x_3 > 0_3(t)\}$, $c_1 = c/q$,

$$q(t) = \int_D c(r,t) dr \;.$$

The variance of the lower part of the cloud is defined analogously.

Assumption. We suppose for the sake of simplicity that the total concentration

$$q(t) = \int_{\mathbf{R}^3_{(+)}} c(r,t) dr \tag{5.97}$$

is a known function of time (e.g., a constant).

Using the notation introduced above in (5.87) we get

$$0_\alpha(t^*) = J_g \tag{5.98}$$

where $g(r,t) = x_\alpha \delta(t - t^*)/q(t)$. It should be noted that (5.98) is linear because we are considering a special case when $q(t)$ is a given determinate function. In the general case, $0_\alpha(t)$ is a non-linear functional.

If the assumption (5.97) holds, then

$$\mu_\alpha(t^*) = J_g(t^*) \;, \tag{5.99}$$

where

$$\mu_\alpha(t^*) = \int_{\mathbf{R}^3_{(+)}} c_1(r,t) x_\alpha^2 dr \;, \quad g(r) = x_\alpha^2 \frac{\delta(t - t')}{q(t)} \;.$$

244

Consequently, from $\sigma_\alpha^2(t^*) = \mu_2(t^*) - 0_\alpha^2(t)$ the statistical characteristics $\sigma_\alpha^2(t^*)$ could be calculated as described above. The fundamental difference between the functionals (5.94, 95) and the functional (5.96) [under the assumption of (5.97)] is that the function q and even the domain D are not known. Therefore it remains only to apply the multiple randomization technique, as described for $\Pr\{J_g > \text{crc}\}$.

5.3.3 Statistical Model of the Turbulent Velocity Field for a Horizontally Homogeneous Boundary Layer

In this section we shall assume that it is sufficient to take into account only the velocity fluctuations on the micrometeorological scale. In this case $u'(x,t)$ can be considered as a stationary horizontally homogeneous random field (i.e., partially homogeneous, Sect. 1.2.4). Let us discuss this briefly.

We denote by $u(x,t) = (u_1, u_2, u_3)^{\mathrm{T}}$ the random field of wind fluctuations. It is generally accepted that $u(x,t)$ can be considered as a vector process with stationary increments. In addition, the six-dimensional vector process $(u(x,t), u(y,t))$ also has stationary increments for arbitrary $x = (x_1, x_2, x_3)$ and $y = (y_1, y_2, y_3)$. We denote by $\psi(\omega_1, x, y)$ the spectral tensor of this process. It is reasonable to assume that

$$\int_{-\infty}^{\infty} \operatorname{Sp} \psi(\omega, x, y) d\omega < \infty . \tag{5.100}$$

For instance, it is sufficient to require that

$$\int_{-\infty}^{\infty} \psi_{ii}(\omega, x, x) d\omega < \infty , \quad i = 1, 2, 3 .$$

Notice that in the inertial interval $\psi_{ii}(\omega, x, x)$ is equal to $c_i \omega^{-5/3}$, so the last integral diverges. To overcome this problem one uses a cut of the spectrum at the point zero.

Thus we define a partial spectral tensor

$$\Phi_{lm}(\omega, x, y) = \frac{1}{2\pi} \int \langle u_l(x, t+\tau) u_m(y, t) \rangle \exp\{-i\omega\tau\} d\tau . \tag{5.101}$$

Assuming that the six-dimensional vector

$$v(x, y, t) = (u(x, y, z_1, t), u(x, y, z_2, t))^{\mathrm{T}}$$

is a local homogeneous field (with respect to the variables x, y, z when z_1 and z_2 are fixed) we can define a spectral tensor $\psi(\omega, k_x, k_y, z_1, z_2)$. Assuming again, if necessary, that

$$\int_{\mathbf{R}^3} \operatorname{Sp} \psi(\omega, x, y) d\omega \, dk_x \, dk_y < \infty ,$$

we find that the field $v_{z_1,z_2}(x, y, z)$ is homogeneous (for arbitrary fixed z_1 and z_2). Consequently, the field $u(x, y, z, t)$ is homogeneous with respect to x, y and t, i.e., it is partially homogeneous. Thus we shall assume in what follows that $u(x, y, z, t)$ is stationary and homogeneous with respect to x, y.

Usually, it is not possible to obtain full information about Φ_{lm} from experiments. However, the rapidly expanding knowledge of Eulerian turbulence statistics in the planetary boundary layer [5.8, 9, 21] can be incorporated, for example, the property of the functions $p_{lm}(\omega, x) = \mathrm{Re}\,[\Phi_{lm}(\omega, x, x)]$, which was recently discovered, that the ratio $p_{lm}(\omega, x)/u_*^2 z$ can be represented as a universal function φ_{lm} of a dimensionless frequency $f = \omega z/U(z)$ and the equilibrium parameter $\zeta = z/L$:

$$p_{lm}(\omega, z) = \frac{u_*^2 z}{U(z)}\, \varphi_{lm}(f, \zeta) . \tag{5.102}$$

Here $z = x_3$, $U(z) = (U_1, 0, 0)$ is the mean velocity, $u_*^2 = -\overline{u_1 u_3}$ is the dynamical velocity, $L = -u_*^2/(H\kappa g/T)$ is the length scale in the Monin–Obuchov theory, g/T is the buoyancy parameter, H is the turbulent flux of the temperature, $\kappa = 0.4$ is Karman's constant.

The form of φ_{lm} is well known in the inertial frequency range [5.11]:

$$\varphi_{lm} = \zeta_m f^{-5/3}[\bar{\varepsilon} z/u_*^3]^{2/3} ,$$

where $\bar{\varepsilon}$ is the mean rate of dissipation energy. For example, in [5.9] $\bar{\varepsilon} = u_*^3/(\kappa z)$ was taken in the case of the neutral stratification of the boundary layer. For stable stratification [5.9] $\bar{\varepsilon}(z) = u_*^3[1 + (\beta_1 - 1)z/L]/(\kappa z)$, where $\beta_1 \approx 5 \pm 0.5$.

The mean wind in the boundary layer is approximated by [5.9]:

$$U(z) = \frac{u_*}{z} \begin{cases} \ln\left(\dfrac{z}{z_0}\right) + 9.9\dfrac{z}{L} , & \dfrac{z}{L} > 0 , \\[2mm] \ln\left(\dfrac{z}{z_0}\right) + 1.45\dfrac{z}{L} , & -0.16 \le \dfrac{z}{L} \le 0 , \\[2mm] 0.24 + 1.25\left(\dfrac{z}{L}\right)^{-1/} - \ln\left|\dfrac{z_0}{L}\right| , & \dfrac{z}{L} < -0.16 \end{cases}$$

in stable, neutral, and nonstable stratification, respectively. From the choice of the coordinate system and the assumed symmetry of u_2 with respect to the plane XZ it follows that there are only the following non-zero entries in the tensor $\{p_{lm}\}$: $p_{ll}(l = 1, 2, 3)$ and $p_{13} = p_{31}$ which can be approximated as described above [5.9 15],

$$p_{ll}(\omega, z) = \frac{c\overline{u_l^2}(z)}{\omega_0^{(l)}(\zeta)} \left[1 + \frac{3}{2}\left(\frac{\omega}{\omega_0^{(l)}}\right)^{5/3}\right]^{-1} , \tag{5.103}$$

$$p_{13}(\omega, z) = \frac{c'\overline{u_*^2}(z)}{\omega_0(\zeta)} \left[1 + \frac{3}{4}\left(\frac{\omega}{\omega_0}\right)^{7/3}\right]^{-1} , \tag{5.104}$$

where $0 \leq \omega < \infty$ and $\omega_0^{(l)}$, ω_0 are characteristic frequencies. The constants c and c' are calculated from the conditions

$$\int_{-\infty}^{\infty} p_{ll}(\omega, z)d\omega = \frac{\overline{u_l^2(z)}}{2} \;, \quad \int_{-\infty}^{\infty} p_{13}(\omega, z)d\omega = \frac{u_*^2(z)}{2} \;,$$

where it is assumed that an even prolongation of the functions (5.103, 104) is used. From this we get [5.18] $c = 1/I = 0.648$, $c' = 1/I' = 0.641$, where

$$I = \int_0^{\infty} [1 + 1.5x^{5/3}]^{-1}dx \;, \quad I' = \int_0^{\infty} [1 + 0.75x^{7/3}]^{-1}dx \;,$$

$$I = \frac{1}{0.648} \;, \quad I' = \frac{1}{0.641} \;.$$

As is seen from (5.103, 104), the characteristic frequencies $\omega_0^{(l)}$, ω_0 are the points of maxima of functions $\omega p_{ll}(\omega, z)$ and $\omega p_{33}(\omega, z)$, respectively. The functions $\overline{u_l^2(z)}$ and $\omega_0^{(2)}(\zeta)$, $\omega_0(\zeta)$ can be approximated by [5.18]

$$\overline{u_l^2(z)} = u_*^2 s_l^2(\zeta) \;, \quad \omega_0^{(l)}(\zeta) = U(z)\frac{f_0^{(l)}(z)}{2} \;, \quad \omega_0(\zeta) = U(z)\frac{f_0(\zeta)}{z} \;,$$

where

$$s_l(\zeta) = \begin{cases} A_l(1 - a_l\zeta)^{1/3} \;, & \zeta \leq 0 \;, \\ A_l(1 + b_l\zeta)^{-d_l} \;, & \zeta \geq 0 \;, \end{cases}$$

$A_1 = 2.2$, $A_2 = 1.8$, $A_3 = 1.42$, $a_1 = 0.75$, $a_2 = 1.4$, $a_3 = 1.7$, $b_1 = 1.5$, $b_2 = 1.0$, $b_3 = 0.5$, $d_1 = 0.75$, $d_2 = 1.0$, $d_3 = 0.5$, $f_0^{(3)}(\zeta) = \exp\{3(2\arctan(\zeta) - 1)/5\}$.

Notice that in the high frequency region of the interval the velocity is well approximated by an isotropic field. Therefore, the spectral tensor $p_{kl}(\omega, z)$ has, in view of the Taylor hypothesis, a diagonal form with

$$p_{11}(\omega, z) = \frac{E_1(\omega/U(z))}{[2U(z)]} \;,$$

$$p_{22}(\omega, z) = p_{33}(\omega, z) = \frac{E_2(\omega/U(z))}{[2U(z)]} \;,$$

where $E_1(k)$ and $E_2(k)$ are the longitudinal and the transversal one-dimensional energy spectra, respectively. Thus, $E_j(k) = 2F_j(k)$, $j = 1, 2$, where F_1, F_2 are the corresponding longitudinal and transversal one-dimensional spectra. In general, they are defined as [5.11]:

$$F_j(k_1) = \int\int_{-\infty}^{\infty} F_{jj}(k_1, k_2, k_3)dk_1 dk_2 dk_3 \;, \quad j = 1, 2 \;.$$

Here $F_{jl}(k)$ are the entries of the spectral tensor of the velocity field $u(x, t)$, t fixed, and $k = (k_1, k_2, k_3)$, i.e.,

$$B_{jl}(r) = \langle u_j(x_1 + r, t)u_l(x, t)\rangle = \int_{\mathbf{R}^3} F_{jl}(k)\exp\{i(k, r)dr\} \ .$$

The one-dimensional spectra are known in our case [5.31]:

$$E_1(k) = c_1 \bar{\varepsilon}^{2/3} k^{-5/3} \ , \qquad E_2(k) = c_2 \bar{\varepsilon}^{2/3} k^{-5/3} \ ,$$

where $c_1 = 18c/55$, $c_2 = 4c_1/3$, and the constant c in

$$\iint\limits_{|k|=k} 0.5 \operatorname{Sp} F(k)ds(k) = E(k) = c\,\bar{\varepsilon}^{2/3} k^{-5/3}$$

is calculated from the condition

$$\int_{k_{\min}}^{k_{\max}} E(k)dk = \alpha^{(*)} \frac{\overline{u^2}(x, t)}{2} \ ,$$

where $[k_{\min}, k_{\max}]$ is a subinterval of the inertial region where the assumption of isotropy holds, α is known ($0 < \alpha \le 1$) which determines the part of the total energy taken over the interval $[k_{\min}, k_{\max}]$.

Now we give a brief derivation of the above representations for p_{ij}. From the Taylor hypothesis $u(x, t) = u(x - U(x_3)t, 0)$, $U = (U(x_3), 0, 0)$ we get

$$\langle u_j(x, t + \tau)u_l(x, t)\rangle = \langle u_j(x - U(x_3)(t + \tau), 0)u_l(x - U(x_3)t, 0)\rangle$$

$$= B_{jl}(-U(x_3)\tau, 0, 0) = \int_{-\infty}^{\infty} p_{jl}(\omega, x)\exp\{i\omega\tau\}d\omega \ .$$

Now, $u(x, t)$ is isotropic, hence

$$B_{jl}(r) = [B_{LL}(r) - B_{NN}(r)]\frac{r_j r_l}{r^2} + B_{NN}(r)\delta_{jl} \ ,$$

where B_{LL} and B_{NN} are the longitudinal and transversal correlation functions, respectively. From the last two equalities we get

$$\int_{-\infty}^{\infty} p(\omega, x)\exp\{i\omega\tau\}d\omega = \begin{pmatrix} B_{LL}(U(x_3)\tau) & 0 & 0 \\ 0 & B_{NN}(U(x_3)\tau) & 0 \\ 0 & 0 & B_{NN}(U(x_3)\tau) \end{pmatrix} .$$

This shows that $p_{33}(\omega, x) = p_{22}(\omega, x)$. Thus we get

$$B_{jj}(-U(x_3)\tau) = \int_{-\infty}^{\infty} \exp\{-iU\tau k_1\} \left\{ \iint_{-\infty}^{\infty} F_{jj}(k_1, k_2, k_3)dk_2 dk_3 \right\} dk_1$$

$$= \int_{-\infty}^{\infty} \exp\{-iU\tau k_1\}F_j(k_1)dk_1$$

$$= \int_{-\infty}^{\infty} \exp\{-i\omega\tau\}F_j(\omega/U(x_3))U^{-1}d\omega$$

and we obtain the desired representations. Thus

$$p_{11}(\omega, x) = c_1(\bar{\varepsilon}U)^{2/3}\omega^{-5/3} \ ,$$

$$p_{22}(\omega, x) = p_{33}(\omega, x) = c_2(\bar{\varepsilon}U)^{2/3}\omega^{-5/3} \ ,$$

$$E^{(t)}(\omega) = 2 \operatorname{Sp} P(\omega, x) = \left[\frac{c_1}{2} + c_2\right](\bar{\varepsilon}U)^{2/3}\omega^{-5/3} \ .$$

Here $E^{(t)}(\omega)$ is the (frequency) energy spectrum. Now we can complete the definition of (5.103 104). From the fact that $p_{33} = p_{22}$ and $p_{11} = 0.75 \, p_{22}$ [which is seen from (5.105, 108)] we conclude that $f_0^{(1)}$ and $f_0^{(2)}$ must be defined as

$$f_0^{(2)}(\zeta) = \left[\frac{s_3(\zeta)}{s_2(\zeta)}\right]^3 f_0^{(3)}(\zeta) \ , \quad f_0^{(1)}(\zeta) = 0.75^{1.5}\left[\frac{s_3(\zeta)}{s_1(\zeta)}\right]^3 f_0^{(3)}(\zeta) \ .$$

In [5.18] it was proposed to take $f_0(\zeta) = f_0^{(2)}(\zeta)$.

Let us return to the general form of the spectral tensor (5.101), and represent it in the form [5.21]

$$\Phi_{lm}(\omega, x_1, x_2) = p_{ll}^{-1/2}(\omega, x_1)\operatorname{Coh}_{lm}(\omega, x_1, x_2)\exp\{-i\theta_{lm}\}p_{mm}^{1/2}(\omega, x_2) \ ,$$

$$(5.105)$$

where

$$\operatorname{Coh}_{lm}(\omega, x_1, x_2) = |\Phi_{lm}(\omega, x_1, x_2)|p_{ll}^{-1/2}(\omega, x_1)p_{mm}^{-1/2}(\omega, x_2)$$

$$\theta_{lm}(\omega, x_1, x_2) = \arctan(-\operatorname{Im}\{\Phi_{lm}\}/\operatorname{Re}\{\Phi_{lm}\})$$

are the coherent spectrum and the phase spectrum, respectively.

Our purpose now is to use the factorization of Φ_{lm} to simulate the random field $u(x, t)$ (Chap. 1). Notice that it is not possible to use this method directly from (5.105), since $\operatorname{Coh}_{lm}(\omega, x_1, x_2)\exp\{-i\theta_{lm}\}$ depends not only on the difference $r = x_1 - x_2$, but also on $R = (x_1 + x_2)/2$. However, the experimental data shows [5.22], that the dependence on R is very slow, so it is reasonable to assume that $\operatorname{Coh}_{lm}(\omega, x_1, x_2)\exp\{-i\theta_{lm}\}$ can be represented in the form

$$\operatorname{Coh}_{lm}(\omega, x_1, x_2)\exp\{-i\theta_{lm}\} = \int_{\mathbf{R}^3} F_{lm}(\omega, k|R)\exp\{i(k, r)\}dk \ ,$$

where

$$F_{lm}(\omega, k|R) = (2\pi)^{-2/3}\int_{\mathbf{R}^3}\operatorname{Coh}_{lm}(\omega, r, R)e^{-i\theta_{lm}(\omega, r, R)}e^{-i(k, r)}dr$$

is a spectral tensor slowly varying with R (R is fixed). Let $f_k = \omega x_k/U(x_3)$, $k = 1, 2, 3$ be the dimensionless frequencies, and let $\Delta f_k = \omega \Delta x_k/U(x_3)$ be the dimensionless coordinate shift ($k = 1, 2, 3$).

In [5.22] a model of vertical coherence ($k = 3$) was considered which was generalized as follows:

$$\text{Coh}_{lm}(\omega, r, R) = r_{lm} \left(\frac{\omega, R_3}{L} \right) \exp \left\{ - \sum_{j=1}^{3} \alpha_{lm}^{(j)} \frac{R_3 |\omega| \, |r_j|}{L\, U(R_3)} \right\} ,$$

$$\theta_{lm}(\omega, r, R) = \sum_{j=1}^{3} \beta_{lm}^{(j)} \frac{R_3 |\omega| \, |r_j|}{L\, U(R_3)} , \tag{5.106}$$

$$r_{lm}(\omega, \zeta) = p_{lm}(\omega, \zeta)[p_{ll}(\omega, \zeta) p_{mm}(\omega, \zeta)]^{-1/2} ,$$

$r = (r_1, r_2, r_3)$, $R = (R_1, R_2, R_3)$. Here r_{lm} is a spectral correlation coefficient ($|r_{lm}| \leq 1$). Experimental data [5.22] led to the choice

$$\{\alpha_{lm}^{(j)}\}_{lm} = \begin{pmatrix} \alpha^{(j)} & 0 & \alpha^{(j)} \\ 0 & \alpha^{(j)} & 0 \\ \alpha^{(j)} & 0 & \alpha^{(j)} \end{pmatrix} , \quad j = 1, 2, 3 ,$$

$$\{\beta_{lm}^{(j)}\}_{lm} = \begin{pmatrix} \beta^{(j)} & 0 & \beta^{(j)} \\ 0 & \beta^{(j)} & 0 \\ \beta^{(j)} & 0 & \beta^{(j)} \end{pmatrix} , \quad j = 1, 2, 3 , \tag{5.107}$$

where $\alpha^{(1)} = 0$, $\alpha^{(2)} = 2$, $\alpha^{(3)} = 3$, $\beta^{(j)} = 1$, $j = 1, 2, 3$.

Now we can construct the simulation formula for the random field with the spectral tensor (5.105) [for the model (5.106)]. We shall take the matrix $\{\beta_{lm}^{(j)}\}_{lm}$ according to (5.107), while the matrix $\{\alpha_{lm}^{(j)}\}_{lm}$ will be taken in the form

$$\{\alpha_{lm}^{(j)}\}_{lm}(\zeta) = \begin{pmatrix} \alpha_{11}^{(j)}(\zeta) & 0 & \alpha_{13}^{(j)}(\zeta) \\ 0 & \alpha_{22}^{(j)}(\zeta) & 0 \\ \alpha_{13}^{(j)}(\zeta) & 0 & \alpha_{33}^{(j)}(\zeta) \end{pmatrix} .$$

We use the factorization method described in Sect. 1.2.4. From (5.105–107) it follows that

$$\Phi(\omega, x_1, x_2) = A(\omega, x_1) \tilde{\Phi}(\omega, r, \zeta) A^*(\omega, x_2) ,$$

where

$$A(\omega, x) = \exp \left\{ -i \sum_{j=1}^{3} \beta^{(j)} \frac{\omega x_j}{U(z)} \right\} p^{1/2}(\omega, x) ,$$

$$p^{1/2}(\omega, x) = \begin{pmatrix} [p_{11}(\omega, x)]^{1/2} & 0 & 0 \\ 0 & [p_{22}(\omega, x)]^{1/2} & 0 \\ 0 & 0 & [p_{33}(\omega, x)]^{1/2} \end{pmatrix} .$$

Here $\zeta = (z_1 + z_2)/2$, $r = x_1 - x_2$.

Notice that, in contrast to the general formulas of Sect. 1.2.4, Φ includes the parameter ζ and has the form

$$\tilde{\Phi}_{lm}(\omega, r, \zeta) = r_{lm}(\omega, \zeta) \exp \left\{ - \sum_{j=1}^{3} \alpha_{lm}^{(j)} \frac{|\omega| \, |r_j|}{|U(\zeta)|} \right\} .$$

To construct the simulation formulas, it is necessary to evaluate the partial spectral tensor

$$F_{lm}(k, \omega, \zeta) = (2\pi)^{-3} \int_{\mathbf{R}^3} e^{-i(k,r)} \tilde{\Phi}_{lm}(\omega, r, \zeta) dr$$

$$= r_{lm}(\omega, \zeta) \prod_{j=1}^{3} \Delta_{lm}^{(j)}(k_j^2, \omega, \zeta) ,$$

(5.108)

where

$$\Delta_{lm}^{(j)}(k_j^2, \omega, \zeta) = \frac{\alpha_{lm}^{(j)} \dfrac{|\omega|}{|U(\zeta)|}}{\pi \left\{ k_j^2 + \left(\dfrac{\alpha_{lm}^{(j)} \omega}{|U(\zeta)|} \right)^2 \right\}} , \quad \text{if} \quad \alpha_{lm}^{(j)} > 0 ,$$

and $\Delta_{lm}^{(j)}(k_j^2, \omega, \zeta) = \delta(k_j)$ if $\alpha_{lm}^{(j)} < 0$. Without loss of generality, we can assume that $\alpha_{lm}^{(j)}(\zeta) > 0$. Then (5.108) defines a spectral tensor of a homogeneous random field, if (and only if) the determinant of (5.108) is non-negative (for all ω and ζ). This leads to some restrictions in the choice of $\alpha_{lm}^{(j)}$. For the model described above in (5.107),

$$F_{lm}(k, \omega, \zeta) = r_{lm}(\omega, \zeta) \delta(k_1) \prod_{j=2}^{3} \frac{\alpha_{lm}^{(j)} \dfrac{|\omega|}{|U(\zeta)|}}{\pi \left\{ k_j^2 + \left(\dfrac{\alpha_{lm}^{(j)} \omega}{|U(\zeta)|} \right)^2 \right\}} .$$

(5.109)

The δ-function in (5.109) just simplifies the simulation formula: we take $k_1 = 0$ when choosing the random vector k. In the general case of (5.108), for $\alpha_{lm}^{(j)} > 0$ we obtain by using the simulation formulas presented in Sect. 1.2.4

$$u(x, t) = [p(\omega)p(k|\omega)]^{-1/2} p^{1/2}(\omega, z) G\left(k, \omega \frac{z}{L} \right) \left\{ \cos\left[\omega t - \left(\beta, \frac{x\omega}{U(z)} \right) \right] \eta_1 \right.$$

$$\left. + \sin\left[\omega t - \left(\beta, \frac{x\omega}{U(z)} \right) \right] \eta_2 \right\} .$$

(5.110)

Here

$$\eta_1 = \xi_{11} \cos(k, x) + \xi_{12} \sin(k, x) ,$$

$$\eta_2 = \xi_{21} \cos(k, x) + \xi_{22} \sin(k, x) , \quad \beta = \left(\beta^{(1)}, \beta^{(2)}, \beta^{(3)} \right)$$

and ξ_{ij} $(i, j = 1, 2)$ are mutually independent three-dimensional random vectors with zero mean and unit covariance matrix; $p(\omega)$ and $p(k|\omega)$ are densities, where $\omega \in \mathbf{R}$, $k \in \mathbf{R}^3$, $G(k, \omega, z/L)$ is a 3×3 matrix defined by

$$F(k, \omega, \zeta) = G(k, \omega, \zeta) G^{\mathrm{T}}(k, \omega, \zeta) .$$

In (5.110) ω is a random scalar number, sampled from $p(\omega)$, $\omega \in \mathbf{R}$, and k is a random vector, sampled from $p(k, \omega)$, $k \in \mathbf{R}^3$, ω fixed.

To choose $p(k, \omega)$ for the model (5.107), we take into account the form of (5.109) and put

$$p(k|\omega) = \delta(k_1) \frac{\alpha^{(2)} \dfrac{|\omega|}{|U(z)|}}{\pi \left\{ k_2^2 + \left(\dfrac{\alpha^{(2)}\omega}{|U(z)|} \right)^2 \right\}} \frac{\alpha^{(3)} \dfrac{|\omega|}{|U(z)|}}{\pi \left\{ k_3^2 + \left(\dfrac{\alpha^{(3)}\omega}{|U(z)|} \right)^2 \right\}} .$$

This means that k is simulated as

$$k_1 \equiv 0 , \quad k_j = \frac{|\omega|}{|U(z)|} \alpha^{(j)} \tan \left[\frac{\pi(2\gamma_j - 1)}{2} \right] , \quad j = 2, 3 . \tag{5.111}$$

Here γ_j $(j = 2, 3)$ are independent random variables uniformly distributed on $[0, 1]$. In this case, (5.110) takes the form

$$u(x, t) = [p(\omega)]^{-1/2} p^{1/2}(\omega, z) G\left(\omega, \frac{z}{L} \right) \left\{ \cos \left[\omega t - \left(\beta, \frac{x\omega}{U(z)} \right) \right] \eta_1 \right.$$

$$\left. + \sin \left[\omega t - \left(\beta, \frac{x\omega}{U(z)} \right) \right] \eta_2 \right\} , \tag{5.112}$$

where

$$G\left(\omega, \frac{z}{L} \right) = \begin{pmatrix} 1 & 0 & 0 \\ 0 & 1 & 0 \\ r_{13}\left(\omega, \frac{z}{L} \right) & 0 & [1 - r_{13}^2] \end{pmatrix}$$

Remark. Notice that the random field constructed is not incompressible and that the field (5.110) has a spectral tensor Φ' which is not coincident with Φ. It is important to estimate the difference $\Phi' - \Phi$. For simplicity, let us consider a complex-valued field. Assume that the partial spectral tensor of a complex-valued partially homogeneous, stationary random vector field $u(x, t)$ can be represented as

$$\Phi(\omega, x_1, x_2) = A(\omega, x_1) \tilde{\Phi}(\omega, r, R) A^*(\omega, x_2) , \tag{5.113}$$

where $R = (x_1 + x_2)/2$, $r = x_1 - x_2$.

Let

$$F(k, |\omega, R) = (2\pi)^{-3} \int_{\mathbf{R}^3} e^{-i(k, r)} \tilde{\Phi}(\omega, r, R) dr$$

be a spectral tensor (with respect to k, R fixed) of a homogeneous field, such that

$$F(k|\omega, R) = G(k|\omega, R) G^*(k|\omega, R) ,$$

where $G(k|\omega, R)$ is a tensor. We introduce the following approximation for the random field $u(x)$

$$u(x, t) = [p(\omega)p(k|\omega)]^{-1/2} e^{i[\omega t + (k, r)]} A(\omega, x) G(k|\omega, x) \eta , \tag{5.114}$$

where ω is a random number distributed with $p(\omega)$, $\omega \in R$, k is a random 3D vector with zero mean and unit covariance matrix. From (5.113, 114) we get

$$\Phi'(\omega, x_1, x_2) - \Phi(\omega, x_1, x_2) = A(\omega, x_1)\delta G(\omega, x_1, x_2)A^*(\omega, x_2) \ ,$$

where

$$\delta G(\omega, x_1, x_2) = \int_{R^3} \left[G(k|\omega, x_1)G^*(k|\omega, x_2) - G(k|\omega, R)G^*(k|\omega, R) \right]$$

$$\times \exp[i(k,r)]dk = \int_{R^3} \left\{ G(k|\omega, x_1) \left[G^*(k|\omega, x_2) - G^*(k|\omega, R) \right] \right.$$

$$+ \left[G(k|\omega, x_1) - G(k|\omega, R) \right] G^*(k|\omega, R) \right\} \exp[i(k,r)]dk \ . \qquad (5.115)$$

This equality shows that for small r, the right-hand side of (5.115) will be small if we require that $G(k|\omega, x)$ is continuous with respect to x. For large r, this follows from the presence of the factor $\exp[i(k,r)]$.

We present now a simulation algorithm based on (5.111, 112). The input parameters are u_*, L and $[\omega_{\min}, \omega_{\max}]$.

1) Sample two independent random numbers uniformly distributed on $[0,1]$: α_1 and α_2; calculate $\omega = \alpha_1 \Delta \omega + \omega_{\min}$; here $\Delta \omega = \omega_{\max} - \omega_{\min}$; if $\alpha_2 < 0.5$ then $\omega = -\omega$.

2) Calculate

$$k_j = \frac{|\omega|}{|U(z)|} \alpha^{(j)} \tan \left[\frac{\pi(2\gamma_j - 1)}{2} \right] \ , \qquad j = 1, 2, 3 \ ,$$

where $\gamma_1, \gamma_2, \gamma_3$ are independent random numbers uniformly distributed on $[0,1]$.

3) Simulate 4 independent standard gaussian random 3D vectors ξ_{ij} ($i = 1, 2; j = 1, 2$).

4) Calculate

$$\eta_i = \xi_{i1} \cos(k, w) + \xi_{i2} \sin(k, x) \ ,$$

$$\theta = \omega t - \left(\beta \frac{x\omega}{U(z)} \right) \ , \qquad \beta = \left(\beta^{(1)}, \beta^{(2)}, \beta^{(3)} \right) \ .$$

5) Calculate

$$u_1(x, t) = \left[\frac{p_{11}(\omega, z/L)}{2\Delta\omega} \right]^{1/2} \left\{ \eta_{11} \cos(\theta) + \eta_{21} \sin(\theta) \right\} \ ,$$

$$u_2(x, t) = \left[\frac{p_{22}(\omega, z/L)}{2\Delta\omega} \right]^{1/2} \left\{ \eta_{12} \cos(\theta) + \eta_{22} \sin(\theta) \right\} \ ,$$

$$u_3(x, t) = \left[\frac{p_{33}(\omega, z/L)}{2\Delta\omega} \right]^{1/2} \left\{ \eta_{13} r_{13} \left(\omega, \frac{z}{L} \right) \cos(\theta) \right.$$

$$+ \eta_{23} \left[1 - r_{13}^2 \left(\omega, \frac{z}{L} \right) \right]^{1/2} \sin(\theta) \right\} \ .$$

The functions $p_{ij}(\omega, z/L)$ and $r_{13}(\omega, z/L)$ are defined in (5.103, 104 106); the constants $\alpha^{(j)}$, $\beta^{(j)}$ are taken from (5.107).

5.4 Applications to Diffusion Problems

In this section we present numerical results obtained by computer simulation of various diffusion processes. Some of these numerical experiments are interesting from a practical viewpoint, e.g., applications to the problem of spreading of clouds of particles of aerosol insecticide, which are described in Sect. 5.4.2, and calculation of the optimal choice of parameters in diffusion batteries which are discussed in Sect. 5.4.3. The numerical results obtained in Sect. 5.4.1 and 5.4.4 lead to new theoretical conclusions in these problems.

5.4.1 Distribution of the First Passage Time for Particles Moving in Classical Isotropic Random Velocity Fields

In applied problems of particle deposition in turbulent velocity fields it is important to evaluate not only the mean time needed to reach a given surface but also, perhaps even more importantly, to calculate the probability distribution of the random time at which a particle reaches the surface (the first passage time). For asymptotically large times (when the characteristic times are large compared to the Lagrangian time T_L) there exists a developed theory based on a Markovian approximation to turbulent diffusion [5.10, 11]. However, when the time is comparable to T_L, it must be expected that this theory will lead to incorrect results. In this section we calculate the distribution of the first passage time for particles moving in classical isotropic random velocity field (Sect. 5.3.1) and compare it with theoretical results [5.10]. Our numerical results have confirmed the hypothesis mentioned above. We have also investigated the relative contributions to the first passage time due to sedimentation and turbulent dispersion.

Statement of the Problem. In the half-space $\mathbb{R}^3_{(+)}$, let us consider a spherical particle with diameter d whose sedimentation rate is described by the Stokes formula: $v_s = cd^2$, where c is a constant related to the density of the particle ϱ and the air viscosity η : $c = \varrho g/18\eta$. Thus, it is assumed that a particle released in the flow acquires the velocity $V = (u'_1, u'_2, u'_3 + v_s)$ so that its motion in Lagrangian coordinates X is given by

$$\frac{dX_1}{dt} = u'_1(X(t))$$

$$\frac{dX_2}{dt} = u'_2(X(t)) \tag{5.116}$$

$$\frac{dX_3}{dt} = u'_3(X(t)) + v_s , \quad X(0) = x_0 .$$

Here u_i' are the components of the turbulent velocity field and the coordinate system is fixed at the release point $x_0 = (0, 0, h)$.

We shall consider two models

1) The classical turbulent diffusion model where the velocity field u' is assumed to be δ-correlated:

$$\langle u_i(t_1) u_j(t_2) \rangle = \delta(t_1 - t_2) .$$

2) The classical isotropic turbulence model where the spatial correlations are described by Kolmogorov's law. [The simulation formulas will be taken from Sect. 5.3.1, e.g. (5.74).]

First, let us present the distribution of the times at which a particle reaches the boundary $\partial G = \Gamma = \{(x, y, z) : z = 0\}$ for model 1 [5.10]. We denote by k the eddy diffusivity; then the distribution density $W(t)$ of the first passage time τ is given by [5.23]

$$W(t) = h[4\pi k t^3]^{-3/2} \exp \left\{ -\frac{(h - v_s t)^2}{4kt} \right\} \tag{5.117}$$

[recall that the release point is $x_0 = (0, 0, h)$]. Note that the mean time needed to reach the boundary Γ

$$\bar{\tau} = \int_0^\infty t W(t) dt = \frac{h}{v_s} \tag{5.118}$$

does not depend on k, while the variance of τ is a linear function of k: $D(\tau) = 2kh/v_s^3$. We introduce a dimensionless time $t' = t/\bar{\tau}$ and a quantity

$$\gamma = \frac{[D(\tau)]^{1/2}}{\bar{\tau}} = \left[\frac{2k}{(v_s h)} \right]^{1/2}$$

which characterizes the relative role of the diffusion and sedimentation in the process of deposition of particles on Γ (in the case $\gamma = 0$ we have pure sedimentation). Following [5.23] we assume that the deposition of particles is practically fully controlled by the diffusion (sedimentation can be neglected) if $\gamma > 0.7$, while in the case $\beta < 0.02$ it can be described only by sedimenation (diffusion can be neglected). In the intermediate region $0.02 < \gamma < 0.7$ both processes must be taken into account. For example, $\gamma = 0.14$ in Fig. 5.10 is a typical value from the intermediate interval.

We now turn to the evaluation of the distribution of the first passage time in the framework of model 2. Thus, u' is constructed as follows:

$$u(X) = u^* \sum_{i=1}^n [(\xi_i \times \omega_i) \cos[\theta_i(X)] + (\eta_i \times \omega_i) \sin[\theta_i(X)]] . \tag{5.119}$$

Here $u^* = \langle u'^2 \rangle / 2$, $\theta_i(X) = \lambda_i(\omega_i, X)$, ξ_i, η_i are mutually independent standard Gaussian random vectors lying in $\mathbf{R}^3$, ω_i are independent isotropic vectors in $\mathbf{R}^3$, and the random numbers λ_i are simulated from

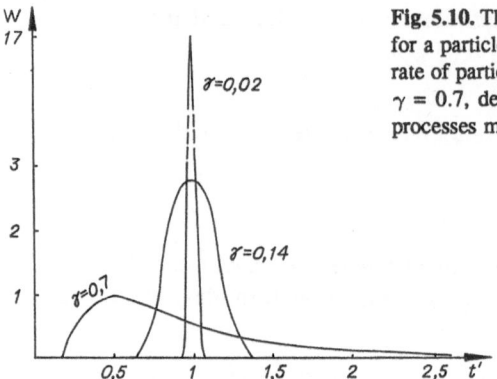

Fig. 5.10. The probability density of the first passage time for a particle to reach a boundary Γ. When $\gamma = 0.02$ the rate of particle deposition is controlled by sedimentation, $\gamma = 0.7$, deposition is diffusion-limited, $\gamma = 0.14$, both processes must be included

$$\lambda_i = 2\pi L^{-1} \left[1 - \frac{i - \alpha_i}{n} \right]^{-3/2} , \quad i = 1, 2, \ldots n ,$$

where n is the number of harmonics (turbulent eddies), L is the external scale of the inertial interval, α_i are mutual independent random numbers uniformly distributed on $[0, 1]$, $\langle u'^2 \rangle / 2$ is the intensity of the velocity fluctuations

$$\frac{\langle u'^2 \rangle}{2} = \int_0^\infty E(\lambda) d\lambda , \tag{5.120}$$

where $E(\lambda)$ is the energy spectrum

$$E(\lambda) = \begin{cases} c \bar{\varepsilon}^{2/3} \lambda^{-5/3} , & \text{if } \lambda > \lambda_{\min} \\ 0 , & \text{if } 0 \leq \lambda \leq \lambda_{\min} . \end{cases}$$

Here $\lambda = 2\pi/L$, $c = 1.4$, $\bar{\varepsilon}$ is the mean rate of dissipation of energy. Thus, the velocity fluctuations are specified by three input parameters: L, $\langle u'^2 \rangle$ and $\bar{\varepsilon}$. Because of (5.120), only two of them are independent. Let us choose L and $\langle u'^2 \rangle$ as the two independent input parameters. It should be noted that n can also be considered as a parameter of our model. Of course, n should be taken as large as possible, but in practice, it suffices to take $n = 16$ or 8, or, sometimes, even 4 – it depends on the functional to be calculated.

Lagrangian statistical characteristics of (5.119) were numerically investigated and we obtained

$$K = \tilde{K} \frac{L}{[\langle u'^2 \rangle]^{1/2}} , \quad T_L = \tilde{T}_L \frac{L}{[\langle u'^2 \rangle]^{1/2}} . \tag{5.121}$$

Here K and T_L are the eddy diffusivity and the Lagrangian time scale, respectively; $\tilde{K}$ and $\tilde{T}_L$ are the corresponding dimensionless values: $\tilde{K} = 0.075$, $\tilde{T}_L = 0.19$.

In Fig. 5.11 the probability density function of the first passage time for a particle moving according to (5.116) in the isotropic velocity field (5.119) is shown for $L = 10$, $\langle u'^2 \rangle = 1$, when $n = 4$ and $n = 9$; N is the number of samples of u'. Here also is shown the function (5.117) where the coefficient K was calculated from (5.121). Note that the cases $n = 4$ and $n = 9$ do not differ noticeably, so it is sufficient to take $n = 9$.

Numerical results for the two models described above are compared further in Fig. 5.12 for $L = 1\,\mathrm{m}$, $h = 0.1\,\mathrm{m}$, and $\langle u'^2 \rangle = 0.01\,\mathrm{m}^2/\mathrm{sec}^2$, and for three particle sizes $d = 1\,\mu\mathrm{m}$, $25\,\mu\mathrm{m}$, and $100\,\mu\mathrm{m}$, with $v_s = 3 \times 10^{-5}\,\mathrm{m}/\mathrm{sec}$, $1.875 \times 10^{-2}\,\mathrm{m}/\mathrm{sec}$, and $0.3\,\mathrm{m}/\mathrm{sec}$, respectively. In the first model, $\gamma = 0.99$, $\gamma = 0.94$, $\gamma = 0.58$, respectively, while the second model results in $\gamma = 0.55$, $\gamma = 0.45$ and $\gamma = 0.17$ for $d = 1\,\mu\mathrm{m}$, $25\,\mu\mathrm{m}$ and $100\,\mu\mathrm{m}$. This implies that model 1 predicts that the role of sedimentation will be negigible for all three diameters. However, the second model shows that for particles with $d = 100\,\mu\mathrm{m}$ ($\gamma = 0.17$) the contributions due to the turbulent scattering and sedimentation process are comparable. Note that in the first model this situation ($\gamma = 0.17$) arises when $d \approx 300\,\mu\mathrm{m}$.

The results shown in Fig. 5.13 are the same as in Fig. 5.12 but for $h = 3\,\mathrm{m}$, $d = 100\,\mu\mathrm{m}$. In the first model $\gamma = 0.13$, and in the second we obtained $\gamma = 0.06$. In this case the first model predicts that we are in the intermediate interval, while the second model shows that sedimentation predominates. A comparison of curves $W(t)$ for both models is given in Figs. 5.14 and 5.15 for $d = 40\,\mu\mathrm{m}$, $h = 1\,\mathrm{m}$, $L = 10\,\mathrm{m}$, where $\langle u'^2 \rangle = 1\,\mathrm{m}^2/\mathrm{sec}^2$ in Fig. 5.14, and $\langle u'^2 \rangle = 10\,\mathrm{m}^2/\mathrm{sec}^2$ in Fig. 5.15. It is seen that at times comparable to T_L model 1 fails, while for $t \gg T_L$ the two models are in good agreement.

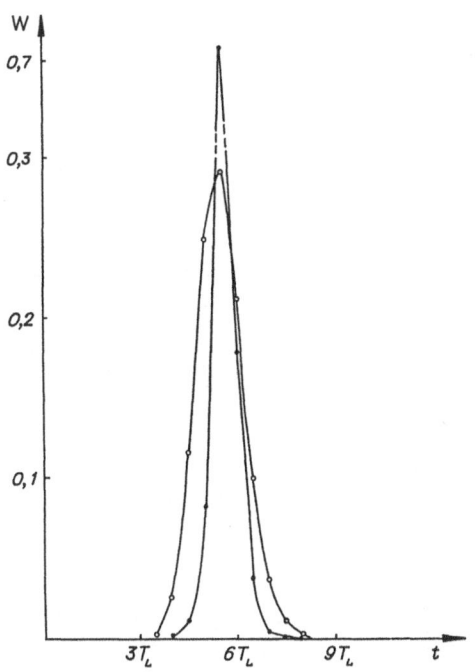

Fig. 5.11. Probability density function W for $n = 4$ (*circles*) and $n = 9$ (*points*)

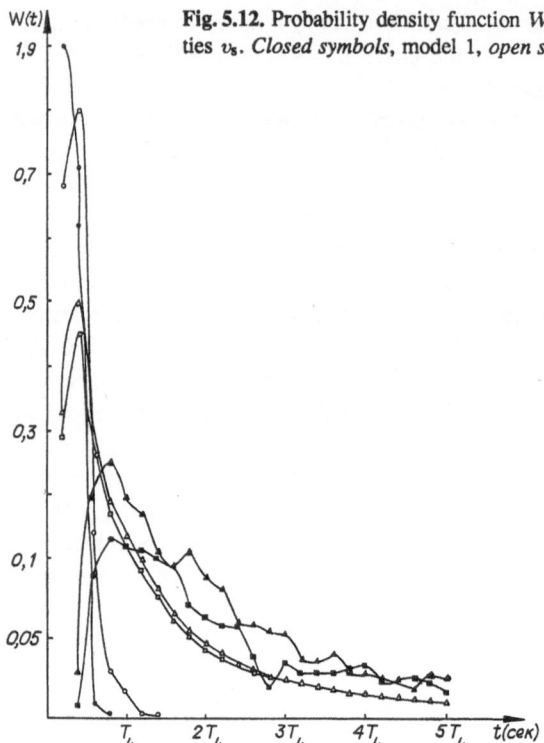

Fig. 5.12. Probability density function W for different values of particle velocities v_s. *Closed symbols,* model 1, *open symbols,* model 2

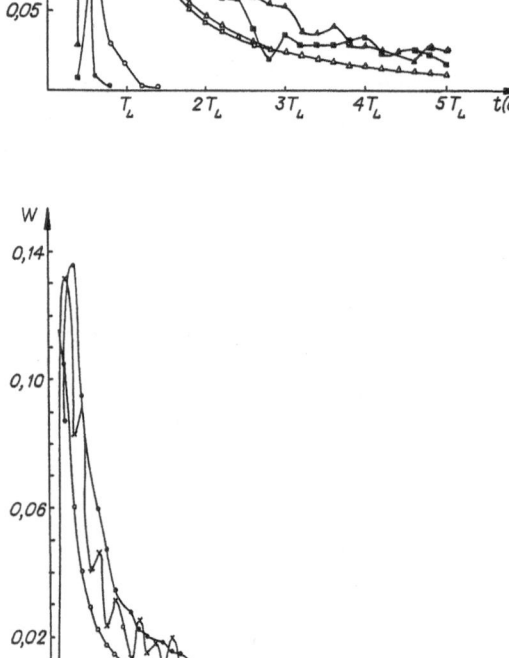

Fig. 5.13. As in 5.12. Comparison of contributions due to diffusion and sedimentation

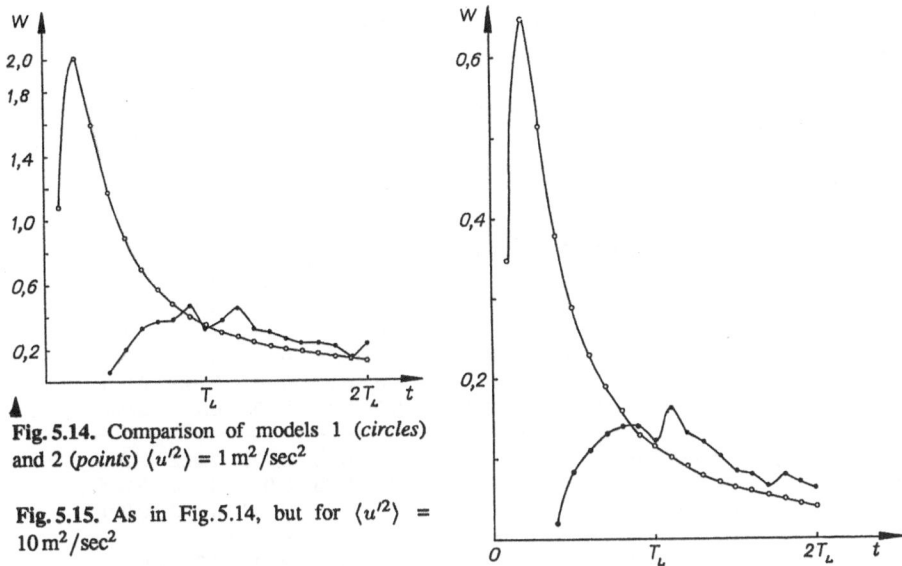

Fig. 5.14. Comparison of models 1 (*circles*) and 2 (*points*) $\langle u'^2 \rangle = 1\,\mathrm{m}^2/\mathrm{sec}^2$

Fig. 5.15. As in Fig. 5.14, but for $\langle u'^2 \rangle = 10\,\mathrm{m}^2/\mathrm{sec}^2$

5.4.2 Spread of Clouds of Particles of Aerosol Insecticide in Arboreal Canopies

In order to improve aerosol technology created to defend arboreal canopies against insects, one is faced with the problems of calculating the concentration of particles absorbed by the canopy or deposited on the ground and estimating the fraction of particles moving in the air (that is, the fraction of particles which is not captured). To apply the algorithm described in Sect. 5.3.2, it is necessary to define the absorption coefficient $\sigma(x)$. Thus we shall consider the canopy as an absorbing layer $G = \{(x, y, z) : 0 < z < H\}$ with absorption coefficient $\sigma(z) = \mathcal{E} S U(z)/H$. Here H is the height of the layer, $U(z)$ is the mean velocity depending only on the height z, S is the total leaf area (leaf index), $\mathcal{E}$ is the probability that a particle reaching a leaf will be deposited on the leaf surface. An empirical dependence $\mathcal{E} = 6.1 \times 10^{-4}(U v_{\mathrm{s}})^{0.65}$, where v_{s} is the sedimentation rate calculated as described in the previous section has been found and was used here. The mean velocity was taken to be constant inside the layer, $U = 1\,\mathrm{m/sec}$, and as a logarithmic function above the layer; the intensity of velocity fluctuations was taken such that $k_z = 0.1\,\mathrm{m}^2/\mathrm{sec}$. The height H was set equal to $15\,\mathrm{m}$, the leaf index $= 7$. The unit particle source was situated at $(0, 0, h)$, where $h = 2\,\mathrm{m}$. Calculations were carried out for 5 particle sizes, $d = 1, 3, 7, 20$ and $30\,\mu\mathrm{m}$. The results of calculations of the mean integral concentrations absorbed by the canopy layer, $\bar{c}_{\mathrm{c}}$, deposited on the ground, $\bar{c}_{\mathrm{g}}$, and suspended in air, $\bar{c}_{\mathrm{s}}$, are given in Table 5.1

Many such numerical experiments and comparisons with experimental data are described in [5.7].

Table 5.1. Distribution of particle concentration in a canopy layer

Particle	Particle diameter	Distance from the source x_1 [m]								
	μm	20	40	80	160	320	640	1280	2560	10240
$\bar{c}_c$		0.0025	0.005	0.009	0.019	0.037	0.07	0.119	0.190	0.398
$\bar{c}_g$	1	0	0	0	0	0	0	0	0.001	0.002
$\bar{c}_s$		0.9975	0.995	0.989	0.973	0.893	0.73	0.545	0.395	0.125
$\bar{c}_c$		0.0104	0.020	0.040	0.078	0.146	0.26	0.404	0.555	0.769
$\bar{c}_g$	3	0	0.001	0.001	0.003	0.006	0.01	0.14	0.023	0.031
$\bar{c}_s$		0.9896	0.978	0.967	0.903	0.781	0.59	0.362	0.158	0.020
$\bar{c}_c$		0.0307	0.059	0.114	0.210	0.366	0.56	0.732	0.822	0.882
$\bar{c}_g$	7	0.0078	0.018	0.027	0.038	0.054	0.06	0.078	0.084	0.086
$\bar{c}_s$		0.9615	0.922	0.858	0.746	0.547	0.29	0.102	0.020	0.001
$\bar{c}_c$		0.1110	0.200	0.339	0.511	0.655	0.71	0.723	0.724	0.725
$\bar{c}_g$	20	0.0528	0.100	0.160	0.220	0.261	0.27	0.275	0.276	0.276
$\bar{c}_s$		0.8366	0.699	0.501	0.267	0.080	0.01	0.000	0.000	0.000
$\bar{c}_c$		0.1875	0.319	0.480	0.607	0.655	0.65	0.660	0.660	0.66
$\bar{c}_g$	30	0.0957	0.176	0.262	0.320	0.338	0.34	0.34	0.34	0.34
$\bar{c}_s$		0.7168	0.504	0.259	0.072	0.006	0.00	0.00	0.00	0.00

5.4.3 Diffuse Deposition of Polydispersed Aerosol Particles in Pipes

When a two-phase material (gas + aerosol particles) is flowing through a pipe, the size distribution of particles is changed because of coagulation, sedimentation, diffusion, and deposition of the particles on the pipe. Such processes occur in many devices, for example, in diffusion batteries, devices for condensative enlargement of particles, generators of aerosols, etc.

Let us introduce the notation $G = \{(x_1, x_2, x_3) : x_2^2 + x_3^2 \leq R^2\}$; $\partial G = \{(x_1, x_2, x_3) : x_2^2 + x_3^2 = R^2\}$, the cylinder of radius R whose axis coincides with x_1; $Q = G \times (0, T]$, $\Gamma = \partial G \times (0, T]$. Let $u(x, t) = (u_1(x, t), u_2(x, t), u_3(x, t))^{\mathrm{T}}$ be the gas velocity. The mean concentration $c(x, t; \mu)$ of aerosol particles with molecular diffusion coefficient $\mu > 0$ is described by

$$\frac{\partial c}{\partial t} - \mu \Delta c(x, t; \mu) + u_i(x, t) \frac{\partial c}{\partial x_i}(x, t; \mu) = f(x, t), \quad (x, t) \in Q$$

$$c(x, 0; \mu) = \varphi(x), \quad x \in G, \quad c(x, t; \mu) = 0, \quad (x, t) \in \Gamma, \tag{5.122}$$

where $f(x, t)$ and $\varphi(x)$ are the densities of the aerosol sources at a time instant $t > 0$, and $t = 0$, respectively. It is assumed that the particles are absorbed on the boundary ∂G.

Divide the boundary ∂G into $m+2$ parts $\partial G_0, \partial G_1, \ldots, \partial G_m, \partial G_{m+1}$ such that

$$\partial G_0 = \{x \in \partial G : x_1 < 0\} ,$$
$$\partial G_j = \{x \in \partial G : (j-1)\delta \le x_1 < j\delta\} , \quad j = 1, 2, \ldots, m ,$$
$$\partial G_{m+1} = \{x \in \partial G : x_1 \ge m\delta\} ,$$

where $\delta > 0$ is a prescribed number. Let

$$\psi_j = \begin{cases} 1 , & x \in \partial G_j , \\ 0 , & x \notin \partial G_j , \end{cases} \quad j = 1, 2, \ldots, m+1 .$$

The functionals (called concentration impulses) are of practical interest

$$q_j(\mu) = \int_\Gamma \left[-\mu \frac{\partial c}{\partial n_x}(x, t; \mu) \right] \psi_j(x) d\sigma_x dt , \quad j = 0, 1, \ldots, m+1 \qquad (5.123)$$

since they describe the number of particles (of fixed size) deposited on ∂G_j during the time T. Note that these functionals could not be calculated by the scheme based on direct calculation of the derivative $\partial c/\partial n_x$ in (5.123), since the algorithms constructed in Sect. 5.1 do not work when it is necessary to calculate the derivatives at a boundary point. However, the relation to the adjoint problem (utilized already in Chap. 3) and using the homogeneity of the boundary conditions permit us to construct an effective algorithm for calculating the quantities (5.123).

We denote by $\bar{c}_j(x, t; \mu)$ the solution of

$$\frac{\partial \bar{c}_j}{\partial t} - \mu \Delta \bar{c}_j(x, t; \mu) - \bar{u}_i(x, t) \frac{\partial \bar{c}_j}{\partial x_i}(x, t; \mu) = 0 , \quad (x, t) \in Q ,$$
$$\bar{c}_j(x, 0; \mu) = 0 , \quad x \in G , \quad \bar{c}_j(x, t; \mu) = \psi_j(x) , \quad (x, t) \in \Gamma , \qquad (5.124)$$
$$(j = 0, 1, \ldots, m+1) ,$$

where $\bar{u}_i(x, t) = u_i(x, T - t)$; $\partial u/\partial x_i \equiv 0$. Then utilization of the connection between the direct and adjoint problems (Sect. 5.2.1) yields

$$q_j(\mu) = \int_\Gamma f(x, T - t) \bar{c}_j(x, t; \mu) dx dt + \int_G \varphi(x) \bar{c}_j(x, t; \mu) dx . \qquad (5.125)$$

Then using (5.124, 125) and the double randomization technique it is possible to calculate $q_j(\mu)$. In this scheme it is necessary to use some Monte Carlo estimate for (5.124). We apply the following method. Assume that $u_2(x, t) \equiv 0$, $u_3(x, t) \equiv 0$, i.e., the velocity field is oriented parallel to the axis of the cylinder. We choose the parameter α in the scheme presented in Sect. 5.2.1 as

$$\alpha(x, t) = \min\{t, e(R - \varrho)^2/6\mu\} ,$$

where $\varrho^2 = x_2^2 + x_3^2$. The transition $(x_k, t_k) \to (x_{k+1}, t_{k+1})$ is sampled in two steps: (1) An auxiliary transition $(x_k, t_k) \to (x_{k+1}, t_{k+1})$ [or $(x, t) \to (x', t')$] is sampled by calculating

$$\tau = \alpha(x,t)\beta^{2/3}\exp\left\{-\frac{\gamma_{2/3}}{1+3/2}\right\},$$

$$r = \left[\frac{4\tau\gamma_{3/2}}{1+3/2}\right]^{1/2}, \quad t' = t - \tau, \quad x' = x + r\omega$$

with probability $q = (5/3)^{-5/2}$, where β is a random number sampled from a β-distribution with parameters $(2, 2/3)$, ω is a 3D vector distributed isotropically. With probability $1 - q$ calculate $\tau = \alpha(x,t)\exp(-2\theta/3\}$, $t' = t - \tau$, $x' = x + R(\tau)\omega$, where θ is a random number whose distribution is proportional to $[1 - \exp(-2\theta/3)]\theta^{2/3}\exp(-\theta)$, $\theta \in (0, \infty)$ which can be simulated by the Neumann rejection method with the majorant function $(1 - q)^{-1}\theta^{2/3}\exp(-\theta)$. (2) We set $t_{k+1} = t_{k+1/2}$, $x_{k+1} = x_{k+1/2} - \bar{u}(x_k, t_k)(t_k - t_{k+1}) = x_{k+1/2} - u(x_k, T - t_k)(t_k - t_{k+1})$. Then the random variable

$$\xi_\varepsilon^{(j)}(x_0, t_0; \mu) = Q^{(N_\varepsilon)}\psi_\varepsilon^{(j)}(x_{N_\varepsilon}, t_{N_\varepsilon})$$

is an ε-biased estimate for $\bar{c}_j(x_0, t_0; \mu)$. Here N_ε is the index of the last state of the chain $\{(x_k, t_k)\}_{k=0}^{N_\varepsilon}$, and $Q^{(0)} = 1$, $Q^{(k)} = Q^{(k-1)}q(x_{k-1}, t_{k-1}; x_k, t_k)$, $k = 1, 2, \ldots$;

$$q(x,t;x't') = \begin{cases} 1 - [\alpha(x,t) - (t - t')k(x,t;x't')], & (x',t') \in B_\alpha \\ 1, & (x',t') \in \partial B_\alpha(x,t); \end{cases}$$

$$k(x,t;x',t') = \frac{u_1(x',t') - u_1(x,t)}{2\mu(t-t')}[x_1' - x_1 - u_1(x,t)(t - t')];$$

$$\psi_\varepsilon^{(j)}(x,t) = \begin{cases} \psi_j(x^*,t), & (x,t) \in \Gamma_\varepsilon, \\ 0, & (x,t) \notin \Gamma_\varepsilon, \end{cases}$$

x^* is a boundary point nearest to $x \in \Gamma_\varepsilon$. In our case the velocity field in (5.122) was taken as

$$u_1(x,t) = V[1 - \varrho^2/R^2]; \quad u_2 = u_3 = 0;$$

$$\varphi(x) = 0; \quad f(x,t) = I_{[0,1]}\delta(x_1 - x_1^{(0)}, x_2 - x_2^{(0)}, x_3 - x_3^{(0)}).$$

In Fig. 5.16 $q_j(j = 1, \ldots, 9)$ is shown as a function of L, the distance from the source position $(x_1^{(0)}, x_2^{(0)}, x_3^{(0)}) = (0, 0, 0)$; $\mu = 0.1\,\text{cm}^2/\text{sec}$ (clusters of molecular size), $R = 1\,\text{cm}$. (The percentage of q_j to the total number of particles is shown.) Curve 1 was obtained for $v = 0.5\,\text{cm/sec}$, curve 2 for $V = 1\,\text{cm/sec}$, and curve 3 for $V = 3\,\text{cm/sec}$. All computations were carried out in 9 min of computer time. The results of Fig. 5.16 show that when $V = 0.5\,\text{m/sec}$, the main fraction of the particles is absorbed in the first steps (from 0.5 cm to 2 cm). If $V = 3\,\text{cm/sec}$, the particles are absorbed more uniformly, which is often desired in practice.

We denote by L_{max} the distance where q_j reaches its minimum (maximal absorption). Calculations showed (Fig. 5.17) that L_{max}, as a function of V, is well approximated by a straight line.

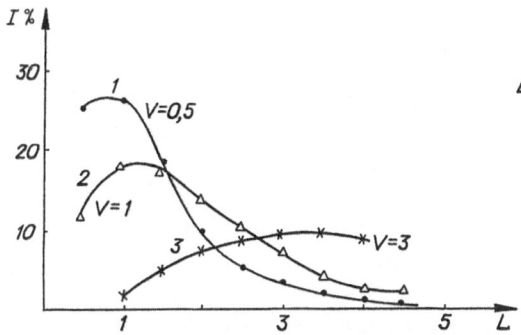

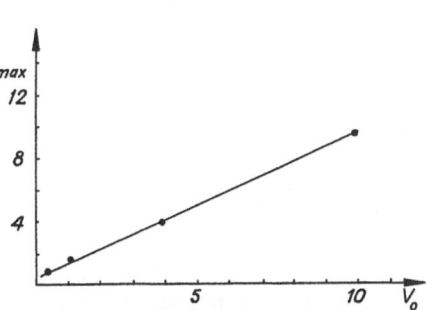

Fig. 5.16. Deposition of particles as a function of distance for different values of V curve 1, 0.5 cm/sec, curve 2, 1 cm/sec, curve 3, 3 cm/sec

Fig. 5.17. Dependence of the maximum particle deposition on the velocity at the molecular cylinder axis

Finally, let us describe the results of calculation of transformation of the particle size distribution. We define the transmission coefficient $p_j(\mu)$ on a part $\partial G_j[j = 1, \ldots, m]$ as

$$p_j(\mu) = 1 - \sum_{i=1}^{j} q_i(\mu) \, .$$

We divide the segment $[\mu_{\min}, \mu_{\max}]$ of given values of the molecular diffusion coefficient in l subintervals $\Delta\mu_k$ and choose a fixed value $\mu_k \in \Delta\mu_k$ [$k = 1, 2, \ldots, l$]. If the matrix $P = \{p_j(\mu_k)\}_{j=1,\ldots,m}^{k=1,\ldots,l}$ is known, then the calculation of the transformation of size distribution is solved easily. Indeed, let us assume that the initial distribution of μ_k is defined by a vector $\pi^{(0)} = (\pi_1^{(0)}, \ldots, \pi_l^{(0)})$, where $\pi_k^{(0)}$ is the index of the fraction of particles with $\mu \in \Delta\mu_k$ ($k = 1, \ldots, l$). Then the distribution for j-th region $\pi^{(j)}$ is calculated as $\pi_k^{(j)} = p_j(\mu_k)\pi_k^{(0)}$, $j = 1, \ldots, m$; $k = 1, \ldots, l$. The initial distribution was taken to be lognormal. As calculations show, this distribution can be strongly changed. In Fig. 5.18 the transmission coefficient ξ is shown as a function of distance L for different values of μ. The error of the calculations lies between 1 and 5%, the computer time was ~ 10 min.

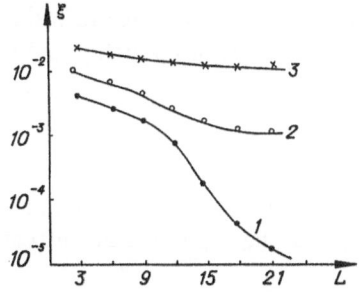

Fig. 5.18. Transmission curves for different values of diffusion coefficient μ. Curve 1, 4×10^{-2} cm^2/sec, curve 2, 1.3×10^{-3} cm^2/sec, curve 3, 0.4×10^{-4} cm^2/sec

5.4.4 Simulation of the Growth
of Nuclei on Highly Dispersed Aerosol Particles

Currently there is great interest in highly dispersed aerosols (HDA) with particle diameters $d < 0.5\,\mu m$, and with complicated coatings [5.24, 25]. This kind of particle is found in combustion processes, in formation of atmospheric aerosols, in heterogeneous catalysis, etc. [5.26]. Here we present the results of numerical and experimental investigations carried out in cooperation with physicists of the Department of Dispersed Media, Institute of Chemical Kinetics and Combustion of the Siberian Branch of the USSR Academy of Science.

For the heterogeneous condensation nuclei, we used latex particles which are ideal spheres. These particles with density $\varrho = 10^2\,kg/cm^3$ enable us to predict and experimentally observe bismuth islands with diameters larger than 1.5–2 nm [5.26]. In the process of heterogeneous nucleation a flux of atoms of a condensed material given by

$$I = \alpha \frac{4\pi D R n_1}{1 + \lambda_0 l/R} \tag{5.126}$$

falls on a spherical particle $S(0, R)$ of radius R, centered at the origin. Here D is the molecular diffusion coefficient in the gas phase, n_1 is the concentration of the condensed atoms, l is the free path length of a molecule in the gas phase, λ_0 is a tabulated constant of order 1, α is the accomodation factor (equal to probability that an atom reaching the particle surface will not immediately reevaporate). Thus the incident atom will remain on the surface with probability α and start to diffuse on $S(0, R)$. We denote by a the atomic free path length. The surface diffusion coefficient is related to a by

$$D = (a^2 \nu) \exp\{E_D/kT\} \tag{5.127}$$

where E_D is the activation energy of the surface diffusion, ν is the frequency ($\approx 10^{13}\,sec^{-1}$). The atom is diffusing on $S(0, R)$ until it collides with another atom. The result of the collision is a fixed dimer island. This island is then built up by direct condensation of atoms from the gas phase [the source (5.126)] and by capture of atoms diffusing on $S(0, R)$ (Fig. 5.19). The experiments show that the shape of the growing island of j atoms is well approximated by a spherical segment; let θ be the angle formed by the tangents to the two surfaces of this segment (Fig. 5.19). We assume that θ is the same for all growing islands. This is also confirmed by classical thermodynamics [5.27] and is in good agreement with our experiments. As mentioned above, the island will capture the free diffusing atoms on $S(0, R)$; the radius of capture $0L$ (Fig. 5.19) is given through j and θ. $0L \equiv a_1 = B\pi R/180°$, where B is obtained from

$$\tan B = \frac{r \sin \theta}{R - r \cos \theta}.$$

The dependence of $0L$ on the number of atoms in the island for different values of θ is shown in Fig. 5.20. From this we obtained that the capture radius a_1 can be approximated (to within 10%) by

$$a_1 = a_2 j^{1/3} \tag{5.128}$$

if $\theta \approx 20°–60°$ and j is not large (from 50 to 100). In (5.128) a_2 depends only on θ. This simple approximation simplifies the calculations in the direct Monte Carlo simulation of the process under study.

The expressions (5.126, 127) depend on many parameters, for example, the temperature and concentration of the condensed material, the radius of the particle, etc. Thus, it is important to reformulate the problem to minimize the number of parameters. To this end, we make the following assumptions:

a) the adatom's lifetime on $S(0, R)$ is much longer than the time of the experiment
b) the atom's diffusion on $S(0, R)$ is assumed to be homogeneous, not depending on the atom's position
c) a collision of two adatoms results, with probability equal to 1, in the appearance of a fixed island.

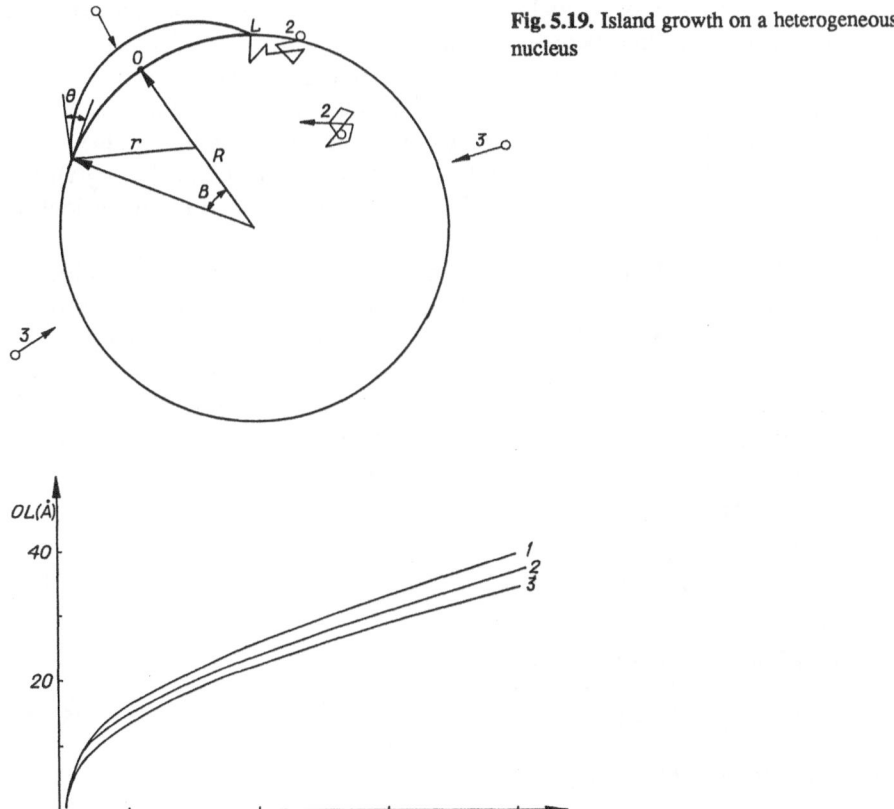

Fig. 5.19. Island growth on a heterogeneous nucleus

Fig. 5.20. Dependence of the capture radius OL of the island on the number of atoms in the island, for different values of the angle θ. Curve 1 30°; curve 2 45°, curve 3 60°

Assumption (a) is satisfied since the temperature in the condensation process is usually less than 100°C. Experiments with materials such as AgI, Bi, BiI$_3$ showed that the lifetime of adatoms on $S(0, R)$ is many orders of magnitude larger than the time of the experiment. Assumption (b) is satisfied by the choice of the condensation nuclei: as experiments show, latex and photoaerosol particles have ideal spherical form with homogeneous surfaces. Finally, assumption (c) that the appeared island is stable and fixed on $S(0, R)$ is justified by theoretical [5.13, 14] and experimental results [5.26]. It should be noted that the assumptions are satisfied if the temperature is less than 100°C. Otherwise it is necessary to take into account the diffusion of the islands over $S(0, R)$.

We now introduce the time T_1 which is the average time interval between successive depositions of two atoms on the sphere, $T_1 = 1/I$, where I is given by (5.126). We define T_2 as the average time between two successive atom collisions while diffusing on $S(0, R)$, $T_2 = \nu^{-1} \exp(E_D/kT)$, where E_d is the activation energy of the surface diffusion. Then $\delta = T_1/T_2$ describes the average number of moves that an atom makes on $S(0, R)$ between two successive depositions of new atoms. Formally, δ varies between 0 and ∞. It is convenient to take $R = 1$. Then the change in R is equivalent to a change of the length of atom step on $S(0, 1)$. We define a dimensionless diffusion length by $c_0 = a_0/R$ where a_0 is this length on the surface of $S(0, R)$. The "capture radius" in dimensionless units is defined as $c_2 = a_1/R$. Thus, the input parameters of our model are δ, c_0, c_2.

We have calculated the following quantities: $N_n(t)$, the total number of islands on the sphere $S(0, 1)$; (x_i, y_i, z_i), the coordinates of the nuclei; $n_j(t)$, the number of atoms in the j-th nucleus. The time t was measured in dimensionless units δ.

It should be noted that the direct Monte Carlo simulation of this process is efficient only for the first non-stationary period until the number N_n becomes stable. But stabilization of N_n implies that our problem becomes linear, since there remain, in fact, only interactions between the atoms and the islands. Thus, we are then faced with a linear diffusion problem on the surface $S(0, R)$. Hence, the walk on spheres (in fact, on circles lying on our sphere) can be applied. Generally, let us consider a diffusion equation in an m-connected domain $G \subset \mathbf{R}^3$ with a boundary

$$\Gamma = \bigcup_{i=1}^{m} \Gamma_i : \quad \frac{\partial c}{\partial t} = D\Delta c + f \tag{5.129}$$

with initial and boundary conditions

$$c(x, 0) = \varphi, \quad c|_\Gamma = 0. \tag{5.130}$$

Consider also a problem which is adjoint to (5.129, 130):

$$\frac{\partial c^*}{\partial t} = D\Delta c^* + f, \quad c^*(x, T) = \varphi^*, \quad c^*|_\Gamma = \psi^*. \tag{5.131}$$

From

$$\int_G (c^* D \Delta c - cD\Delta c^*) dx = \int_\Gamma \left(c^* D \frac{\partial c}{\partial n} - cD \frac{\partial c^*}{\partial n} \right) ds_\Gamma$$

we obtain for $f^* = 0$, $\psi^* = 1$ on Γ_i, and $\psi^* = 0$ on $\Gamma \backslash \Gamma_i$

$$\int_{\Gamma_i} D \frac{\partial c}{\partial n} ds_\Gamma = \int_G \left\{ \frac{\partial}{\partial t} (cc^*) - c^* f \right\} dx .$$

Hence

$$\int_G \left(cc^* |_0^T \right) dx - \int_0^T dt \int_G c^* f dx = J_i \tag{5.132}$$

where

$$J_i = \int_0^T dt \int_{\Gamma_i} D \frac{\partial c}{\partial n} ds_\Gamma . \tag{5.133}$$

From (5.132) we get for $\varphi^* = 0$

$$J_i = \int_G \varphi(x) c^*(x, 0) dx + \int_0^T dt \int_G c^* f dx . \tag{5.134}$$

Thus, the flux J_i on Γ_i can be calculated from (5.134) using Monte Carlo estimates for solving (5.131) and the double randomization technique. In the stationary case analogous arguments yield for the flux

$$J_i = \int_{\Gamma_i} D \frac{\partial c}{\partial n} ds_\Gamma$$

the relation

$$J_i = \int_G c^* f dx \tag{5.135}$$

where

$$D\Delta c = f , \quad c|_\Gamma = \varphi = 0 , \quad D\Delta c^* = f^* = 0 ,$$
$$c^*|_\Gamma =)\psi^* = \begin{cases} 1 & \text{on } \Gamma_i \\ 0 & \text{on } \Gamma \backslash \Gamma_i . \end{cases}$$

Indeed, from

$$\int_G (c^* D \Delta c - cD\Delta c^*) dx = D \int_\Gamma \left(c^* \frac{\partial c}{\partial n} - c \frac{\partial c^*}{\partial n} \right) ds_\Gamma$$

we get

$$\int_G (c^* f - cf^*) dx = \int_\Gamma \psi^* D \frac{\partial c}{\partial n} ds_\Gamma .$$

Taking now $f^* = 0$, $\psi^* = 1$ on Γ_i and $\psi^* = 0$ on $\Gamma \backslash \Gamma_i$ we obtain (5.135). (Derivation of (5.135) for the case when the equation describes surface diffusion is more complicated but follows exactly the same scheme.) From this we can formulate the following walk on circles algorithm. Let Γ_i be the boundary of the i-th island. According to the density f/D [where f is determined from (5.126), and D from (5.127)] a random point on the sphere is simulated. Then a walk on circles starting from this point and absorbed in Γ_{i_e} (in this case the contribution is $\psi^* = 1$) is constructed.

We now describe the results of simulation. We focus our attention on two quantities: N_n, the number of islands, and n_a, the number of adatoms moving on the sphere ("free" adatoms). At the initial moment $N_n = 0$, $n_a = 0$. After a given time interval, we have some fixed islands and free adatoms. Then, when the relative size of the surface occupied by the islands becomes larger, n_a decreases and a stable state arises where $N_n \approx \tilde{N}_n$ does not change, and $n_a \to 0$. Two examples are shown in Fig. 5.21 in which after $\tilde{t} = \tilde{t}_0$ the process stabilizes. The dependence of the mean number of atoms per island in the stable state on the total number of incident atoms is shown in Fig. 5.22. In this example stabilization occurred when the number of atoms in islands is about 10^3. Thus, the non-stationary part of the process cannot be observed in real experiments. However, from the practical viewpoint it is important to note that the island configuration becomes stable when the number of condensed atoms in each island is quite

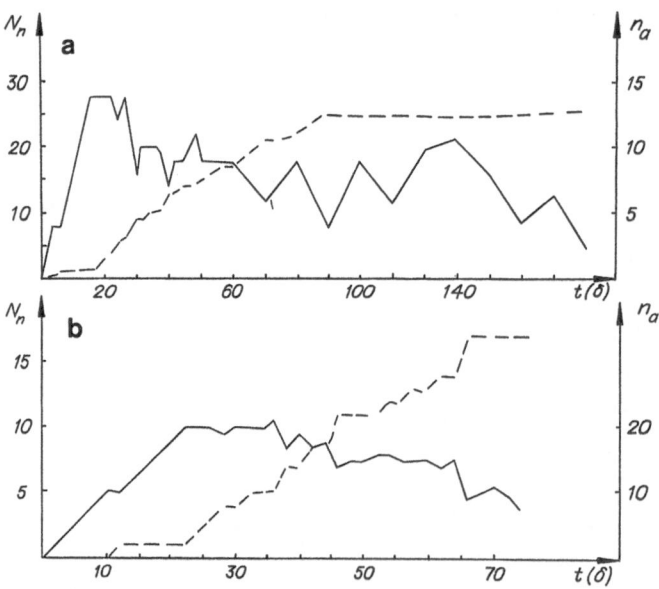

Fig. 5.21. Formation of stable nuclei on a sphere: Dependence of $\tilde{N}_n$ the number of islands (*solid line*) and total number of free atoms n_a (*dashed line*) on the time; (a): $\delta = 3$, $c_0 = 0.1$, $c_2 = 0.05$; (b): $\delta = 100$, $c_0 = 0.05$, $c_2 = 0.05$

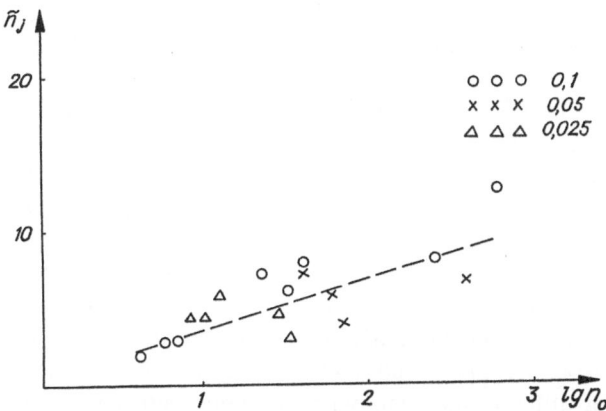

Fig. 5.22. Dependence of the mean number of atoms in an island on the total number of incident atoms (at the time instant $t = \tilde{t}$, i.e., a stationary regime is reached): 1 : $c_0 = 0.1$, 2 : $c_0 = 0.05$, 3 : $c_0 = 0.025$

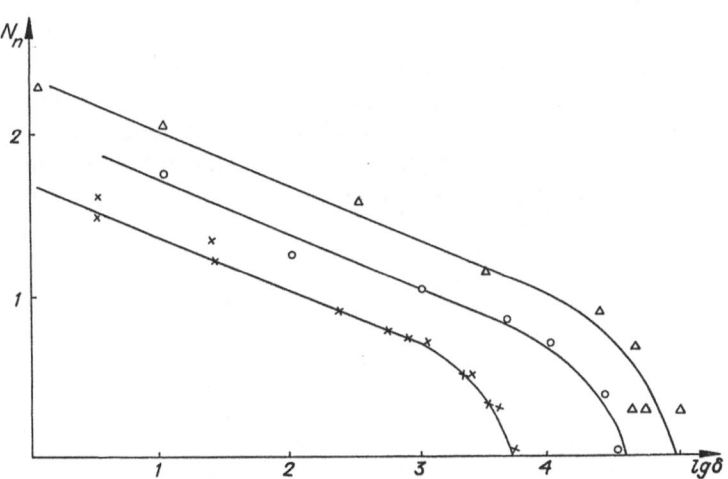

Fig. 5.23. Dependence of the number of islands (in stationary regime) on the dimensionless time δ: 1: $c_0 = 0.1$, 2: $c_0 = 0.05$, 3: $c_0 = 0.025$

small. In Fig. 5.23 we show the dependence of $\tilde{N}_n$ on the dimensionless time δ for different values of c_0. Two regions of different behavior of $\tilde{N}_n$ are easily seen: a linear (on a logarithmic scale), the slope is close to 0.30–0.35, and a region where $\tilde{N}_n$ is rapidly changing.

It should be noted that the concentration n_1 of the condensed material, that is, the source of the incident atoms, is one of the important parameters in this problem. The dependence of $\tilde{N}_n$ on n_1 is shown in Fig. 5.24 for different sizes of the sphere. In the linear region the slope is close to 0.5 which corresponds to

269

$\tilde{N}_n \approx n_1^{1/2}$. Note that analogous results were obtained for nucleation on a plane surface [5.24, 25]. In our calculations, we obtained nonlinear behavior (Fig. 5.22) beginning from $\tilde{N}_n \approx 4$.

Let us discuss what the values of n_1 in actual experiments are. The concentration of the condensed material must be higher by at least $10^8 \, \text{cm}^{-3}$, since, to investigate the steady state by electron microscopy, the size of the islands must be on the order of 20–30 Å. On the other hand, n_1 must not be higher than $10^{12} \, \text{cm}^{-3}$ since, otherwise, homogeneous nucleation will begin and completely change the process under study. So, there is an interval of only 4 orders of magnitude where n_1 can be verified. As is seen from Fig. 5.24, in situations close to realistic experimental conditions there is always a possibility of obtaining one island if $R \leq 50 \, \text{Å}$. Note that this situation often arises in practice when simultaneous condensation of different materials with different vapor pressures is carried out to obtain island structures on latex solutions. The most favorable situation occurs at $R \approx 200 \, \text{Å}$. Here, varying of the vapor pressure between 10^{10} and $10^{12} \, \text{cm}^{-3}$ may result in 1 to 10 islands appearing on the sphere.

The next important parameter in this problem is the temperature in the condensation cell. Variation of this temperature strongly affects the surface diffusion coefficient D, which in turn leads to the change of N_n. For example, changing T_1 from 250 to 500 K in some cases results in decreasing N_n from 100 to 5 (for realistic values of n_1). The dependence $N_n(T_1)$ is shown in Fig. 5.25 for different values of n_1. When the number of nuclei becomes larger, the measurement error rapidly increases, since the derivative $dT_n/d\tau_1$ is large. It should be noted that the quantity $N_n(T_1)$ is only slightly dependent on the size. This effect shifts the

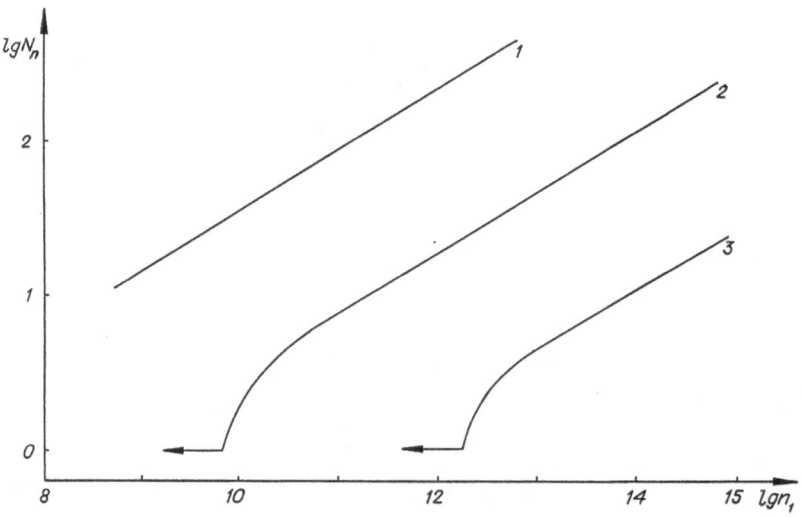

Fig. 5.24. Dependence of N_n on the concentration of condensed vapor (n_1 in cm^{-3}) for different values of the nucleus size; 1: $R = 1000 \, \text{Å}$; 2: $R = 200 \, \text{Å}$; 3: $R = 50 \, \text{Å}$ ($T_1 = 300 \, \text{K}$, $E_D = 0.3 \, \text{eV}$)

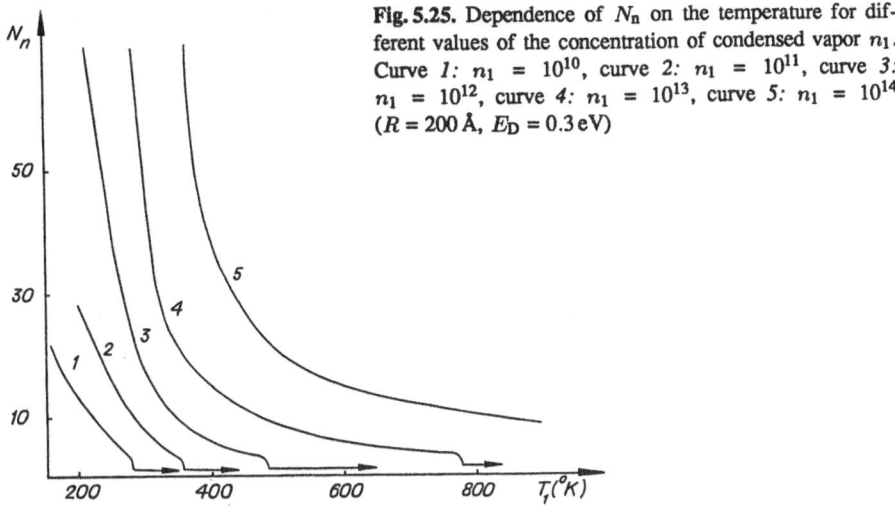

Fig. 5.25. Dependence of N_n on the temperature for different values of the concentration of condensed vapor n_1. Curve 1: $n_1 = 10^{10}$, curve 2: $n_1 = 10^{11}$, curve 3: $n_1 = 10^{12}$, curve 4: $n_1 = 10^{13}$, curve 5: $n_1 = 10^{14}$ ($R = 200\,\text{Å}$, $E_D = 0.3\,\text{eV}$)

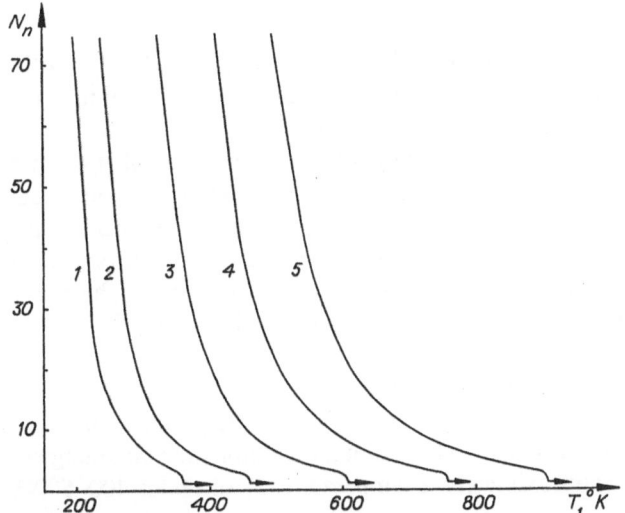

Fig. 5.26. Same as Fig. 5.25 for different values of E_D. Curve 1: $E_D = 0.25\,\text{eV}$, curve 2: $E_D = 0.35\,\text{eV}$, curve 3: $E_D = 0.45\,\text{eV}$, curve 4: $E_D = 0.5\,\text{eV}$, curve 5: $E_D = 0.6\,\text{eV}$; $n_1 = 10^{12}$, $R = 200\,\text{Å}$

critical temperature at which the first nucleus appears by 10–20 K. However, the size effect strongly affects the critical concentration n_1 (by more than 1 order of magnitude, Fig. 5.24).

Now, let us discuss the parameter E_D. This parameter is important but practically not controllable. The dependence $N_n(T_1)$ is shown in Fig. 5.26 for different values of E_D. The number of nuclei (for the case $t = 300\,\text{K}$, $R \approx 200$–$400\,\text{Å}$) for various condensed materials varies between 10 and 40. This enables us to find

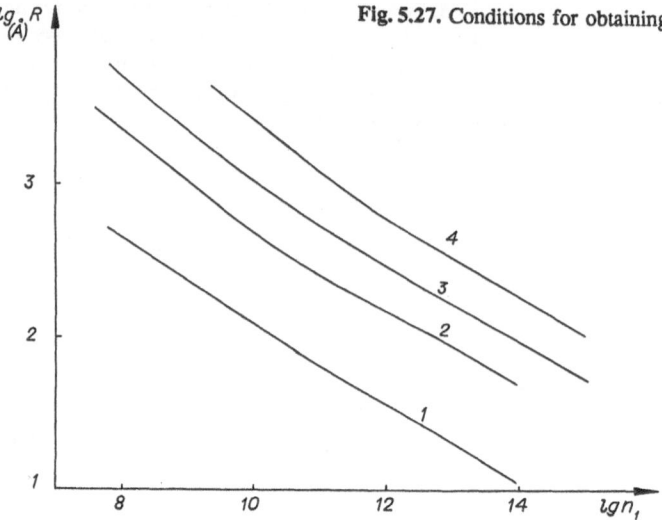

Fig. 5.27. Conditions for obtaining a single nucleus

the value of $E_D \approx 0.25$–$0.35\,\mathrm{eV}$ (compare curves 1, 2, 3 at $T = 300\,\mathrm{K}$). Notice that N_n strongly depends on E_D so that a $0.2\,\mathrm{eV}$ change in E_D leads to a two orders of magnitude change in N_n.

In conclusion, let us consider two limiting situations: (1) the spherical particle is fully coated by a thick layer of nuclei, (2) there exists only one small nucleus. Both cases are important from the practical viewpoint: the first case is interesting for problems of light scattering, and the second one is relevant to the heterogeneous processes. In Fig. 5.27 we show the curves separating the region where one nucleus occurs (to the left of curves) and the region where $N_n \geq 4$, on the scales $\log(R)$, $\log(n_1)$. As can be seen, it is difficult to obtain one nucleus on the sphere with radius $R = 1000\,\text{Å}$ if the concentration n_1 is high ($n_1 \geq 10^{12}$), since the needed temperature must be about $700\,\mathrm{K}$. But this high temperature may lead to evaporation of the nuclei. The optimal conditions for obtaining one nucleus at $R = 10^3\,\text{Å}$ are approximately a temperature T_1 of about 400–$500\,\mathrm{K}$, and $n_1 \approx 10^{10}\,\mathrm{cm}^{-3}$.

Now, let us discuss the second case where it is desired to obtain a fully coated particle. Assume that θ is between 60 and 90° and that it is necessary to obtain a layer of height equal to $10\,\text{Å}$. The corresponding conditions are shown in Fig. 5.28: the desired coating is obtained if $T < \log(n_1)$ (right region). These conditions are satisfied practically for all sizes ($R > 20\,\text{Å}$). The results show that for $n_1 \leq 10^{12}$, $E_D = 0.3\,\mathrm{eV}$, the desired coating can be obtained only if $T \leq 200\,\mathrm{K}$. Even if $E_D \approx 0.6\,\mathrm{eV}$, it is difficult to obtain such a coating at room temperature.

It was discovered in the experiments that the islands are uniformly distributed over the surface of the particle. In addition, the sizes of the islands do not vary strongly. In Fig. 5.29 and in Table 5.2 we present the results of numerical simulation for $R = 200\,\text{Å}$, $\theta = 90°$. The total number of adatoms was 6.4×10^3. We obtain 43 stable nuclei. The mean radius of the nucleus is $\bar{r} = 13\,\text{Å}$, as the results of Fig. 5.29 show. The relative variance of the radius $\Delta r / \bar{r}$ is 0.46. This condensation regime corresponds to an intense flux of the incident atoms ($\delta \approx 1$). In another variant (Table 5.3) the flux was 10^3 times less ($\delta \approx 10^3$). Here we obtained 5 stable islands with sizes 12, 10, 13, 13.3, and $10\,\text{Å}$. In this case, $\Delta r / \bar{r} = 0.25$. Of course, the size distribution of the islands and the spatial

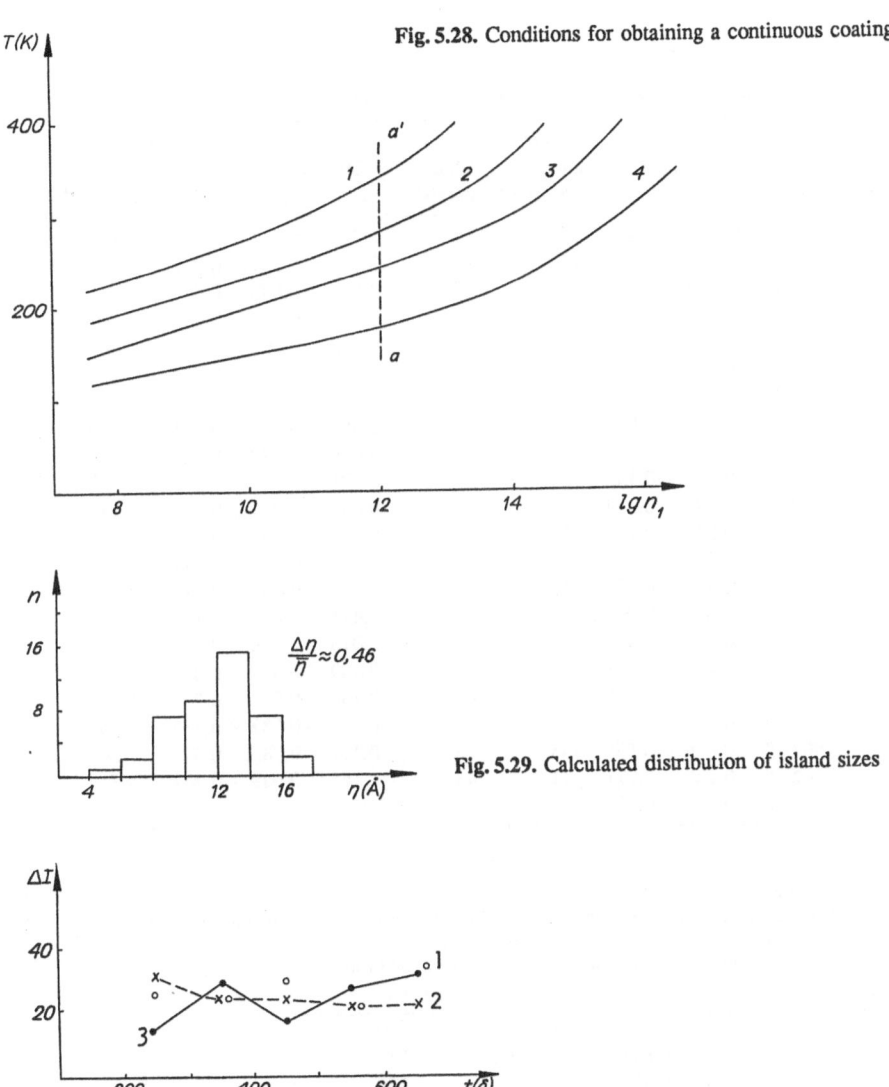

Fig. 5.28. Conditions for obtaining a continuous coating

Fig. 5.29. Calculated distribution of island sizes

Fig. 5.30. Flux of atoms on the islands. Curve *1*: island which appeared first, curve *2*: last island, curve *3*: an island which appeared at an intermediate time

distribution of islands over the surface of the particle are closely related. It is therefore important to estimate the fluxes of diffusing adatoms on each island. This problem is difficult to solve by conventional numerical methods [5.28]. These fluxes were calculated by the Monte Carlo method, the corresponding results are shown in Fig. 5.30 for the five large islands. The fluxes are shown for the island which appeared first (curve 1), last (curve 2), and at an intermediate moment (curve 3). These results show that the average fluxes do not differ by more than a factor of 2. This implies that the sizes of islands differ by not more than $2^{1/3}$ (≈ 1.25) times.

Table 5.2. Calculated island positions and sizes with high incident atom flux $\delta \sim 1$

Island number	Island coordinates x	y	z	Number of atoms per island	Island number	Island coordinates x	y	z	Number of atoms per island
1	−0.023	0.234	−0.971	153	23	0.249	0.652	0.716	167
2	0.579	−0.742	0.335	150	24	−0.625	0.321	−0.710	199
3	−0.530	0.594	−0.604	15	25	−0.358	−0.833	−0.417	128
4	0.168	−0.898	−0.403	23	26	−0.815	0.324	0.478	147
5	−0.490	−0.608	0.622	149	27	0.972	0.199	−0.122	87
6	0.322	0.892	0.318	95	28	0.483	−0.208	0.850	289
7	−0.182	−0.296	0.937	180	29	−0.400	−0.907	0.119	148
8	−0.818	−0.565	0.096	49	30	−0.659	0.688	−0.301	166
9	0.290	−0.627	0.722	125	31	0.076	−0.979	0.189	314
10	0.967	−0.213	−0.138	63	32	−0.271	0.936	0.223	38
11	0.516	0.182	0.837	41	33	0.856	−0.079	−0.509	111
12	−0.949	−0.244	0.192	163	34	0.916	0.093	0.391	392
13	0.156	−0.974	−0.161	37	35	−0.743	0.644	0.177	231
14	−0.833	−0.383	−0.394	166	36	0.715	−0.469	−0.516	195
15	0.769	0.510	−0.384	49	37	0.471	0.864	−0.178	306
16	0.385	0.753	−0.533	58	38	0.016	−0.350	−0.935	178
17	0.592	0.053	−0.803	213	39	−0.409	0.383	0.828	257
18	−0.924	0.175	−0.335	124	40	−0.138	0.828	0.553	175
19	0.152	0.069	0.986	6	41	−0.615	−0.232	−0.752	253
20	−0.025	0.844	−0.535	135	42	0.411	−0.745	−0.524	260
21	−0.025	0.238	0.971	71	43	−0.145	0.979	−0.142	219
22	−0.062	−0.629	−0.773	75					

Table 5.3. Calculated island positions and sizes under low incident atom flux $\delta \sim 10^3$

Island number	Island coordinates x	y	z	Number of atoms per island
1	−0.287	−0.191	−0.937	110
2	−0.495	0.615	−0.613	171
3	−0.650	−0.174	0.738	169
4	−0.826	−0.536	−0.167	68
5	0.542	−0.807	−0.231	132

References

Introduction

0.1 G.I. Marchuk, G.A. Mikhailov, M.A. Nazaraliev, R.A. Darbinjan, B.A. Kargin, B.S. Elepov: *Monte Carlo Methods in Atmospheric Optics* (Springer, Berlin, Heidelberg 1980)

0.2 S.M. Yermakov, G.A. Mikhailov: *Statistical Simulation* (Nauka, Moscow 1982) [in Russian]

0.3 I.M. Sobol: *Numerical Monte Carlo Methods* (Nauka, Moscow 1973) [in Russian]

0.4 G.A. Mikhailov: *Optimization of Weighting Monte Carlo Methods* (Nauka, Novosibirsk 1980) [in Russian]

0.5 A.N. Kolmogorov: "Über die analytischen Methoden in der Wahrscheinlichkeitsrechnung", Math. Ann. **104**, 415–458 (1931)

0.6 N.Wiener: "Differential Space" J. Math. Phys. **2**, 131–174 (1923)

0.7 I.G. Petrovsky: "Über das Irrfahrtproblem" Math. Ann. **109**, 425–444 (1934)

0.8 A.Ya. Khinchin: *Asymptotische Gesetze der Wahrscheinlichkeitsrechnung* (Berlin 1933)

0.9 R. Courant, K. Friedrichs, H. Lewy: Math. Ann. **100**, 1 (1928)

0.10 J.L. Doob: Trans. Amer. Math. Soc. **77** (1954)

0.11 E.B. Dynkin: *Markov Processes* (Fizmatgiz, Moscow 1963) [in Russian]

0.12 R.H. Cameron, W.T. Martin: Ann. Math. **45**(1) (1944)

0.13 I.M. Kovalchik: Dokl. Akad. Nauk SSSR **138**(6) (1961) [in Russian]

0.14 A.L. Ventsel: Usp. Mat. Nauk. **15**(2), 92 (1960) [in Russian]

0.15 K.K. Sabelfeld: Sov. J. Num. Anal. Math. Model. **1**(3) (1986)

0.16 R.H. Cameron, W.T. Martin: Ann. Math. **45**(2) (1944)

0.17 V.S. Vladimirov: Usp. Mat. Nauk. **15**(4) (1960) [in Russian]

0.18 B.S. Elepov, A.A. Kronberg, G.A. Mikhailov, K.K. Sabelfeld: *Solution of Boundary Value Problems by Monte Carlo Methods* (Nauka, Novosibirsk 1980) [in Russian]

0.19 L.A. Yanovich: *Approximate Evaluation of Continual Gaussian Integrals* (Nauka, Minsk 1976) [in Russian]

0.20 I.M. Gelfand, A.S. Frolov, N.N. Chentsov: Izv. Vuzov. Matematika No. 5, 32–45 (1958)

0.21 M.E. Müller: Ann. Math. Statistics **27**(3) (1956)

0.22 M. Motoo: Ann. Inst. Stat. Math. XI (1959)

0.23 S.M. Yermakov, V.V. Nekrutkin, A.S. Sipin: *Random Processes for Solving Classical Equations of Mathematical Physics* (Nauka, Moscow 1984) [in Russian]

0.24 K.K. Sabelfeld: *Monte Carlo Methods in Boundary Value Problems* (Nauka, Novosibirsk 1989) [in Russian]

0.25 K.K. Sabelfeld: Dokl. Akad. Nauk SSSR **262**(5) (1982) [in Russian]

Chapter 1

1.1 L.W. Kantorowitsch, G.P. Akilow: *Funktionalanalyse in normierten Räumen* (Akademie, Berlin 1964)

1.2 S.M. Yermakov, G.A. Mikhailov: *Statistical Simulation* (Nauka, Moscow 1982) [in Russian].

1.3 J.M. Hammersley, D.C. Handscomb: *Monte Carlo Methods* (Methuen, London 1964)

1.4   E. Goursat: *Cours D'Analyse Mathematique III* (Gauthier–Villars, Paris 1930)
1.5   V.N. Kublanovskaya: Trudy Matem. Inst. Steklov. **53**, 145–185 (1959) [in Russian]
1.6   L.W. Kantorowitsch, V.I. Krylov: *Approximate Methods of Superior Analysis* (Fizmatgiz, Leningrad 1962) [in Russian]
1.7   K.K. Sabelfeld: Sov. J. Num. Anal. Math. Model. **2**(3), 201–218 (1987)
1.8   B.S. Elepov, A.A. Kronberg, G.A. Mikhailov, K.K. Sabelfeld: *Solution of Boundary Value Problems by Monte Carlo Methods* (Nauka, Novosibirsk, 1980) [in Russian]
1.9   C. Miranda: *Partial Differential Equations of Elliptic Type* (Springer, Berlin, Heidelberg 1970)
1.10 S. Mizohata: *Theory of Partial Differential Equations* (Cambridge University Press 1973)
1.11 B.W. Schulze, G. Wildenhein: *Methoden der Potentialtheorie für elliptische Differentialgleichungen beliebiger Ordnung* (Akademie, Berlin 1977)
1.12 N. Wiener: Proc. Nat. Acad. Sci. **7**(10) (1921)
1.13 R.H. Cameron: Duke Math. J. **18**(1) (1951)
1.14 R.H. Cameron, W.T. Martin: Ann. Math. **45**(2) (1944)
1.15 L.A. Yanovich: *Approximate Evaluation of Continual Gaussian Integrals* (Nauka, Minsk 1976)
1.16 R. Hersh: Springer Lect. Notes Math. **446**, 283–300 (1975)
1.17 M.E. Müller: Ann. Math. Stat. **27**(3) (1956)
1.18 A.S. Sipin: in *Monte Carlo Methods in Numerical Mathematics and Mathematical Physics*, ed. by G.A. Mikhailov (Nauka, Novosibirsk 1976)
1.19 S. Hoshino, K. Ichida: Numer. Math. **18**, 61–72 (1971)
1.20 K.K. Sabelfeld, O. Kurbanmuradov: Computing Center, Novosibirsk Preprint No. 506 (1984) [in Russian]
1.21 M. Motoo: Ann. Inst. Stat. Math. **11**, 49–54 (1959)
1.22 A.A. Borovkov: *Probability Theory* (Nauka, Moscow 1976) [in Russian]
1.23 K.K. Sabelfeld: Sov. J. Num. Anal. Math. Model. **1**(3) (1986)
1.24 K.K. Sabelfeld, N.A. Simonov: Chisl. Metody Mekh. Splosh. Sredy, **14**(1) (1983) [in Russian]
1.25 I.M. Sobol: *Numerical Monte Carlo Methods* (Nauka, Moscow 1973) [in Russian]
1.26 A.J. Chorin: Math. Comp. **27**(121), 1–15 (1973)
1.27 A.S. Monin, A.M. Jaglom: *Statistical Fluid Mechanics of Turbulence*, Vols. 1, 2 (MIT, Cambridge, MA 1967)
1.28 I.A. Malyshev, Yu.I. Palagin: Tekhn. Kibernetika, **6**, 13–21 (1980) [in Russian]
1.29 Yu.I. Palagin: Avtomatika i Telemekhanika, **2**, 35–42 (1981) [in Russian]
1.30 K.K. Sabelfeld, O. Kurbanmuradov: Sov. J. Num. Anal. Math. Model. **4**(1), (1989)
1.31 S.M. Ponamaryeva, A.S. Gavrilov: Gidrometeorologia, Ser. Meteorologia, No. 1 (Obninsk, 1984) [in Russian]
1.32 S.M. Rytov, Yu.A. Kravtsov, V.I. Tatarskii: *Principles of Statistical Radiophysics 3, Elements of Random Field Theory* (Springer, Berlin, Heidelberg 1987)

Chapter 2

2.1   S.M. Yermakov, G.A. Mikhailov: *Statistical Simulation* (Nauka, Moscow 1982) [in Russian]
2.2   G.A. Mikhailov: *Optimization of Weighting Monte Carlo Methods* (Nauka, Novosibirsk 1980) [in Russian]
2.3   I.M. Sobol: *Numerical Monte Carlo Methods* (Nauka, Moscow 1973) [in Russian]
2.4   L.W. Kantorowitsch, V.I. Krylov: *Approximate Methods of Superior Analysis* (Fizmatgiz, Leningrad 1962) [in Russian]
2.5   C. Miranda: *Partial Differential Equations of Elliptic Type* (Springer, Berlin, Heidelberg 1970)
2.6   S. Mizohata: *Theory of Partial Differential Equations* (Cambridge University Press, 1973)
2.7   B. W. Schulze, G. Wildenhein: *Methoden der Potentialtheorie für elliptische Differentialgleichungen beliebiger Ordnung* (Akademie, Berlin 1977)
2.8   V.N. Kublanovskaya: Trudy Matem. Inst. Steklov **53**, 145–185 (1959) [in Russian]
2.9   K. Jörgens: *Linear Integral Operators* (Pitman, Boston 1970)

2.10 L.W. Kantorowitsch, G.P. Akilow: *Funktionalanalyse in normierten Räumen* (Akademie, Berlin 1964)
2.11 E. Goursat: *Cours D'Analyse Mathematique III* (Gauthier–Villars, Paris 1930)
2.12 V.D. Kupradze, T.G. Gegelia, M.O. Basheleishvili, T.V. Burkhuladze: *Three Dimensional Problems in Mathematical Elasticity Theory* (Nauka, Moscow 1976) [in Russian]
2.13 G.E. Lewis: in *Methods in Computational Physics* (Academic, New York 1965)
2.14 R.D. Richtmayer: AEC Computing and Applied Mathematics Center, Courant Institute of Mathematical Sciences, New York University Report No. 40 (1973)
2.15 M.M. Lavrentyev, V.G. Romanov, S.P. Shishatskii: *Ill-posed Problems in Mathematical Physics and Analysis* (Nauka, Novosibirsk 1973) [in Russian]
2.16 G. Polya, G. Szego: *Aufgaben und Lehrsätze aus der Analyse* (Springer, Berlin, Heidelberg 1964)
2.17 D.K. Faddeev, V.N. Faddeeva: *Numerical Methods of Linear Algebra* (Fizmatlit, Moscow 1963) [in Russian]
2.18 C.P. Rao: *Linear Statistical Inference and its Applications* (Wiley, New York 1960)
2.19 V.Ya. Skorobogatko: *Theory of Continued Fractions and Applications to Computational Mathematics* (Nauka, Moscow 1983) [in Russian]
2.20 W. Fair: Rocky Mountains J. Math. **4**, 357–260 (1974)
2.21 G.H. Hardy: *Divergent Series* (Oxford University Press, 1949)
2.22 E. Hille: *Functional Analysis and Semigroups*, Amer. Math. Soc. Colloq., Vol. 31 (Providence, 1957)
2.23 G. Cooke: *Infinite Matrices and Sequence Spaces* (London 1950)
2.24 R.P. Agnew: Amer. J. Math. **66**(2) (1944)
2.25 G.S. Salekhov: *Evaluation of Series* (Gostekhizdat, Moscow 1956)
2.26 D. Shanks: Math. Phys. **34**, 1–42 (1955)
2.27 L. Wuytack (ed.): *Padé Approximation and its Applications* Springer Lect. Notes Math. Vol. 765 (1979)
2.28 J. Gilewicz: *Approximants de Padé.* Springer Lect. Notes Math. Vol. 667 (1978)
2.29 *Padé Approximants Method and Its Applications to Mechanics* Lect. Notes Phys. Vol. 47 (1976)
2.30 V.M. Friedman: Usp. Mat. Nauk. **11**(1), 67 (1956)

Chapter 3

3.1 C. Miranda: *Partial Differential Equations of Elliptic Type* (Springer, Berlin, Heidelberg 1970)
3.2 W. Sternberg: Math. Z. **21** (1924)
3.3 W.I. Smirnov: *Lehrgang der höheren Mathematik Bd. IV, V* (Berlin 1961, 1962)
3.4 M.V. Keldysh: Usp. Mat. Nauk, alte ser. **8** (1941)
3.5 N.M. Günter: *Die Potentialtheorie und ihre Anwendung auf Grundprobleme der mathematischen Physik* (Leipzig, 1957)
3.6 W.M. Rvachev: *Methods of Algebraic Logic in Mathematical Physics* (Naukova Dumka, Kiev 1975)
3.7 O. Kurbanmuradov, K.K. Sabelfeld, N.A. Simonov: *The Walk on Boundary Algorithms* (Nauka, Novosibirsk 1989) [in Russian]
3.8 V.N. Kublanovskaya: Trudy Matem. Inst. Steklov **53**, 145–185 (1959) [in Russian]
3.9 M.E. Müller: Ann. Math. Statistics, **27**(3) (1956)
3.10 S.M. Yermakov, G.A. Mikhailov: *Statistical Simulation* (Nauka, Moscow 1982) [in Russian]
3.11 B.S. Elepov, A.A. Kronberg, G.A. Mikhailov, K.K. Sabelfeld: *Solution of Boundary Value Problems by Monte Carlo Methods* (Nauka, Novosibirsk 1980) [in Russian]
3.12 H. Amann: Z. angew. Math. Mech. **46**(5) (1967)
3.13 B. Aggarwala: Z. angew. Math. Mech. **54**(3) (1974)
3.14 K. Gopalsamy, B. Aggarwala: Z. angew. Math. Mech. **53**(6) (1973)
3.15 H.J. Kushner: J. Math. Anal. Appl. **58**, 644–668 (1976)

3.16 R. Courant: *Methods of Mathematical Physics*, Vols. I, II (Interscience, New York 1962)
3.17 K.K. Sabelfeld: Sov. J. Num. Anal. Math. Model. **2**(3), 201–218 (1987)
3.18 V.S. Vladimirov: *Equations of Mathematical Physics* (Nauka, Moscow 1981) [in Russian]
3.19 B.W. Schulze, G. Wildenhein: *Methoden der Potentialtheorie für elliptische Differentialgleichungen beliebiger Ordnung* (Akademie, Berlin 1977)
3.20 V.Ya. Ivanov: Computing Center, Novosibirsk, Preprint No. 8 (1976)
3.21 V.M. Voloshuk: *Introduction to Hydrodynamics of Dispersed Media* (Gidrometeoizdat, Leningrad 1971) [in Russian]
3.22 L.M. Levin: *Investigations in Physics of Low Dispersed Aerosols* (Izd. AN SSSR, Moscow 1961) [in Russian]
3.23 Yu.M. Zyrkunov: Izv. AN SSSR. MDJG, 1 (1982)
3.24 A.G. Sutugin, Z.I. Kotsev, N.A. Fuks: Colloid J. **33**(4) (1971)

Chapter 4

4.1   S.P. Timoshenko: *Vibrations in Engineering* (Nauka, Moscow 1967) [in Russian]
4.2   R. Courant: *Methods of Mathematical Physics*, Vols. I, II (Interscience, New York 1962)
4.3   B. Aggarwala: Z. angew. Math. Mech. **54**(3) (1974)
4.4   S.L. Sobolev: *Introduction to Cubic Formulas* (Nauka, Moscow 1974) [in Russian]
4.5   H. Poritsky: Trans. Amer. Math. Soc. **43**, 199–225 (1938)
4.6   B.E. Pobedrya: *Numerical Methods in Elasticity and Plasticity* (Moscow University Press 1981) [in Russian]
4.7   S. Roux: J. Stat. Phys. **48**(1/2) (1987)
4.8   V.D. Kupradze, T.G. Gegelia, M.O. Basheleishvili, T.V. Burkhuladze: *Three Dimensional Problems of Mathematical Elasticity Theory* (Nauka, Moscow 1976) [in Russian]
4.9   V. Vagner: Z. Vychisl. Mat. Mat. Fiz. **22**(3) (1982) [in Russian]
4.10 I.B. Diaz, L.E. Payne: "Mean Value Relations in the Theory of Elasticity" Proc. Third U. S. Nat. Congr. Appl. Mech. (1958)
4.11 K.K. Sabelfeld, I.A. Shalimova: Sov. J. Num. Anal. Math. Model. **3**(3) (1988)
4.12 V.V. Bolotin: *Random Vibrations of Elastic Systems* (Nauka, Moscow 1979) [in Russian]
4.13 G. Fairweather, A. Mitchell: Comput. J. **7**(3) (1964)

Chapter 5

5.1   O.A. Ladyshenskaya, V.A. Solonnikov, N.N. Uraltseva: *Linear and Quasilinear Equations of Parabolic Type* (Nauka, Moscow 1967) [in Russian]
5.2   K.K. Sabelfeld: Sov. J. Num. Anal. Math. Model. **1**(3) (1986)
5.3   M.L. Rasulov: *Application of Contour Integrals to Solving Problems for Parabolic Systems of Second Order* (Nauka, Moscow 1975) [in Russian]
5.4   A.M. Ilyin, A.S. Kalashnikov, O.A. Oleinik: Usp. Mat. Nauk, **17**(3) (1962) [in Russian]
5.5   K.K. Sabelfeld: Sov. J. Num. Anal. Math. Model. **2**(3), 201–218 (1987)
5.6   R. Bellman, R. Kalaba, I. Lockett: *Numerical Inversion of Laplace Transforms* (Elsevier, New York 1966)
5.7   K.K. Sabelfeld: *Monte Carlo Methods in Boundary Value Problems* (Nauka, Novosibirsk 1989) [in Russian]
5.8   B.L. Sawford: Quart. J. Roy. Meteor. Soc. **108**, 207 (1982)
5.9   F.T.M. Niewstadt, Kh. van Don (eds.): *Atmospheric Turbulence and Air Pollution Modelling* (Reidel, Amsterdam 1985)
5.10 F.A. Gifford: J. Met. **1**, 14 (1957)
5.11 A.S. Monin, A.M. Jaglom: *Statistical Fluid Mechanics of Turbulence* Vols. 1, 2 (MIT, Cambridge, MA 1967)
5.12 R.H. Kraichnan: Phys. Fluids. **13**(1) (1970)

5.13   B.S. Elepov, A.A. Kronberg, G.A. Mikhailov, K.K. Sabelfeld: *Solution of Boundary Value Problems by Monte Carlo Methods* (Nauka, Novosibirsk 1980) [in Russian]

5.14   K.K. Sabelfeld, O. Kurbanmuradov: Sov. J. Num. Anal. Math. Model. 4(1), (1989)

5.15   G.K. Batchelor: Quart. J. Roy. Meteor. Soc. 76(328) 1 (1950)

5.16   R.V. Ozmidov: Izv. Akad. Nauk SSSR, Ser. Geofiz. No. 8, (1960) [in Russian]

5.17   V.I. Tatarsky: Izv. Vuzov. Radiofizika, 23(4) (1960) [in Russian]

5.18   V.N. Ivanov: Izv. Akad. Nauk SSSR, Ser. Geofiz. No. 10 (1963) [in Russian]

5.19   A.A. Khanonyan: Izv. Akad. Nauk SSSR, Ser. FAO, 4, No. 1 (1988) [in Russian]

5.20   G.A. Mikhailov: *Optimization of Weighting Monte Carlo Methods* (Nauka, Novosibirsk 1980) [in Russian]

5.21   S.M. Ponamaryeva, A.S. Gavrilov: in *Gidrometeorologia*, Ser. Meteorologia (Obninsk 1984) [in Russian]

5.22   J.C. Kaimal, J.C. Wyngaard, D.A. Haugen, O.R. Cote, Y. Izimi, S.J. Caughey, C.J. Readings: J. Atmos. Sci. 33, (1976)

5.23   N.A. Fuks: *Mechanics of Aerosols* (Moscow University Press, 1955) [in Russian]

5.24   B. Lewis, D.S. Campbell: J. Vac. Sci. Technol. 4, 209–211 (1967)

5.25   B. Lewis: Surface Sci. 21, 273–276 (1970)

5.26   S.E. Pashenko, A.V. Bubnov, K.P. Kutsenogii, K.K. Sabelfeld: in *Thirteenth Symposium on Rarefied Gas Dynamics* (Novosibirsk 1982) 345–346

5.27   A. Poppa: Appl. Phys. 38 (1967)

5.28   R. Hrach, V. Stary: Czech. J. Phys. 28, 1382–1388 (1978)

Subject Index

G. A. Mikhailov

Optimization of Weighted Monte Carlo Methods

1991. Approx. 220 pp.
Hardcover in prep.
ISBN 3-540-53005-3

D. W. Heermann

Computer Simulation Methods
in Theoretical Physics

2nd ed. 1990. IX, 145 pp. 30 figs.
Softcover DM 39,–
ISBN 3-540-52210-7

Computational methods pertaining to many branches of science, such as physics, physical chemistry and biology, are presented. The text is primarily intended for third-year undergraduate or first-year graduate students. However, active researchers wanting to learn about the new techniques of computational science should also benefit from reading the book. It treats all major methods, including the powerful molecular dynamics method, Brownian dynamics and the Monte-Carlo method. All methods are treated equally from a theoretical point of view. In each case the underlying theory is presented and then practical algorithms are displayed, giving the reader the opportunity to apply these methods directly. For this purpose exercises are included. The book also features complete program listings ready for application.

K. Binder, D. W. Heermann

Monte Carlo Simulation in Statistical Physics
An Introduction

1988. VIII, 127 pp. 34 figs.
(Springer Series in Solid-State Sciences, Vol. 80) Hardcover DM 54,–
ISBN 3-540-19107-0

Contents: Introduction: Purpose and Scope of This Volume, and Some General Comments. – Theoretical Foundations of the Monte Carlo Method and Its Applications in Statistical Physics. – Guide to Practical Work with the Monte Carlo Method. – Appendix. – References. – Subject Index.

Springer-Verlag
Berlin
Heidelberg
New York
London
Paris
Tokyo
Hong Kong
Barcelona
Budapest

S. M. Rytov, Y. A. Kravtsov, V. I. Tatarskii

Principles of Statistical Radiophysics 1

Elements of Random Process Theory

1987. X, 253 pp. 28 figs.
Hardcover DM 124,– ISBN 3-540-12562-0

Principles of Statistical Radiophysics is a four-volume series that introduces the newcomer to the theory of random functions. It aims at providing the background necessary to understand papers and monographs on the subject and to carry out independent research in fields where fluctuations are of importance, e.g. radiophysics, optics, astronomy, and acoustics.
"Elements of Random Process Theory", the first volume, contains the essential mathematical prerequisites and definitions related to this topic. It deals in particular with the physics of random pulse processes, shot and flicker noises, fluctuations in self-oscillatory systems, random actions on linear and nonlinear discrete dynamical systems, Markov processes and stochastic differential equations.

S. M. Rytov, Y. A. Kravtsov, V. I. Tatarskii

Principles of Statistical Radiophysics 2

Correlation Theory of Random Processes

Translated from the Russian by A. P. Repyev
1988. X, 234 pp. 54 figs.
Hardcover DM 148,– ISBN 3-540-16186-4

Contents: Fundamentals of Correlation Theory. – Applications of Correlation Theory. – Spectral Theory of Random Actions on Dynamic Systems. – Certain Kinds of Nonstationary Processes. – References. – Subject Index.

S. M. Rytov, Y. A. Kravtsov, V. I. Tatarskii

Principles of Statistical Radiophysics 3

Elements of Random Fields

1989. X, 239 pp. 36 figs.
Hardcover DM 168,– ISBN 3-540-17829-5

Contents: Fundamentals. – Radiation and Diffraction of Random Wave Fields. – Thermal Electromagnetic Fields. – Single Scattering Theory. – References. – Subject Index.

S. M. Rytov, Y. A. Kravtsov, V. I. Tatarskii

Principles of Statistical Radiophysics 4

Wave Propagation Through Random Media

Translated from the Russian by A. P. Repyev
1989. X, 188 pp. 43 figs.
Hardcover DM 136,– ISBN 3-540-17828-7

This Volume is concerned with Markov random fields, Feyman diagrams in perturbation theory, wave propagation in media with large random inhomogeneities, multiple scattering and scattering by rough surfaces.

Springer-Verlag
Berlin
Heidelberg
New York
London
Paris
Tokyo
Hong Kong
Barcelona
Budapest

N. G. Chetaev

Theoretical Mechanics

Translated from the Russian by I. Aleksanova

1989. 407 pp. 190 figs. Hardcover DM 68,–
ISBN 3-540-51379-5

This university-level textbook reflects the extensive teaching experience of N. G. Chetaev, one of the most influential teachers of theoretical mechanics in the Soviet Union. The mathematically rigorous presentation largely follows the traditional approach, supplemented by material not covered in most other books on the subject. To stimulate active learning numerous carefully selected exercises are provided. Attention is drawn to historical pitfalls and errors that have led to physical misconceptions.

Extensive appendices contain material from additional lectures on optics and mechnics analogies, Poincaré's equation and the special theory of elasticity.

**Distribution rights for the socialist countries, India and Iran:
V/O "Mezhdunarodnaya Kniga", Moscow**

D. Park

Classical Dynamics and Its Quantum Analogues

2nd enl. and updated ed. 1990. IX, 333 pp. 101 figs.
Hardcover DM 78,– ISBN 3-540-51398-1

The primary purpose of this textbook is to introduce students to the principles of classical dynamics of particles, rigid bodies, and continuous systems while showing their relevance to subjects of contemporary interest. Two of these subjects are quantum mechanics and general relativity. The book shows in many examples the relations between quantum and classical mechanics and uses classical methods to derive most of the observational tests of general relativity. A third area of current interest is in nonlinear systems, and there are discussions of instability and of the geometrical methods used to study chaotic behaviour. In the belief that it is most important at this stage of a student's education to develop clear conceptual understanding, the mathematics is for the most part kept rather simple and traditional.

This book devotes some space to important transitions in dynamics: the development of analytical methods in the 18th century and the invention of quantum mechanics.

A. Hasegawa

Optical Solitons in Fibers

2nd enl. ed. 1990. XII, 79 pp. 25 figs.
Softcover DM 48,– ISBN 3-540-51747-2

Already after six months high demand made a new edition of this textbook necessary. The most recent developments associated with two topical and very important theoretical and practical subjects are combined: **Solitons** as analytical solutions of nonlinear partial differential equations and as lossless signals in dielectric **fibers**. The practical implications point towards technological advances allowing for an economic and undistorted propagation of signals revolutionizing telecommunications. Starting from an elementary level readily accessible to undergraduates, this pioneer in the field provides a clear and up-to-date exposition of the prominent aspects of the theoretical background and most recent experimental results in this new and rapidly evolving branch of science. This well-written book makes not just easy reading for the researcher but also for the interested physicist, mathematician, and engineer. It is well suited for undergraduate or graduate lecture courses.

Springer-Verlag
Berlin
Heidelberg
New York
London
Paris
Tokyo
Hong Kong
Barcelona
Budapest

B. N. Zakhariev, A. A. Suzko

Direct and Inverse Problems

Potentials in Quantum Scattering

1990. XIII, 223 pp. 42 figs.
Softcover DM 48,–
ISBN 3-540-52484-3

This textbook can almost be viewed as a "how-to" manual for solving quantum inverse problems, that is, for deriving the potential from spectra or scattering data and also, as somewhat of a quantum "picture book" which should enhance the reader's quantum intuition. The formal exposition of inverse methods is paralleled by a discussion of the direct problem. Differential and finite-difference equations are presented side by side. The common features and (dis)advantages of a variety of solution methods are analyzed. To foster a better understanding, the physical meaning of the mathematical quantities are discussed explicitly. Wave confinement in continuum bound states, resonance and collective tunneling, energy shifts and the spectral and phase equivalence of various interactions are some of the physical problems covered.

A. G. Sitenko

Scattering Theory

1991. XI, 294 pp. 32 figs.
(Springer Series in Nuclear and Particle Physics)
Hardcover DM 88,–
ISBN 3-540-51953-X

This book is an introduction to non-relativistic scattering theory. The presentation is mathematically rigorous, but is accessible to upper level undergraduates in physics. The relationship between the scattering matrix and physical observables, i. e. transition probabilities, is discussed in detail. Among the emphasized topics are the stationary formulation of the scattering problem, the inverse scattering problem, dispersion relations, three-particle bound states and their scattering, collisions of particles with spin and polarization phenomena. The analytical properties of the scattering matrix are discussed. Problems round off this volume.

Springer-Verlag
Berlin
Heidelberg
New York
London
Paris
Tokyo
Hong Kong
Barcelona
Budapest